Welding Technology Fundamentals

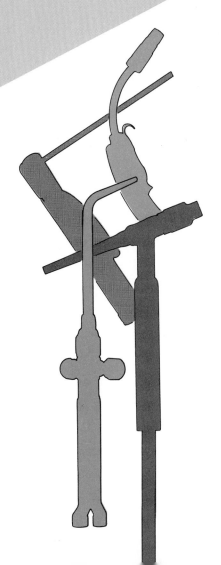

by

William A. Bowditch

Technical-Vocational Education Consultant
Gaylord, Michigan
Life Member of the American Welding Society
Member of the American Vocational Association

and

Kevin E. Bowditch

Welding Engineer
Production Manager
Vista Industrial Products, Inc.
Vista, California
Member of the American Welding Society
Member of the American Vocational Association

Publisher
The Goodheart-Willcox Company, Inc.
Tinley Park, Illinois

Library of Congress Catalog Card Catalog Number 96-3968
International Standard Book Number 1-56637-314-X

6 7 8 9 10   97   03 02

William A. Bowditch has an extensive teaching and welding background. He has been a teacher, department head, and supervisor of special needs and vocational programs. In addition to his formal college training in preparation for teaching, Bill has taken several specialized courses in industry, such as the Hobart Welding School and American Welding Society courses. He is a member of the American Vocational Association and a life member of the American Welding Society. As a coauthor of *Modern Welding,* he has guided that Goodheart-Willcox book through many revisions to keep it up-to-date and technically correct to maintain its value as an authoritative welding text.

Kevin E. Bowditch is the production manager of a precision sheet metal firm that produces fabricated metal parts and weldments. His welding experience includes working for an automotive firm, a construction company building two nuclear plants, and two aerospace firms. While working for one aerospace firm, Kevin designed resistance welding and soldering equipment, special equipment for custom applications, and worked to develop correct welding and soldering schedules for customers. He has a bachelors degree in welding engineering from Ohio State and has attended specialized conferences and courses sponsored by the American Welding Society, American Society of Mechanical Engineers, and American National Standards Institute. Kevin joined his father as a co-author of *Modern Welding,* beginning with the 1984 edition, and has been a coauthor of *Welding Technology Fundamentals* since its first edition was published in 1991.

Weldment on cover courtesy of F.H. Ayers Manufacturing Co. **Photo Credits:** American Welding Society, Fibre-Metal Co., GE Fanuc Robotics.

Library of Congress Cataloging in Publication Data

Bowditch, William A.
    Welding technology fundamentals / by William A.
Bowditch and Kevin E. Bowditch.

    p. cm.
Includes index.
ISBN 1-56637-314-X
    1. Welding. I. Bowditch, Kevin E. II. Title
TS227.B69 1997
671,5'2--dc20   96-3968
                                                CIP

Introduction

Welding Technology Fundamentals is written for secondary and postsecondary students, apprentices, journeymen, and individuals who wish to learn to weld.

This book covers the equipment and techniques used for the welding and cutting processes most often employed in industry today. These processes are: oxyfuel gas welding and cutting, shielded metal arc welding, gas metal arc welding, flux cored arc welding, gas tungsten arc welding, and resistance welding.

It contains information about welding careers and the physics of welding. Technical information regarding weld inspection and testing, welder qualification, drawing interpretation, and welding symbols is also included.

General welding safety is covered prior to Chapter 1. Safety information and cautions are also written into the text wherever they apply. Safety information and cautions are printed in color, so that they will stand out.

The text is organized into eight sections. Each covers a group of related welding or cutting processes. A section can be studied independently or in sequence with other sections.

You may begin your study of welding with any section and progress in any desired sequence from section to section. However, when you use the *Laboratory Manual for Welding Technology Fundamentals,* we recommend that Chapter 31, Welding Symbols, be studied early. Welding symbols are used in the manual to describe the sample joints and welds for each job assignment.

The authors recognized that oxyfuel gas welding is not widely used in industry today. However, it is a slow and relatively inexpensive way of learning to manipulate a weld pool. This slower speed makes it possible for you to study the weld pool, and learn to manipulate it. In other processes, the actions of the weld pool occur at a much faster rate of speed and they are harder to observe. Oxyfuel gas welding is also an inexpensive method of preparing for gas tungsten arc welding.

Welding Technology Fundamentals is written in an easy-to-read and understandable form. All welding terms used are those approved by the American Welding Society (AWS). In cases where nonstandard terms are used by some people in the trade, such terms are often given in parentheses after the correct AWS term. The book is extensively illustrated with drawings and photographs to show the various processes or welding techniques.

Many tables and charts are provided to help you select the variables required to make a good weld. Photographs of industrial welding applications have been used, along with photographs of practice welds in progress. Equivalent SI metric measurement units are shown in parentheses following US conventional measurements.

You should read the caption accompanying each illustration, since the caption often gives information that is not covered in the text. Review questions are provided at the end of each chapter to test your knowledge of the information covered. In most chapters, practice exercises are provided to test your skills as a student welder in completing various welding tasks.

It is our sincere hope that *Welding Technology Fundamentals* will help you to progress in an organized manner toward a mastery of the essential welding skills.

William A. Bowditch
Kevin E. Bowditch

Contents

SAFETY IN THE WELDING SHOP

Working and moving about in a welding shop or welding environment can present many dangers such as heat, sparks, fumes, radiation, high voltage, hot metal, moving vehicles, hazardous machinery, and moving overhead cranes and their loads. However, through the use of training and the use of adequate safety equipment all these safety hazards can be controlled.

Three of the most important factors in safety on the job are:
• Staying healthy in mind and body.
• Becoming well trained in the required job or task and its possible hazards.
• Having a good attitude toward safety rules, equipment, and training on the job.

Personal Safety and Clothing

Welders should wear work clothes or coveralls. The shirt or coveralls should have covered pockets and have buttons at the neck. Trousers and coveralls should not have cuffs which could catch hot metal spatter. A cap of some kind should be worn to protect the hair from hot metal spatter. Gloves should be worn to protect against hot and sharp metal. Leather gloves with gauntlet-type cuffs are recommended for welders. Safety glasses should be worn by everyone working in an eye hazard area.

Steel-toed safety shoes are recommended for welders and other workers who handle heavy articles. Oxyfuel gas welders and cutters should wear approved goggles with the correct shade lens.

Anyone performing arc welding should wear an arc welding helmet with the correct shade lens for the process and amperage being used. See Fig. S-1. Some arc welders also wear a pair of goggles under their arc welding helmet to protect them from arc rays and metal spatter that might reflect off the inside of the helmet.

Shop *housekeeping* is one of the most important factors in shop safety. The floors and work benches should be kept clean of dirt, scrap metal, grease, oil, and anything that is not essential to the job at hand. *Combustibles* such as wood, paper, rags, and flammable liquids must be kept clear of all areas where sparks or hot metal may fly. Aisles must be kept clear of hoses and electric cables, which can cause tripping accidents. Hoses and cables also could be run over and possibly damaged.

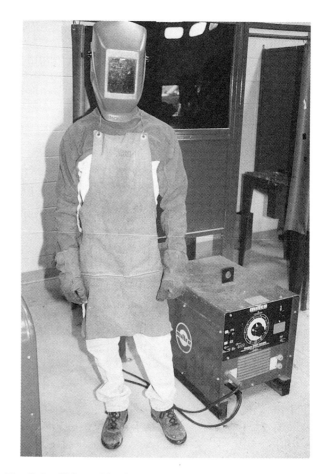

Fig. S-1. This welder is wearing an arc helmet, cap, gloves, leather apron, leather cape, cuffless trousers, and high-top shoes as protection from hot metal and harmful arc rays.

Fire Hazards

Paint, oil, cleaning chemicals, and other possible combustibles must be kept in steel cabinets designed for such storage in accordance with the local fire codes.

Fire exits, fire blankets, and extinguishers should be clearly marked. Fire extinguishers and blankets are usually mounted on a bright red surface for visibility. Every worker should know how to use the extinguishers and blankets. They should also know where the exits are and which one to use in an emergency. Periodic fire drills are a good idea. It is also important to know what equipment should be shut down before leaving the work area to reduce fire hazards.

Electrical Hazards

Electrical devices of various kinds are common in welding shops. All electrical devices are hazardous, but some use extremely high and dangerous voltages. All equipment and areas where 220 volts or more are used must be well-marked. The installation and repair of electrical equipment must only be done by well trained and competent technicians.

Machinery Hazards

Machinery must be operated only after thorough training on how the machine operates, its safety hazards, its safety features, the correct placement of hands and feet, and the proper sequence of operation.

The pedestal grinder, a simple piece of equipment, can be one of the most dangerous pieces of machinery in the shop! For example, failing to adjust the tool rest to its closest safe point and securely tighten it can result in an accident.

Fumes and Ventilation

Dust, fumes, and metal particles can be a hazard to health. Adequate and approved ventilation is required in welding and cutting shops to ensure that toxic fumes, dust, and dirt are removed properly. In addition to ventilation, the air may need to go through filters and cleaners before it is recirculated. Ventilation pickups should be located so that fumes are picked up below the level of the welder's nose, before he or she has to inhale them.

Toxic fumes may be created when fluxes and metals containing such chemicals as cadmium, chromium, lead, zinc, and beryllium are heated as when welding or brazing. When these chemicals are present or suspected to be present, welding or brazing work must be done with excellent ventilation or while wearing a purified air breathing apparatus. Air-purifying respirators may also be worn to filter dust-like particles, See Fig. S-2.

Fig. S-2. An air-purifying respirator filters dust and other particles from the air. It will not protect against toxic fumes, however.

Lifting

Lifting is always a hazard to the body, but training in how to lift objects safely can reduce the chance of injury. Limits should be set on how much weight a worker may lift. The maximum weight to be lifted may vary from shop to shop and by worker classification. In many shops, any worker who may have to lift anything is required to wear a lifting belt or back brace.

Hazardous Obstacles

Safety hazards or obstacles in the shop should be well-marked. Signs, fences, or barriers should be erected while *temporary hazards* are present, so that all workers are fully aware of them. *Permanent hazards* are often painted with yellow and black stripes to create high visiblity.

Hand and Power Tools

All hand and power tools should be examined prior to use for loose parts that may come off and injure the operator or a worker nearby. Power cords should be checked for cut or frayed insulation and exposed wires. Return and report any defective equipment to the shop supervisor or to maintenance.

Designated Welding and Cutting Areas

All welding and cutting should be done in designated areas of the shop, if possible. These areas should be made safe for welding and cutting operations with concrete floors, filter screens, protective drapes or curtains, and fire extinguishers. No combustibles should be stored nearby.

Suffocation Hazards

Gases that are heavier than air or lighter than air can be extremely dangerous to welders in closed tanks or confined spaces. Argon and carbon dioxide are examples of heavier-than-air gases. Helium is an example of a lighter-than-air gas. These gases are colorless and odorless and will displace the oxygen in a closed space. Argon, for example, can asphyxiate a person in about seven seconds. Closed spaces must be well-ventilated when heavier-than-air or lighter-than-air gases are to be used. If proper ventilation cannot be obtained, the welder must go into the space using an air-supplied purifier.

Welding on Hazardous Containers

Unless you are trained in the proper procedures for doing so, never try to weld or cut a container that has held flammable or hazardous materials. Such a container may explode. Proper procedures are described in the American Welding Society (AWS) Standard F4. 1-88, *Recommended Safe Practices for the Preparation for Welding and Cutting of Containers That Have Held Hazardous Substances.* You should also consult with the local fire marshal before welding or cutting such containers.

SECTION 1

INTRODUCTION TO WELDING

Fibre-Metal Products Co.

8

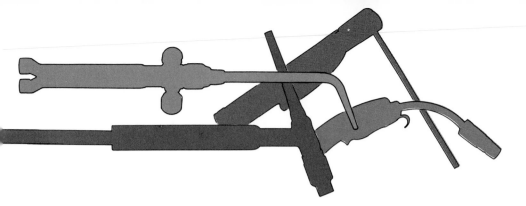

Chapter 1

Welding and Cutting Processes

After studying this chapter, you will be able to:
☐ Cite advantages of welding over other joining processes.
☐ List the significant developments in the history of welding.
☐ Identify recent developments in welding and cutting processes.
☐ Identify several occupations in the welding industry and list the recommended amount of education for each.

TECHNICAL TERMS

Force welding, fushion welding, occupation, process, prototype parts, weld, welding, weldment.

THE WELDING PROCESS

A *process* is an operation used to produce a product. Riveting, forging, cutting, turning, bending, and welding are processes used in industry. Metalworking processes have been used for thousands of years. Metals have been joined by riveting and welding for thousands of years. Blacksmiths used their skills and knowledge of metalworking processes to work metal into many desired products. Hinges, nails, cooking pots, farm implements, wagon wheels, and swords were produced by forging and forge welding.

Blacksmiths are employed today to make prototype parts for cars, airplanes, and other equipment. *Prototype parts* are the first models of parts that later may be mass produced. Prototype parts are generally handcrafted. After the prototype parts are approved, machines and equipment are designed to mass produce the parts.

Forge welding is a process used by blacksmiths to produce welded metal parts. In forge welding, the parts are heated until they are very hot and in a softened state. The heated ends are then placed together on an anvil and struck repeatedly with a hammer. See Fig. 1-1. This hammering creates a *weld* by forcing the hot softened metal from the pieces to intermix and form a single unit. If the welding process is performed properly, the weld is stronger than the original metal.

Welding is defined as a process of "joining metallic or nonmetallic material in a relatively small area by heating the area to the welding temperature. Pressure may or may not be applied; welds may be made by applying pressure only. Welds may be made with or without the addition of filler material." Not all materials can be welded, but most metals and plastics are weldable.

Fig. 1-1. A blacksmith is joining the ends of a chain link by the forge welding method.
(Henry Ford Museum and Greenfield Village)

9

Most welding processes require the addition of heat to the weld area. In *fusion welding,* heat is applied to a small area until it becomes molten (liquid). The molten areas of the weld joint then flow together, forming a single until after they cool.

Welds are also made with the metal at or near room temperature. Cold welding and explosion welding are examples of welding processes that are performed at room temperature. Extremely high pressure is applied to the area to be welded. Metal at the surface of one part is driven into the surface of the adjoining part by high pressure. The pieces are thus joined or welded at the surface.

ADVANTAGES OF WELDING AND CUTTING PROCESSES

Welding and cutting processes offer a number of advantages over other joining and cutting methods.

Most welding and cutting equipment is portable. It can be operated inside boilers, furnaces, large containers, or pipes to make repairs. Welding and cutting can be done under water to repair ships or to cut up and remove water hazards. Welding and cutting equipment can be transported into the field to repair farm machinery, trucks, and earth-moving equipment. See Fig. 1-2. Welding is used to construct pipelines, buildings, and ships.

Because of modern welding techniques, engineers are able be design strong parts that are lightweight, complex, and often less expensive. As late as 1920, many steel structures were built using rivets. Automobile and truck chassis also were constructed using rivets. Handles were attached to cooking pots using rivets. Holes are drilled into parts to be riveted. This weakens the parts in that area. Thicker and heavier metal is used to compensate for the weakness. See Fig. 1-3. Today, many parts are designed for welding. The welded

A

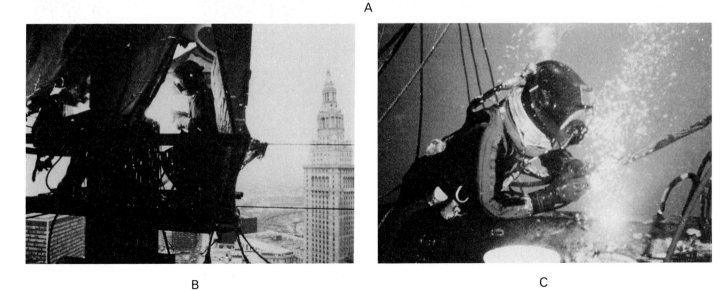

B   C

Fig. 1-2. Portable welding and cutting equipment. A—Wheel housing of a piece of earth-moving equipment being repaired in the field. (Hobart Brothers Co.) B—Two welders on a skyscraper. The welding machine is kept as close as possible to the welders. (Lincoln Electric Co.) C—Diver on the ocean floor using underwater welding on a large pipe.

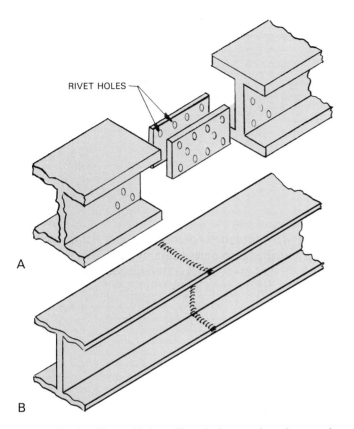

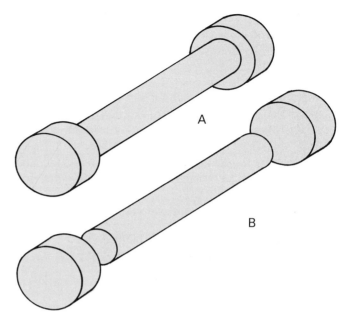

Fig. 1-4. A—Part machined from one large-diameter rod. Metal removed from the middle section is wasted. B—Part produced by welding three individual pieces of stock together. Machining is not required.

Fig. 1-3. A—Riveted joints. Rivet holes weaken the metal at the joint. Thicker metal must be used. B—Welded joints. Rivet holes are not required. Thinner metal may be used.

parts are lighter and generally cheaper to produce than riveted parts. Parts joined by welding are known as **weldments.**

Many complex shapes are cast or forged and then machined into their final shape. Cast or forged parts are heavy and require large, costly equipment to produce them. A large amount of equipment and machining would be required to produce the part in Fig. 1-4A from 3″ (76 mm) solid stock. The same part can be produced from 3″ (76 mm) and 1 1/2″ (38 mm) stock by cutting the stock to length and welding it at both ends. See Fig. 1-4B. The weldment would be generally faster, use less material, and be cheaper to produce.

Applications of welding and cutting processes are unlimited. For example, stones can be welded together to build or repair a statue. A layer of weldable metal can be added to the stone surfaces by thermal spraying. The metal can then be welded to join the stones together. Fig. 1-5 shows a layer of harder material being applied to a soft metal shaft using the thermal spraying process.

Fig. 1-5. Thermal spraying used to add a layer of harder material to a soft metal shaft. The harder material is superheated and sprayed onto the surface of the shaft. (Wall Colmonoy Corp.)

HISTORY OF WELDING

The use of welding dates back to 2000 B.C., but the development of modern forms of welding began in 1881. Some of the major dates and developments are as follows:

2000 B.C.—Forge welding used to join copper and bronze.

1881—Electric arc welding performed by Auguste de Meritens. A carbon electrode was used.

1883-1885—Electric resistance welding developed by Elihu Thomson. It was used to weld the ends of wires together.

1894—Carbon arc welding used commercially to produce steel barrels.

1902—Arc cutting demonstrated.

1903—Oxyfuel gas welding and cutting torches developed by Fouché and Picard.

1907—Covered steel electrode developed by Kjellborg in Sweden.

1917—Strength of welded joints was first tested by the British and United States governments and by Lloyd's of London (a ship insurance company). Test results proved that welded joints were as strong as riveted joints.

Welding became an accepted practice for repair and construction. When the U.S. entered World War I, the Germans severely damaged 105 of their own ships that were in U.S. ports. This was done to prevent their use by the United States. Repairs to boilers and pumps were performed using welding, saving $20,000,000.

1918—First all-welded ship was launched.

1920—First all-welded building was constructed. Built by Electric Welding Company of America, it measured 40′ x 60′.

1940—Submerged arc welding process developed in Russia.

1942—Gas tungsten arc welding (GTAW) developed, using two tungsten electrodes and helium as a shielding gas. Also known as heliarc welding.

1948—Gas metal arc welding (GMAW) developed.

1990s—A total of 94 welding and cutting processes are in use with a variety of energy sources and shielding gases. American Welding Society (AWS) approved names and abbreviations are shown in Fig. 1-6.

MASTER CHART OF WELDING AND ALLIED PROCESSES

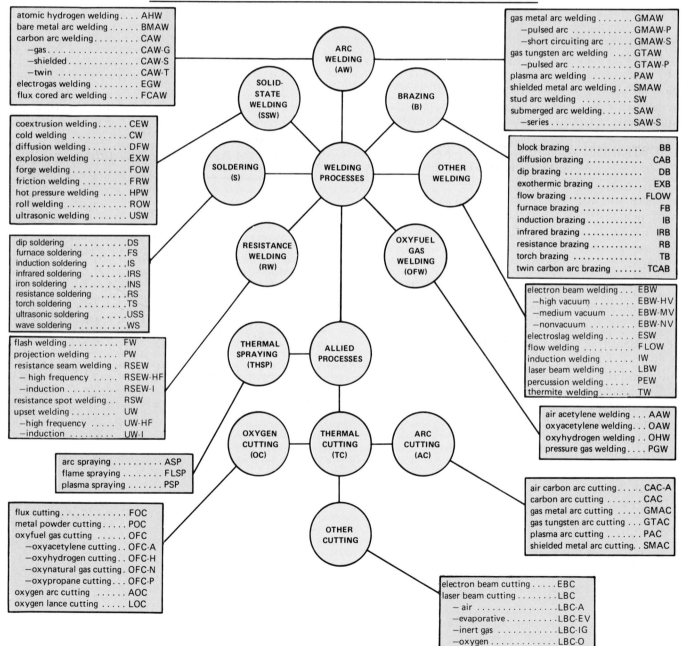

Fig. 1-6. Welding and cutting processes as defined by the American Welding Society (AWS).

RECENT DEVELOPMENTS IN WELDING AND CUTTING PROCESSES

Engineers have developed new welding and cutting processes in recent years. New processes were developed because new metals and alloys were being used in newly developed products such as supersonic aircraft, nuclear plants, submarines, and spacecraft. Welding and cutting processes which have been recently developed include:
- Plasma arc welding and cutting
- Electron beam welding and cutting
- Electroslag welding
- Electrogas welding
- Laser beam welding and cutting
- Plasma spraying
- Electric arc spraying
- Flame spraying
- Explosion welding

Although the future of the welding industry is unknown, it is certain that it will continue to grow. New welding and cutting processes will have to be developed as new metals, alloys, plastics, and ceramics are created.

OBTAINING AND HOLDING A JOB IN THE WELDING INDUSTRY

There are, and will be, many jobs available in the welding industry. Most of these jobs will require a high school education or an apprenticeship. Other jobs may require a junior college (associates), four-year (bachelors) or advanced degree.

A number of welding and welding-related jobs, together with the educational level usually required and where this education may be obtained, are listed below:

High School or Technical School
- Gas welder
- Assembler brazer
- Gun welder
- Arc welder
- Fitter welder
- Tack welder
- Resistance welder
- Assembler welder
- Combination welder
- Solderer

Technical School or Community College
- Ultrasonic welding machine operator
- Welding inspector
- Laser beam machine operator
- Welding supervisor
- Experimental welder
- Welding machine repair person
- Robotic welding machine programmer
- Tool-and-die welder

Military Training Schools
- Combination welder
- Diver/welder
- Repair welder
- Specialty welder

Trade Apprenticeship
- Sheet metal worker
- Pipe and steam fitter
- Structural iron worker
- Ornamental iron worker
- Blacksmith for experimental parts

College or University
- Welding engineer
- Metallurgist
- Metallurgical engineer

Getting a job is sometimes difficult, but often, the job seeker has control over the reasons for that difficulty. To get the job you want, you must take the required courses in school. You must have and demonstrate the personal traits that employers are looking for in an employee.

SUGGESTED SCHOOL SUBJECTS FOR SUCCESS

Welding is a technical trade, but to succeed an employee should know more than how to weld. Some of the school subjects suggested for greater success in finding and holding a job in the welding industry are: *Print Reading, Mechanical Drafting, Electricity, Electronics, Metals, Physics, Math, Algebra, Geometry, Trigonometry, Calculus, and Welding.*

PERSONAL TRAITS SOUGHT BY EMPLOYERS

Employers want the following personal traits in an employee:
- Dependability.
- The ability to follow directions.
- The ability to get along with peers.
- Thoroughness.
- Self-confidence.
- An ability to accept responsibility.
- Initiative.
- An ability to get along with supervisors.
- The ability to communicate written ideas.
- An ability to communicate ideas orally.

Since school is a student's first job, you should learn and demonstrate the traits required by a future employer while still in school. All of the traits can be demonstrated in class, by participating in school sports and clubs, and by doing good work in a timely manner.

ACADEMIC SKILLS SOUGHT BY EMPLOYERS

The academic skills which employers are seeking include the ability to:
- Read and understand written materials.
- Write and understand the technical terms and language of the trade or business in which you desire to work.
- Read and understand graphs and charts.
- Understand basic math.
- Use mathematics to solve problems.
- Use research and library skills.
- Use the tools and equipment involved in the business.
- Speak the technical language in which the business is conducted.
- Use the scientific method of solving problems.

FACTORS THAT CAN LEAD TO REJECTION FOR EMPLOYMENT

Employers have identified several personal factors or traits which would cause a person to be rejected for a job. These factors include:
- A poor scholastic record.
- Inadequate personality.
- Lack of goals.
- Lack of enthusiasm.
- Inability to express yourself.
- Unrealistic salary demands.
- Poor personal appearance.
- Lack of maturity.
- Excessive interest in security and benefits.
- A poorly completed job application.

FACTORS THAT MAY LEAD TO TERMINATION FROM A JOB

Employers have also identified several reasons for a person being overlooked for promotion or even being

terminated. These reasons include:
- Poor attendance without cause.
- Habitually coming in late.
- Alcoholism.
- Illegal drug use.
- Inability to perform the tasks assigned.
- Inability to work as a team member.
- Fighting and threats to peers.
- Insubordination to directions from a supervisor.
- Talking with others too much and too often.
- Lack of respect for others.
- Lack of respect for others' property.
- Always making excuses.
- Constant complaining.

REVIEW QUESTIONS

Write your answers on a separate sheet of paper.
1. Welding has been performed for _____ of years.
2. Parts are heated until they are molten when forge welding. True or False?
3. Most welding and cutting equipment is portable. True or False?
4. Thicker metal is used for riveted parts to compensate for strength lost due to rivet holes. True or False?
5. Welding is performed on metallic and nonmetallic material. True or False?
6. In what year did welding become an accepted practice for repair and construction?
7. In what year were the first oxyfuel gas welding and cutting torches developed?
8. Refer to Fig. 1-6. Give the correct name for the SMAW process.
9. Refer to Fig. 1-6. What AWS abbreviation is used for oxyacetylene welding?
10. Identify one welding occupation that you would like to pursue at each educational level. Discuss your choices with a career counselor. Write a brief summary of each.

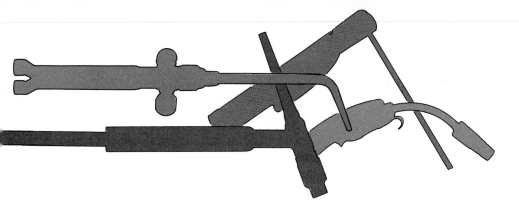

Chapter 2

The Physics of Welding

After studying this chapter, you will be able to:
- ☐ Identify the three general methods by which a weld is achieved.
- ☐ Describe the difference between chemical and mechanical properties and give examples of each.
- ☐ Explain the effects of welding on metal.
- ☐ Identify processes used to heat-treat metal.
- ☐ Describe the relationship between voltage and current.
- ☐ Give examples of US conventional units of measurement.
- ☐ Give examples of SI metric units of measurement.
- ☐ Convert US conventional units of measurement to SI metric units.
- ☐ Convert SI metric units of measurement to US conventional units.

TECHNICAL TERMS

Amperes, annealing, base metal, brittleness, chemical composition, chemical properties, compressive strength, contraction, corrosion resistance, current, density, ductility, electrode, expansion, full anneal, grain size, hardness, interpass heating, normalizing, ohms, open circuit voltage, oxidation resistance, physical properties, preheating, quenching, resistance, root opening, shielded metal arc welding (SMAW), SI metric system, strength, stress relieving, tempering, tensile, strength, toughness, US conventional system, volts, voltage.

WELDING THEORY

Welding is a group of processes used to join metallic or nonmetallic materials. Welding is often done using

heat. It can also be done using pressure or a combination of heat and pressure. A filler material may or may not be added to the weld joint.

WELDING WITH HEAT

Heat is used to create welds in many welding processes. Filler material is commonly used when welding with heat only. *Shielded metal arc welding (SMAW)* is a process that uses heat and filler material. In the SMAW process, heat is created by an arc that is struck between an *electrode* and *base metal* (metal to be welded). See Fig. 2-1. The heat causes the end of the electrode and

Fig. 2-1. Shielded metal arc welding (SMAW) being used to make repairs on a ship. (Miller Electric Mfg. Co.)

an area of the base metal to melt. When two pieces of base metal are placed together, both are heated by the arc. A portion of each piece melts and the liquid areas flow together. The molten filler material combines with the molten base metal. The molten material cools and becomes solid, creating a weld.

WELDING WITH HEAT AND PRESSURE

Some welding processes use both heat and pressure. Filler material is generally not used in these processes. In resistance spot welding, pressure is applied through opposing electrodes. Electrical current flows from one electrode through the base metal to the other electrode. Resistance to the electrical current provides the heat required to join the pieces. A resistance spot welding machine is shown in Fig. 2-2.

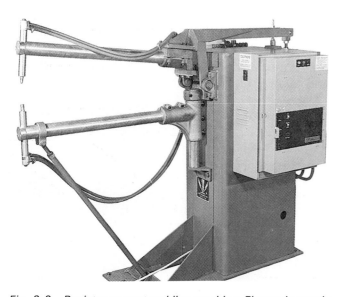

Fig. 2-2. Resistance spot welding machine. Electrodes apply pressure to pieces to be welded. Resistance to electrical current between the electrodes results in heat.
(LORS Machinery, Inc.)

WELDING WITH PRESSURE

Metal or other material also may be welded together using pressure alone. Heat and filler material are not required. Cold welding is an example of a welding process that does not require the use of heat or filler material. In the cold welding process, very clean pieces of metal are forced together under considerable pressure. The pressure forces the atoms of the materials together to create a weld.

PROPERTIES OF METALS

The properties of a metal determine how it is used. Properties can be grouped into three categories— physical, chemical, and mechanical. These properties are largely determined by the chemical composition of a metal.

The *chemical composition* of a metal alloy is a list of the different metals or elements that are combined to produce that metal alloy. For example, the chemical composition of low carbon steel is a combination of iron and carbon. The chemical composition of stainless steel includes iron, chromium, nickel, manganese, and carbon.

Properties of metals are affected by the method of processing. Processing includes bending, rolling, forming, heat treating, and welding.

PHYSICAL PROPERTIES

Physical properties are used to identify or describe a metal. These include color, melting temperature, and density. *Density* is the weight of a particular metal per unit volume.

CHEMICAL PROPERTIES

Chemical properties determine the way a material reacts in a given environment. Corrosion resistance and oxidation resistance are chemical properties that determine how a material will withstand the effects of the environment. *Corrosion resistance* is the ability of a material to withstand corrosive attack. Acids and salt water are both corrosive. A material resistant to attack by an acid, however, may not be resistant to the effects of salt water. *Oxidation resistance* is the ability of a material to resist the formation of oxides. Metal oxides occur when oxygen combines with the metal.

MECHANICAL PROPERTIES

Mechanical properties determine how a material reacts under applied loads or forces. Testing procedures for mechanical properties are discussed in Chapter 32.

Strength is the ability of a material to withstand applied loads without failing.

Tensile strength is the ability of a material to resist pulling forces.

Compressive strength is the ability of a material to resist pressing or crushing forces. Tensile strength and compressive strength are exact opposites, with respect to the direction of the applied load. Fig. 2-3 shows the difference between tensile and compressive forces acting on a material.

Ductility is the ability of material to stretch or bend without breaking.

Brittleness is the inability of a material to resist fracturing. Brittleness is the exact opposite of ductility. A material that is brittle has very low ductility and breaks easily.

Toughness is the ability of material to resist cracking and prevent a crack from progressing. Materials with a high amount of ductility are usually very tough.

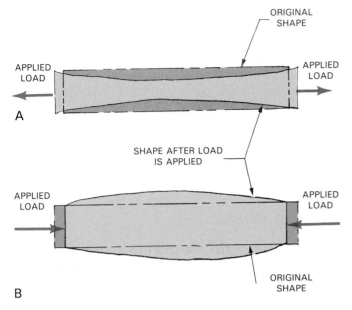

Fig. 2-3. A—Tensile strength. Thickness of the material is reduced and length is increased. B—Compressive strength. Thickness of material is increased and length is decreased.

Hardness is the ability of a material to resist identation or scratching. Hardness testing is done by forcing a steel ball or pointed diamond into the surface of a material. Hardness of some metals can be improved by quenching and tempering. Quenching and tempering are discussed later in the chapter.

Grain size is important in determining the mechanical properties of a material. Large or coarse-grained materials have a high amount of brittleness and a low amount of ductility. Fine-grained materials have a low amount of brittleness and a high amount of ductility.

EFFECTS OF WELDING

The chemical and mechanical properties of a metal are often affected when welded. Heat produced by welding creates stress in the metal. Heat also affects the ductility and toughness of the metal. If the metal was heat treated before being welded, the effects of heat treating are lost in the area around the weld.

A weld that is made correctly is usually stronger than the base metal. An incorrect welding procedure may result in serious problems. A stainless steel pipe, for example, that is corrosion resistant to certain chemicals, may lose its resistance at an improperly welded joint. Chemicals within the pipe may attack the weld area, causing the pipe to leak.

Another example where the correct welding procedure is critical is armor plate for the military. Armor plate is very tough. It resists cracks and prevents cracks that do start from getting larger. The metal near the weld area may lose its toughness if an improper welding procedure is used. If a shell hits the armor, a crack may develop at the weld and continue along the weld joint.

EXPANSION AND CONTRACTION OF METAL

When heat is applied to metal, the metal *expands* (increases in size). When heat is removed, the metal cools and *contracts* (reduces in size). Expansion and contraction create stress in the metal. The metal may relieve the stress by changing shape, or warping. In many welding applications, deformation or movement is not acceptable. Welding jigs or fixtures are commonly used to keep parts from moving. However, stress remains in parts that are clamped while being welded.

When welding a butt joint, the root opening may be reduced in size toward the end of the weld. The *root opening* is the space between the pieces to be welded. This occurs because the weld metal contracts and pulls the pieces together, as in Fig. 2-4A. To prevent this, the pieces should be tack welded as shown in Fig. 2-4B. The outer edges of the base metal may also bend toward the weld bead. To prevent this, the pieces may be clamped into a welding fixture or positioned with a reverse angle to compensate for the movement. Fig. 2-5 shows the effects of an unrestrained butt joint and a method used to compensate for the movement.

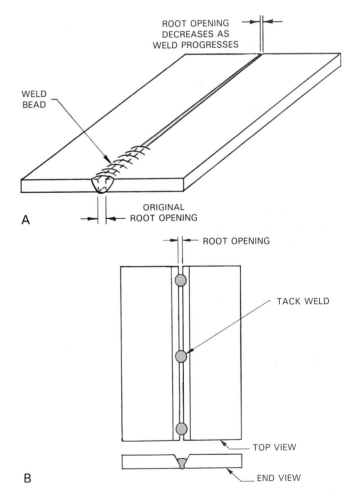

Fig. 2-4. A—Butt joint that is not tack welded. Root opening decreases as weld progresses. In extreme cases, the pieces may overlap. B—Two views of a V-groove butt joint that is tack welded. Root opening is uniform.

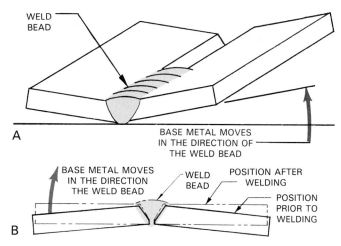

Fig. 2-5. A—Unrestrained butt joint. Outer edges bend toward the weld bead. B—Butt joint in which pieces are tack welded with a reverse angle. The amount of offset is determined by experimentation.

When welding a T-joint, the vertical piece may be pulled toward the weld bead. To prevent this, place the pieces in a welding fixture. Another solution is to tack weld the vertical piece a few degrees from the perpendicular, as shown in Fig. 2-6.

In general, six techniques may be used to reduce movement and/or stress when welding.

1. Tack weld the parts.
2. Align the parts to allow for contraction during welding.
3. Use welding jigs or fixtures.
4. Preheat the parts.
5. Heat treat the welded parts.
6. Use the proper welding procedure.

Straightening

The principles of expansion and contraction can be used to straighten parts. If a flat plate or weldment is bent or warped, heat can be used to straighten the piece. An oxyfuel gas torch is used to locally heat the metal, causing it to expand. This heating must be done on the correct side of the metal, and in the correct location, to be effective. This metal will not contract to its original shape or position when it cools. Although straightening a part is not easily performed, it is a skill that can be acquired through practice.

HEAT TREATING

Various heat treating processes are used in industry. Different heat treating processes are used to accomplish different purposes. Typical processes include annealing, stress relieving, and quenching and tempering. In addition to heat treating of a part after welding, heat may be applied before or during the welding process. *Preheating* is done just before the welding operation. *Interpass heating* is done while the weld is being completed.

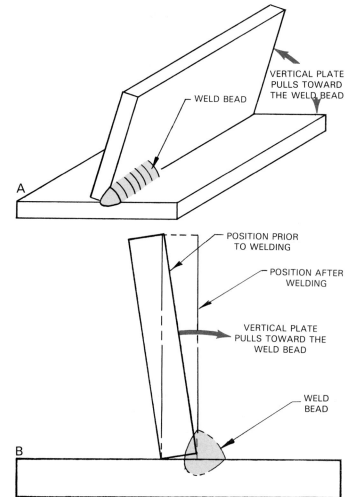

Fig. 2-6. A—Unrestrained T-joint. Vertical piece is pulled toward the weld bead. B—T-joint that is tack welded a few degrees from vertical. Offset compensates for movement of vertical piece.

Preheating

Preheating is used to raise the temperature of the metal before welding. Preheating causes less local expansion of the part during welding. When preheating, the entire part is heated rather than one specific area. When the part cools, less contraction occurs and less stress is developed.

Interpass heating

Interpass heating is a method of heating metal while it is being welded or between weld passes. Interpass heating is commonly used on thick plates and pipes. Preheating is used to heat the metal to the desired temperature. Interpass heating is used to maintain an elevated temperature. Preheating and interpass heating are both used to reduce or minimize the amount of expansion, contraction, and stress resulting from welding.

Annealing

Annealing is a heat treating process in which the welded metal is heated and allowed to cool slowly. The thickness of the metal determines the amount of time that the part must be held at a constant temperature. Thicker metal requires more time than thinner metal. The type of metal determines the temperature. Steel, for example, requires a higher temperature than aluminum. Mild steel is heated to about 1650°F (900°C) and must be held at that temperature for a few hours. The metal is then allowed to cool to room temperature. Slow cooling is very important to achieve a *full anneal.* A faster cooling rate is possible if the metal is cooled in air. This process is called *normalizing.*

Stress relieving

Stress relieving is similar to annealing, except that lower temperatures are used. Mild steel is heated to about 1200°F (650°C) and is kept at that temperature for a few hours. The metal is then allowed to air-cool. This process relieves some of the stress in the weldment.

Quenching and tempering

Quenching and tempering are processes used to harden steel and steel alloys. In the *quenching* process, metal is heated to a fairly high temperature. The temperature is maintained for a given period of time. The metal is then cooled quickly by immersing it in a bath of water, oil, or other liquid. This produces a very hard and brittle metal. The metal is then tempered by reheating it to several hundred degrees and cooling it. After *tempering,* the metal is less hard and no longer brittle. A tempered metal has good toughness.

ELECTRICAL PRINCIPLES

Electricity is produced by the movement of electrons within a circuit. Electricity is measured in terms of voltage and current. *Voltage* is the force that causes electrons to flow through a circuit. Voltage is measured in *volts*. It can be compared to water pressure—when your kitchen sink faucet is turned off, water pressure is still present. See Fig. 2-7A. When the faucet is opened, water begins to flow because the pressure is forcing the water out of the faucet. Likewise, voltage is always present in an electrical circuit. Although an arc may not be struck (started) between the base metal and electrode, voltage is still present, Fig. 2-7B. This is known as *open circuit voltage.* When an arc is struck, voltage forces electrons across the arc.

The air gap between the electrode and base metal offers resistance to the flow of electrons. *Resistance* is the opposition to the flow of electrons. Resistance is measured in ohms. A higher voltage setting on electric arc welding equipment allows the arc length to be longer. The arc stops if the arc length is longer than the voltage allows.

Current is the flow of electrons in an electrical circuit. Current is measured in *amperes,* or amps. Current can be compared to the flow of water from a faucet. When the faucet is turned off, water is unable

to flow. When the faucet is opened, water begins to flow, as in Fig. 2-7C. Likewise, when there is no arc, there is no current flowing. When an arc is struck, current is produced as electrons flow across the arc. See Fig. 2-7D.

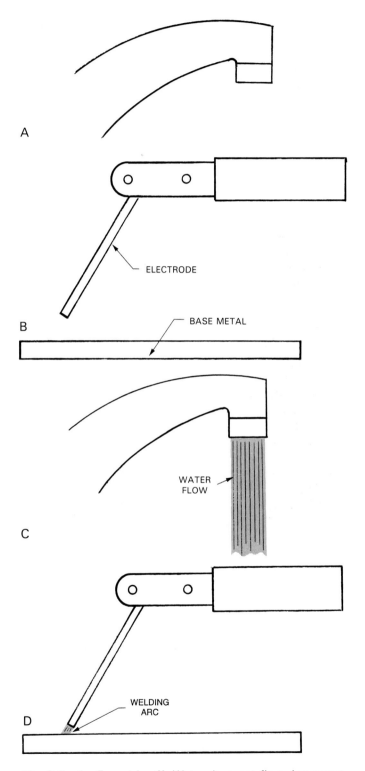

Fig. 2-7. A—Faucet is off. Water does not flow, but water pressure exists in the pipe. B—Welding arc has not been struck. Electrons do not flow, but voltage exists in the circuit. C—Faucet is on. Water flows and the water pressure drops slightly. D—Arc is struck. Electrons flow and voltage drops slightly.

19

Sometimes, when you turn a faucet all the way on, water flows very fast with a great deal of pressure. This is due to high water pressure and a high flow rate. If you turn the faucet half-way off, the same pressure is there, but the flow rate is reduced. At other times when a faucet is turned all the way on, only a small amount of water flows and it does not have much force. This is due to low water pressure and a low flow rate. If you place your thumb over part of the faucet opening, water is delivered with much greater force. The pressure is higher, although the flow rate has not changed. Similar principles apply to voltage and current in an electrical circuit. In some welding applications, a higher voltage and/or current is required. In other applications, a lower voltage and/or current may be required.

In later chapters, instructions will be given on the setup of different types of welding machines. Most machines require that the welder set the voltage and/or the current that will be used.

UNITS OF MEASUREMENT

There are two different measurement systems in common use—the US conventional system and the SI metric system. The *US conventional system* is used primarily in the United States. The *SI metric system* is the standard in much of the rest of the world, and is also used in the United States. Some of the basic units used in the US conventional system are inches, feet, gallons, pounds, and pounds per square inch. Some of the basic units in the SI metric system are millimeters, meters, liters, kilograms, and pascals.

In the US conventional system, the basic unit of measurement is commonly converted and renamed. For example, 12 inches are converted to 1 foot, 3 feet are converted to 1 yard, and 1760 yards are converted to 1 mile. Fig. 2-8 lists common conversions in the US conventional system.

The SI metric system adds a prefix to the basic unit of measurement to increase or decrease the value. For example, there are 1000 millimeters in 1 meter, and 1000 meters in 1 kilometer. In these examples, "meter" is the basic unit of measurement. The prefixes milli- and kilo- are added to it to change the value. See Fig. 2-9 for a list of common prefixes used in the SI metric system.

Conversions between the US conventional system and the SI metric system may need to be made occasionally. Common conversions are listed in Fig. 2-10.

REVIEW QUESTIONS

Write your answers on a separate sheet of paper.
1. A weld can be achieved through what three methods?

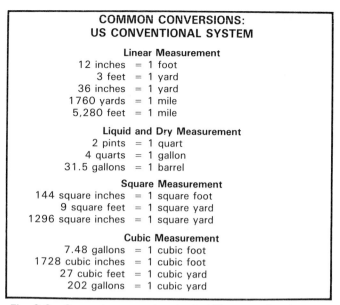

COMMON CONVERSIONS: US CONVENTIONAL SYSTEM	
Linear Measurement	
12 inches	= 1 foot
3 feet	= 1 yard
36 inches	= 1 yard
1760 yards	= 1 mile
5,280 feet	= 1 mile
Liquid and Dry Measurement	
2 pints	= 1 quart
4 quarts	= 1 gallon
31.5 gallons	= 1 barrel
Square Measurement	
144 square inches	= 1 square foot
9 square feet	= 1 square yard
1296 square inches	= 1 square yard
Cubic Measurement	
7.48 gallons	= 1 cubic foot
1728 cubic inches	= 1 cubic foot
27 cubic feet	= 1 cubic yard
202 gallons	= 1 cubic yard

Fig. 2-8. Common conversions in the US conventional system.

COMMON SI METRIC PREFIXES

micro	= 1/1 0000 000 or .000001
milli	= 1/1 000 or .001
centi	= 1/100 or .01
deci	= 1/10 or .1
kilo	= 1 000
mega	= 1 000 000

THE UNIT OF LENGTH IS THE **METER**
Examples of prefixes: millimeter (mm) = .001 or 1/1000 of a meter
centimeter (cm) = .01 or 1/100 of a meter
kilometer (km) = 1000 meters

THE UNIT OF WEIGHT IS THE **GRAM**
Examples of prefixes: milligram (mg) = .001 or 1/1 000 or a gram
centimeter (cm) = 1 000 grams
megagram (Mg) = 1 000 000 grams

THE UNIT OF VOLUME IS THE **LITER**
Examples of prefixes: milliliter (mL) = .001 or 1/1000 of a liter
kiloliter (kL) = 1000 liters

THE UNIT OF TIME IS THE **SECOND**
Examples of prefixes: millisecond = .001 or 1/1 000 sec.
microsecond = .000001 or 1/1 000 000 sec.

Fig. 2-9. Common prefixes used in the SI metric system. Examples of basic units of measurement and prefixes are also shown.

2. Briefly define the following terms.
   a. tensile strength
   b. ductility
   c. toughness
   d. hardness
3. Welding changes the properties of the metal being welded. True or False?
4. What happens to the size of base metal when it is heated?

5. What can happen when welding a T-joint that is not tack welded or clamped in a welding jig or fixture?
6. List four ways to reduce stress in a weld.
7. What type of heat treatment process is used to harden steel and steel alloys?
8. Define the following terms.
   a. voltage
   b. current
   c. resistance
9. The two measurement systems in common use are the _____ system and the _____ system.
10. List the unit from the other measurement system that corresponds to the given unit of measurement.

Length	inch	_____
Mass	pound	_____
Temperature	degree Celsius	_____
Volume	liter	_____
Tensile strength	pounds per square inch (psi)	_____

PROPERTY	TO CONVERT FROM	TO	MULTIPLY BY
AREA	in.²	mm²	645.16
	in.²	m²	.00064516
	ft.²	mm²	92903
	ft.²	m²	.092903
CURRENT DENSITY	A/in.²	A/mm²	.00155
	A/mm²	A/in.²	645.16
DEPOSITION RATE	lb./h	kg/h	.045
	kg/h	lb./h	2.2
ELECTRODE FORCE	lb. (force)	N	4.4482
	kg (force)	N	9.8067
	N	lb. (force)	.22481
FLOW RATE	cfh	Lpm	.47195
	gallons/h	Lpm	.06309
	gallons/min.	Lpm	3.7854
	Lpm	cfh	2.1188
HEAT INPUT	J/in.	J/m	39.37
	J/m	J/in.	.0254
LENGTH	in.	mm	25.4
	in.	m	.0254
	ft.	mm	304.8
	ft.	m	.3048
	mm	in.	.03937
	mm	ft.	.0032808
MASS	lb.	kg	.45359
PRESSURE (GAS AND LIQUID)	psi	Pa	6894.8
	psi	kPa	6.8948
	N/mm²	Pa	1,000,000
	kPa	psi	.14504
	kPa	lb./ft²	20.885
	kPa	N/mm²	.0001
TENSILE STRENGTH	psi	kPa	6.8948
	N/mm²	MPa	1.000
	MPa	psi	145.04
TORQUE	in.•lb.	N•m	.11298
	in.•lb.	N•m	1.3558
TRAVEL SPEED	in./min.	mm/s	.42333
	mm/s	in./min.	2.3622
VOLUME	in.³	mm³	16387
	in.³	m³	.000016387
	ft.³	mm³	28316850
	ft.³	m³	.028317
	in.³	L	.016387
	ft.³	L	28.317
	gallons	L	3.7854

To convert from °F to °C, use °C = (°F − 32) ÷ 1.8.
To convert from °C to °F, use °F = (1.8 × °C) + 32.

Fig. 2-10. Common conversions for US conventional and SI metric systems. To convert from units in Column 3 to Column 2, divide by the value in Column 4.

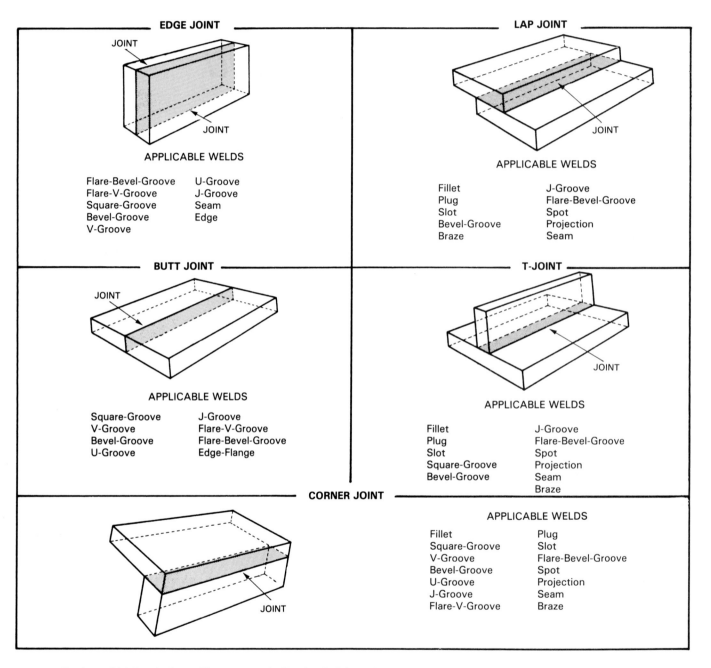

Basic weld joint designs. There are only five basic joints, but many types of welds for each joint. (AWS)

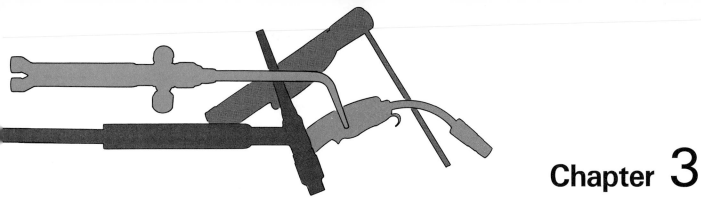

Chapter 3

Weld Joints and Positions

After studying this chapter, you will be able to:
☐ Identify the five basic weld joints.
☐ Identify the types of welds that can be made on each joint.
☐ Identify the parts of a fillet weld.
☐ Identify the parts of a groove weld.
☐ Describe a stringer bead and a weaving bead.
☐ List four welding positions.
☐ State the conditions for welding in the four welding positions.

TECHNICAL TERMS

Bead, bevel angle, butt joint, corner joint, cover pass, edge joint, edge preparation, effective throat, face of the weld, face reinforcement, filler material, filler passes, fillet weld, fillet, fillet weld size, flange joint, flare-groove joint, flat (1G) position, groove angle, groove face, groove weld, horizontal (2G) position, inside corner joint, joint geometry, joint penetration, lap joint, leg, multiple-pass weld, outside corner joint, overhead (4G) position, pass, penetration, root face, root of the weld, root pass, root reinforcement, stringer bead, tack weld, T-joint, toe of the weld, vertical (3G) position, weaving bead, weld axis, weld bead, weld face, weld size, welding positions.

BASIC WELD JOINTS

A weld joint refers to how the parts to be joined are assembled prior to welding. There are five basic types of joints used in welding: butt, lap, corner, T-, and edge joint. See Fig. 3-1.

The metal to be joined is the base metal. It is also known as the workpiece or work. The edges of the base

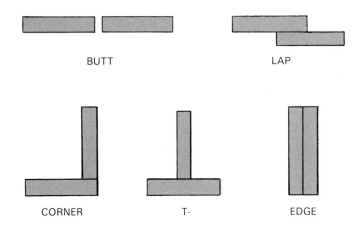

Fig. 3-1. The five basic weld joints.

metal are often machined, sheared, gouged, flame cut, or bent to prepare them for welding. Weld joint design and metal thickness usually determine how the joint is prepared. Generally, the weld joint design is determined by an engineer.

BUTT JOINT

Butt joints are used when parts are joined end-to-end, as in a pipeline or a ship's deck plates. Some type of groove configuration is specified on the ends of the pieces being joined. The edges of the base metal may require preparation before welding. *Edge preparation* refers to how the edges of the joint are shaped prior to welding. If the base metal is thin, the edges may just be squared without additional machining or cutting. The edges of thin metal may also be bent to form flare-groove or edge-flange joints, as shown in Fig. 3-2.

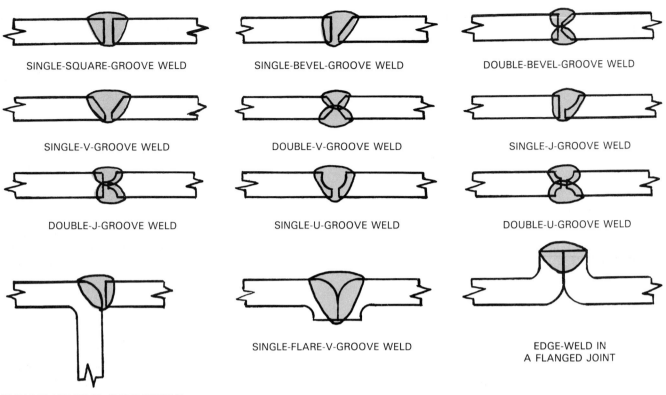

SINGLE-SQUARE-GROOVE WELD

SINGLE-BEVEL-GROOVE WELD

DOUBLE-BEVEL-GROOVE WELD

SINGLE-V-GROOVE WELD

DOUBLE-V-GROOVE WELD

SINGLE-J-GROOVE WELD

DOUBLE-J-GROOVE WELD

SINGLE-U-GROOVE WELD

DOUBLE-U-GROOVE WELD

SINGLE-FLARE-V-GROOVE WELD

EDGE-WELD IN
A FLANGED JOINT

SINGLE-FLARE-BEVEL-GROOVE WELD
IN A FLANGED JOINT

Fig. 3-2. Various methods of preparing the edges of a butt joint. The completed weld is shown in red. Double grooves are used on thick metal that is welded from both sides. The base metal is bent to form the bottom three joints.

Generally, when base metal over 3/16″ (4.8 mm) thick is used, edges are beveled by machining or flame cutting. Edge preparation is required to allow the weld to penetrate to the required depth. Thick base metal may be machined, gouged, or flame cut along the upper or lower edges of the joint, or both, to form a double-bevel, V-, J-, or U-groove. A butt joint may be prepared using any of the edge preparations shown in Fig. 3-2.

A welder should know the names of the various parts of a groove joint, as shown in Fig. 3-3A. The *groove face* is the surface formed on the edge of the base metal after it has been machined or flame cut. The total angle formed between the groove face on one piece and the groove face on the other piece is the *groove angle*. The *bevel angle* is the angle from the root face to the groove face on one piece. The *root of the weld* is the bottom edge of the base metal. The distance from the root of the weld to the point where the bevel angle begins is the *root face*. A root opening is the distance between the two pieces at the root of the weld.

Fig. 3-3B shows a cross section of a completed weld. The *face of the weld* is the outer surface of the weld bead on the side of the weld. *Face reinforcement* is the distance from the top of the weld face to the surface of the base metal. The *toe of the weld* is the point where the weld bead contacts the base metal surface. It occurs twice on each weld bead. *Root reinforcement* is

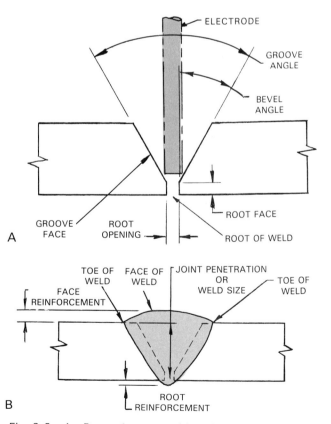

Fig. 3-3. A—Parts of a grooved butt joint. The groove angle should be just large enough to allow the torch or electrode to reach the root opening. B—Parts of a groove weld.

24

the distance that the penetration projects from the root side of the joint. *Joint penetration* or *weld size* is the depth that a weld extends into the joint from the surface.

LAP JOINT

A *lap joint* is formed by two overlapping pieces of base metal. The top surface of one piece is in contact with the bottom surface of the other, as shown in Fig. 3-1. Special edge preparation is not required. However, the edges of the pieces may be prepared to form a bevel-groove, J-groove, or flare-bevel-groove joint.

CORNER JOINT

A *corner joint* is formed by placing one piece of base metal along the outer edge of another piece. As shown in Fig. 3-4, at least one edge of the two pieces is exposed. The pieces may be joined at any angle, but are commonly welded at a 90° angle. Corner joints may be welded as inside corners, outside corners, or a combination of both. *Inside corner joints* are welded along the inside of the intersection of the two pieces. *Outside corner joints* are welded along the outside edge of the joint. The edges may be squared, beveled, grooved, flared, or edge-flanged. See Fig. 3-4.

T-JOINT

A *T-joint* is formed by two pieces of base metal that are at an angle of approximately 90° to one another. The main difference between a corner and T-joint is that a corner joint is formed along the edge of one piece, while a T-joint is formed anywhere but along the edge. The edges of the base metal may be prepared as a square, beveled, grooved, or flared-bevel-groove joint, as shown in Fig. 3-5. Both edges of the base metal may be prepared to form a double-bevel-groove joint.

EDGE JOINT

An *edge joint* is formed when the surfaces of two pieces are in contact and their edges are flush (even). The pieces are joined by welding along at least one of the flush edges. Fig. 3-6 shows the edge preparation for an edge joint.

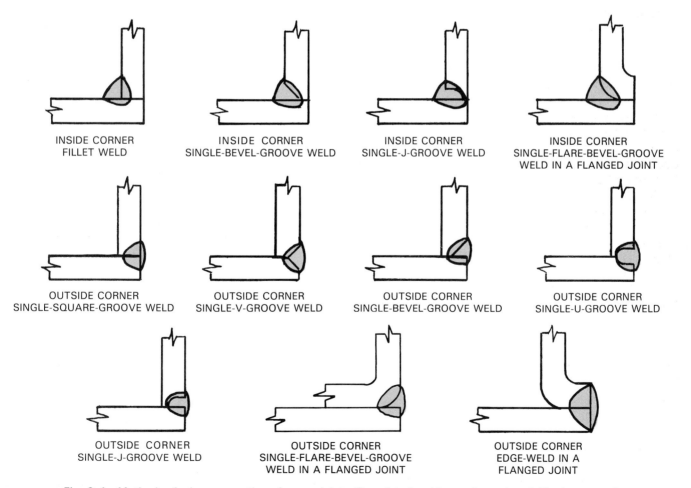

INSIDE CORNER
FILLET WELD

INSIDE CORNER
SINGLE-BEVEL-GROOVE WELD

INSIDE CORNER
SINGLE-J-GROOVE WELD

INSIDE CORNER
SINGLE-FLARE-BEVEL-GROOVE
WELD IN A FLANGED JOINT

OUTSIDE CORNER
SINGLE-SQUARE-GROOVE WELD

OUTSIDE CORNER
SINGLE-V-GROOVE WELD

OUTSIDE CORNER
SINGLE-BEVEL-GROOVE WELD

OUTSIDE CORNER
SINGLE-U-GROOVE WELD

OUTSIDE CORNER
SINGLE-J-GROOVE WELD

OUTSIDE CORNER
SINGLE-FLARE-BEVEL-GROOVE
WELD IN A FLANGED JOINT

OUTSIDE CORNER
EDGE-WELD IN A
FLANGED JOINT

Fig. 3-4. Methods of edge preparation of corner joints. Completed welds are shown in red. The base metal is bent to form three of these joints.

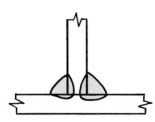

SQUARE-GROOVE WITH
DOUBLE FILLET WELDS

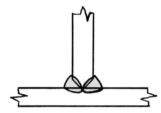

DOUBLE-BEVEL-GROOVE WELD

SINGLE-BEVEL-GROOVE WELD

SINGLE-J-GROOVE WELD

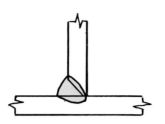

FLARE-BEVEL-GROOVE WELD
IN A FLANGED JOINT

Fig. 3-5. Methods of edge preparation of T-joints. Completed welds are shown in red.

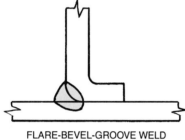

SQUARE-GROOVE WELD   BEVEL-GROOVE WELD

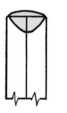

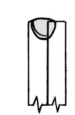

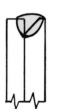

V-GROOVE WELD   J-GROOVE WELD   U-GROOVE WELD

EDGE-WELD IN A
FLANGED BUTT JOINT

Fig. 3-6. Methods of preparing base metal edges for edge-type joints. Completed welds are shown in red.

Flange joint

A *flange joint* is formed when the edge of one or more pieces of the joint is bent to form a flange. The flanged or unflanged edges are aligned and a weld is placed along the specified edges. Figs. 3-2, 3-4, and 3-6 show flange joints.

Flare-groove-joint

Flare-groove joints are formed when the flanged edges of one or both pieces are placed together to form a single-flare-bevel or double-flare-V-groove. The weld is placed in the bevel or V-groove, as shown in Figs. 3-2, 3-4, and 3-5.

TYPES OF WELDS

A weld is defined as "the blending or mixing of two or more metals or nonmetals by heating them until they are molten and flow together. This may be done with or without the addition of a filler material." See Fig. 3-7. Welding is the process of making a weld on a joint. *Fillet* welds are placed into lap, inside corner, and T-joints. Groove welds can be used on all types of weld joints.

When the edges of thicker metal are machined or flame cut, metal is removed from the pieces. *Filler material* must be added to replace the metal that is removed. The addition of filler metal ensures that the completed weld joint is as thick and as strong as the base metal. Edge, flange, or flare-groove joints for thin

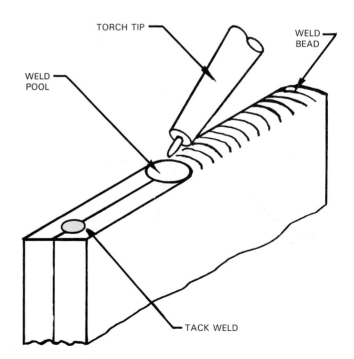

Fig. 3-7. A square-groove weld for an edge joint in progress. The weld pool (puddle) extends to the outer edges of the base metal. Filler metal may not be required on thin pieces of base metal.

metal may be welded without the addition of filler material. Figs. 3-2 through 3-6 show edge, flange, and flare-groove joints.

The parts and dimensions for fillet welds are the same for lap, inside corner, and T-joints. Refer to Fig. 3-8. The *weld face* is the outer surface of the weld bead. The *toe of the weld* is the point where the weld face touches the surface of the base metal. A fillet weld is made up of three primary dimensions. The *fillet weld size* is the length of one side. The *leg* is the shortest distance from the toe to the surface of the other piece of base metal. The *effective throat* is the minimum distance from the weld face to the root of the weld without any convexity.

Fig. 3-8 shows two fillet welds with the same leg dimensions, but different sizes. The size of the weld with a concave bead, Fig. 3-8A, is smaller than the size of the weld with a convex bead, Fig. 3-8B. A fillet weld with a convex bead is stronger than one with a concave bead because of the additional filler metal.

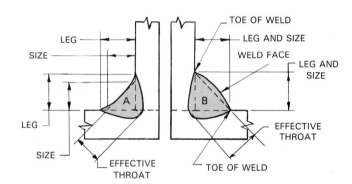

Fig. 3-8. Parts of a fillet weld. Weld A is concave. The weld size is smaller than Weld B, which is a convex weld. Notice that the leg sizes in Weld A and B are the same, but the weld size is larger with a straight or slightly convex bead.

WELD BEADS AND PASSES

A *weld bead* is one thickness of filler metal that is added to a weld joint. A *pass* occurs each time a welder lays one bead across a weld joint. One weld bead or pass is required for thin base metal, Fig. 3-9A. If the metal is thick, more than one pass is required to make a strong joint with complete penetration. See Fig. 3-9B. A *multiple-pass weld* is sometimes required, Fig. 3-10. The first pass is the *root pass.* Intermediate passes are *filler passes.* The final pass is the *cover pass.* Generally, a weld bead should not be thicker than 1/4″ (6.4 mm). A bead may be made as a stringer bead or a weaving bead.

STRINGER BEAD

A *stringer bead* is used when a standard bead width is acceptable. Stringer beads are made by moving the

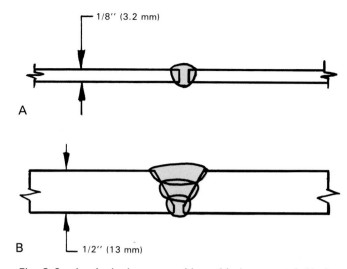

Fig. 3-9. A—A single-pass weld on thin base metal. Notice the build-up of weld metal and complete penetration. B—A multiple-pass weld on thick base metal. The edges have been prepared to form a V-groove joint. Notice the root opening required. Three beads were used with each bead measuring less than 1/4″ (6.4 mm) thick. A weaving bead may be used for the wider, upper bead.

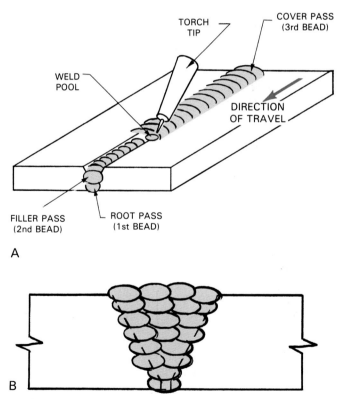

Fig. 3-10. A multiple-pass weld. A—Three weld passes are used in this example. Any number of passes may be used. B—Twenty passes were made in this weld.

torch or electrode holder along the weld without any side-to-side motion. See Fig. 3-11A. A deep and wide joint can be filled using several stringer beads.

WEAVING BEAD

A *weaving bead* is used to create a wider weld pool. A weaving bead is formed by moving the torch or electrode holder from side-to-side as the pass progresses along the weld joint. See Fig. 3-11B.

Various torch or electrode movement patterns can be used when making a weaving bead. The crescent motion shown in Fig. 3-11C, is one of the most popular patterns.

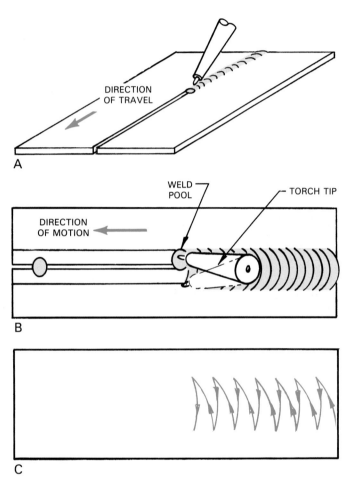

Fig. 3-11. A—A stringer bead in progress on a square-groove butt joint. Bead width is about 2-3 times the metal thickness. B—A weaving bead in progress. As the weld moves in the direction of the arrow, the torch tip and weld pool are moved from side to side. C—Suggested motion for a weaving bead. The bead width is seldom greater than 3/4"-1" (19 mm-25 mm).

JOINT GEOMETRY

The American Welding Society defines *joint geometry* as, "the shape and dimensions of a (weld) joint, in cross section, prior to welding." Joint geometry is generally determined by a welding engineer or designer. The assembly design and the dimensions of a joint depend on the metal thickness and shape, and on the load requirements of the parts. The parts are prepared to ensure that the weld will have adequate penetration. The joint geometry design also provides space for the welder to reach near the root of the weld with the torch or electrode.

PREPARATION

The edges of thick metal are prepared for welding by flame cutting, gouging, or machining. Preparation allows the weld to penetrate as deep as required by the engineer or weld designer. A groove joint allows the welder to reach the root (bottom) of the weld. The groove angle must be large enough to allow the torch tip or electrode to reach near the root of the weld. If the groove angle is too large, filler metal and the welder's time are wasted. This increases the cost of making a weld. See Fig. 3-12A. A properly designed J-groove or U-groove joint also decreases the groove dimensions while allowing adequate space for welding. See Fig. 3-12B.

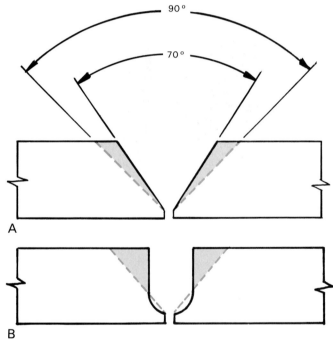

Fig. 3-12. A—A 70° and 90° groove angle comparison. A 70° groove angle is cheaper to weld than a 90° groove angle. The shaded area represents an unnecessary cost in filler metal and welder time. B—A U-groove joint. The root of the weld can be reached easily. Little filler metal and welder time are wasted.

JOINT ALIGNMENT

The alignment of a joint before welding is very important. In the shop, the alignment of the weld joint is often referred to as "fit-up." A ragged edge or an edge that is not cut straight is hard to weld. See Fig. 3-13. Edges to be welded must be straight and cut to exact size.

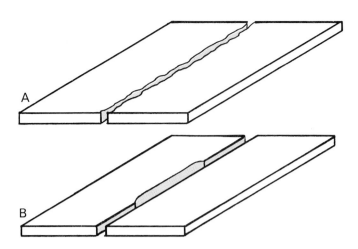

Fig. 3-13. Poorly prepared base metal edges. The edges in A are ragged. One edge in B is not cut straight, which changes the width of the joint. Both joints will be difficult to weld.

Parts of a weldment should be properly aligned and held in position during the welding operation. Tack welding is usually adequate to hold parts during welding. A *tack weld* is a small weld used to hold pieces in alignment. Parts may also be held mechanically during the welding operation because the metal expands, bends, and changes shape when heated. Clamps or other devices, such as jigs and fixtures, are used to hold weldments when welding. See Fig. 3-14.

PENETRATION

A completed weld joint must be at least as strong as the base metal. The weld must penetrate deeply into the base metal to be strong. *Penetration* is the depth of fusion of the weld below the surface. Total (100%) penetration occurs when a weld penetrates through the

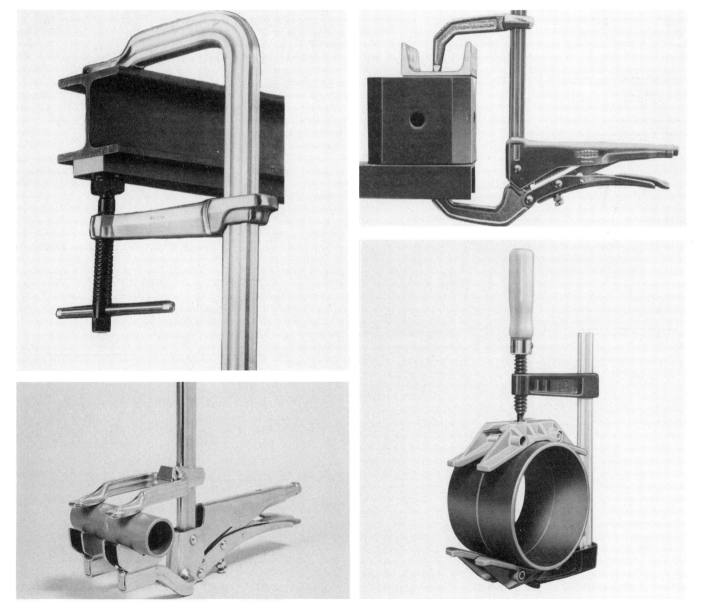

Fig. 3-14. Clamps used to position and hold parts to be welded.   (James Morton, Inc.)

entire thickness of the base metal. Generally, total penetration is required only on a butt joint. The edges of thick metal may need to be machined or flame cut to achieve 100% penetration. Thick metal also may have to be welded from both sides of the joint.

WELDING POSITIONS

Welders often must weld in a variety of positions. Welds may be made in the flat, horizontal, vertical, or overhead welding position. On welding drawings, these positions are often abbreviated at the end of the welding symbol as F, H, V, and O. The American Welding Society refers to *welding positions* with a number and letter combination. Groove joints in the flat, horizontal, vertical and overhead positions are referred to as 1G, 2G, 3G, and 4G, respectively. Fillet joints in the flat, horizontal, vertical, and overhead position are designated as 1F, 2F, 3F, and 4F, respectively.

Welding positions are determined by the positions of the weld axis and weld face. Fig. 3-15 shows the weld axis and weld face. The *weld axis* is an imaginary line running lengthwise through the center of a completed weld. The weld face is the exposed surface of a completed weld on the side which welding is done.

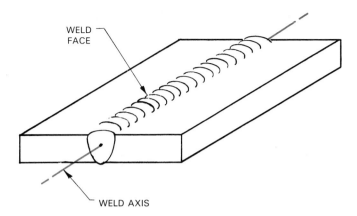

Fig. 3-15. Weld axis and weld face. The weld axis is an imaginary line running lengthwise through the center of the weld. The weld face is the exposed surface of the finished bead.

FLAT (1G) WELDING POSITION

Welds made on a groove joint in the *flat (1G) welding position* must meet these conditions:
1. The weld axis must be within 15° of horizontal. See Fig. 3-16.
2. The weld face is within 30° of horizontal.
3. The weld is made from the upper side of the joint.

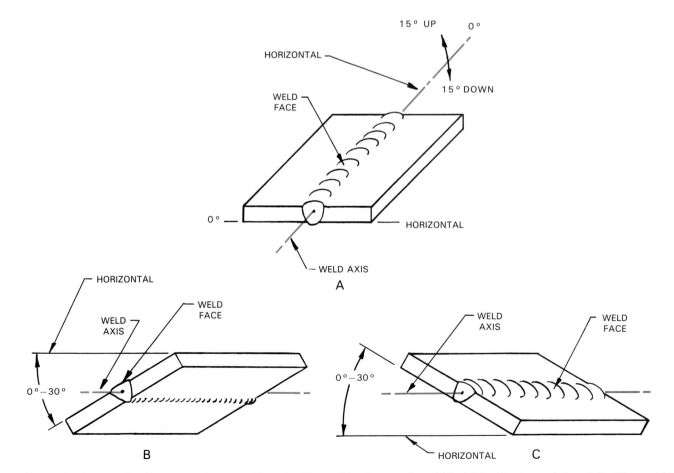

Fig. 3-16. The AWS 1G or flat welding position. A—The weld axis must be within 15° of horizontal. B and C—The weld face just be within 30° of horizontal.

HORIZONTAL (2G) WELDING POSITION

Groove welds made in the *horizontal (2G) welding position* must meet these conditions:
1. The weld axis must be within 15° of horizontal.
2. The weld face must be between 80°−150° or 210°−280°. See Fig. 3-17. Angles are measured clockwise with 0° at the bottom.

VERTICAL (3G) WELDING POSITION

A weld on a groove joint in the *vertical (3G) welding position* must meet either of these sets of conditions:

Condition A
1. The weld axis is between 80°−90° from horizontal.
2. The weld face is between 0°−360° from horizontal. See Fig. 3-18A.

Condition B
1. The weld axis is between 15°−80° from horizontal.
2. The weld face is between 80°−280° from horizontal.
3. The weld is made from the upper side of the joint. See Fig. 3-18B.

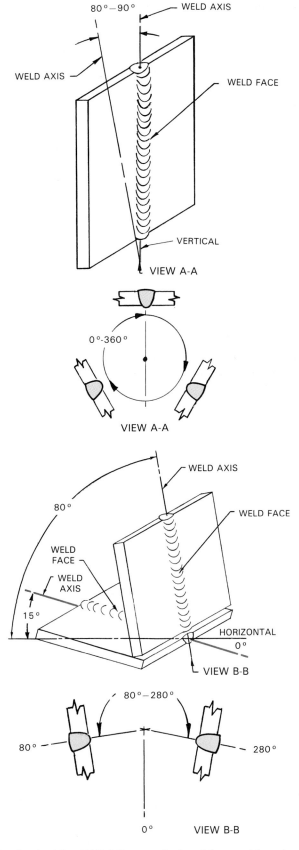

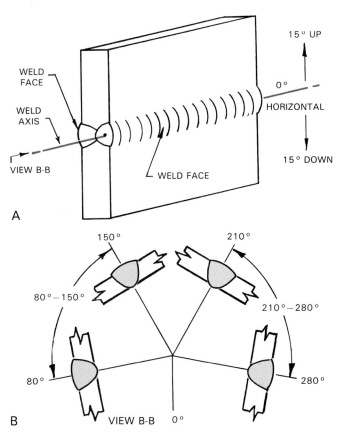

Fig. 3-17. The AWS 2G or horizontal welding position. A— The weld axis must be within 15° of horizontal. View B-B— An end view with the weld shown in red. The weld face must be within 80°−150° or 210°−280°. All angles are measured clockwise with 0° at the bottom.

Fig. 3-18. The AWS 3G or vertical welding position. A ver- tical weld must meet either of the following sets of conditions. A—The weld axis must be between 80° and 90° from horizon- tal. View A-A—The weld face may be rotated from 0°-360°. B—The weld axis is between 15°-80° from horizontal. View B-B—The weld face must be between 80°-280°.

OVERHEAD (4G) WELDING POSITION

Welds made on groove joints in the *overhead (4G) welding position* are made under these conditions:
1. The weld axis is between 0° − 80°.
2. The weld face is between 0° − 80° or 280° − 360°.
3. The weld is made from the lower side of the joint. See Fig. 3-19.

REVIEW QUESTIONS

Write your answers on a separate sheet of paper.
1. List the five basic weld joints.
2. The metal to be welded is the _____ metal.
3. The metal added to a weld joint is the _____ metal.
4. Which joint is always welded without any special edge preparation?
5. Sketch the end view of a groove joint and label the bevel angle and groove angle.
6. _____, _____, and _____ joints are commonly welded without using filler material.
7. What type of weld bead is formed by moving the torch or electrode from side to side as the weld progresses?
8. Name three methods of preparing the edges of metal for welding.
9. List two reasons why it costs more to make a weld if the groove angle is too large.
10. State the conditions of the AWS 2G position.

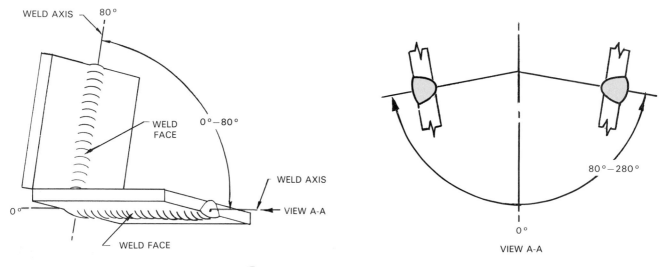

Fig. 3-19. The AWS 4G or overhead welding position. The weld axis must be between 0° and 80°. View A-A—The weld face must be between 80° and 280°. The weld is made from the lower side of the base metal.

OXYFUEL GAS PROCESSES

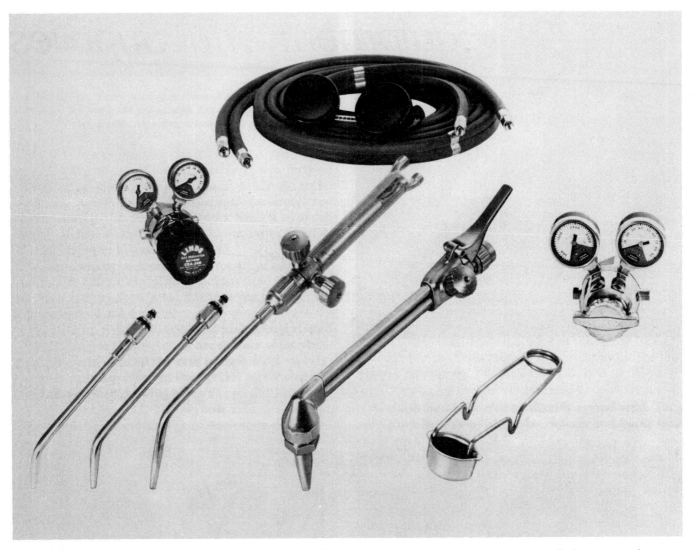

Some of the equipment that makes up an oxyfuel gas welding and cutting outfit. Oxygen and fuel gas cylinders are not shown.
(Linde Div., Union Carbide Corp.)

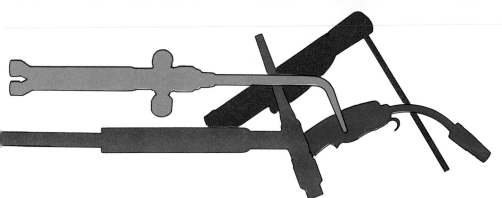

Chapter 4

Oxyfuel Gas Welding Equipment and Supplies

After studying this chapter, you will be able to:
- ☐ Identify the parts of an oxyfuel gas welding outfit.
- ☐ Describe the function of each of the parts.
- ☐ List the steps required to assemble an oxyfuel gas welding outfit.
- ☐ Identify the safety features of an oxyacetylene welding outfit.
- ☐ Describe protective clothing used for oxyacetylene welding.
- ☐ List safety precautions that must be taken when performing oxyfuel gas welding.

TECHNICAL TERMS

Acetone, acetylene, acetylene generator, backfire, check valve, cover plates, cylinder, cylinder pressure gauge, cylinder valve, Dewar flask, distill, filter lenses, fitting, flashback, flashback arrestor, fuel gases, fusible plug, hand truck, hose, infrared rays, injector-type torch, leathers, liquifaction, manifold, mixing chamber, orifice, oxyfuel gas welding, oxygen, positive pressure torch, pressure regulators, regulator adjusting screw, safety cap, safety valve, single-stage regulator, torch, torch tip, torch valves, two-stage regulator, ultraviolet rays, welding outfit, welding station, welding rod, working pressure gauge.

OXYFUEL GAS WELDING

Oxyfuel gas welding is a general term for a group of welding and cutting processes (methods) that use heat produced by a gas flame. Oxyacetylene welding and oxyhydrogen welding are types of oxyfuel gas welding processes. See Fig. 4-1.

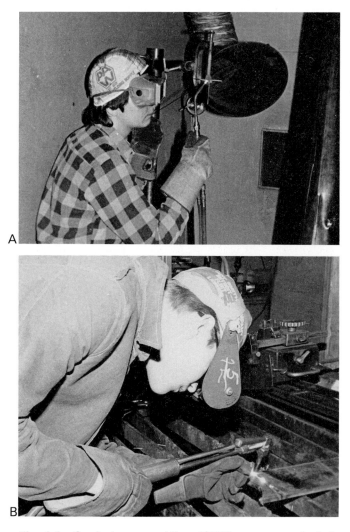

Fig. 4-1. Oxyfuel gas welding (OFW) processes include welding and cutting operations. A—Making an oxyacetylene weld in the overhead position. B—Using an oxyacetylene cutting torch to cut 3/8'' (9.5 mm) steel plate.

Fuel gases are those that will support combustion (burn) when combined with oxygen. They include acetylene, propane, butane, hydrogen, city gas, natural gas, and MPS (methylacetylene-propadiene) gas. Acetylene is commonly used for welding and cutting. When combined with oxygen in a neutral flame, it produces temperatures around 5600°F (3093°C). This is the highest temperature produced by any combination of oxygen and fuel gas. Propane, butane, city gas, and natural gas do not produce enough heat for welding. However, they may be used for soldering and brazing.

An oxyfuel gas *welding outfit*, Fig. 4.2, consists of the following pieces of equipment:

• Fuel gas cylinder.
• Oxygen cylinder.
• Means of securing the cylinders in an upright position.
• Fuel gas regulator and gauges.
• Oxygen regulator and gauges.
• Fuel gas hose and fittings.
• Oxygen hose and fittings.
• Flashback arrestors.
• Check valves.
• Welding torch.

A complete *welding station* includes the welding outfit, a welding table, means of ventilation, goggles, and a sparklighter.

ACETYLENE

Acetylene is a colorless fuel gas with a distinct garlic-like odor. It is produced by adding calcium carbide to water.

An *acetylene generator* is used to produce acetylene in a controlled environment. See Fig. 4-3. Calcium carbide is placed in a hopper and fed into the water at a controlled rate. The chemical combination produces acetylene (C_2H_2).

A low-pressure acetylene generator produces the gas at a gauge pressure of .25 psig (pounds per square inch gauge) (1.72 kPa). A medium-pressure generator produces acetylene at a gauge pressure of 15 psig (103.42 kPa). Pure acetylene is very unstable at a gauge pressure above 15 psig (103.42 kPa). It may explode or burn rapidly.

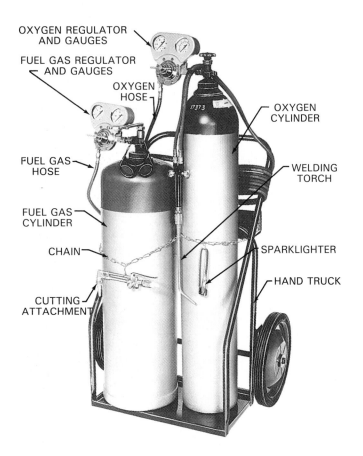

Fig. 4-2. An oxyfuel gas welding outfit. The pressure gauges shown have protective shields around them. Notice the cutting torch attachment and the sparklighter hanging from the security chain. (Modern Engineering Co., Inc.)

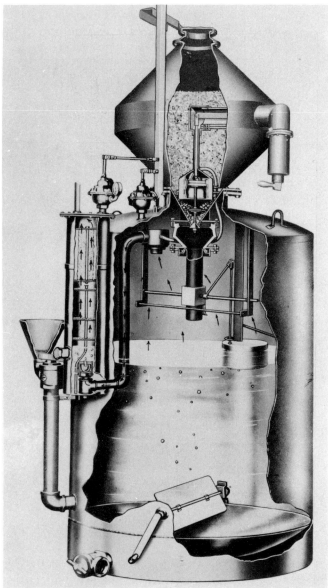

Fig. 4-3. Cut-away view of an acetylene generator. (Rexarc, Inc.)

ACETYLENE CYLINDER

Acetylene can be stored safely as a gas at pressures above 15 psig (103.42 kPa) if it is kept from collecting in large volume. An acetylene *cylinder* is a portable container used to store acetylene. Cylinders are available in sizes to meet the needs of most users, Fig. 4-4.

Fig. 4-5. Cut-away view of an acetylene cylinder. Notice how the porous material completely fills the cylinder.

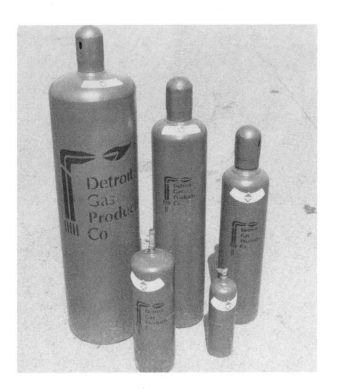

Fig. 4-4. Acetylene cylinders are available in a variety of sizes.

An acetylene storage cylinder is constructed of steel and filled with a porous material. The porous material has many small air pockets. Fig. 4-5 shows a cut-away view of an acetylene cylinder. The cylinder is filled with *acetone,* a colorless and extremely flammable liquid. The acetone fills the air pockets of the porous material. Acetylene is then pumped into the cylinder and absorbed into the acetone. The porous material and acetone prevent the acetylene from collecting in large volume. Acetylene bubbles out of the acetone when the cylinder valve is opened. Acetylene can be safely stored at pressures up to 250 psig (1724 kPa) using this method.

A *safety cap* should be screwed over the cylinder valve when a cylinder is stored or moved. Cylinders should be moved by using a properly designed hand truck or by rolling them along the bottom edge. A *hand truck* is a two-wheel cart with a chain or strap for securing cylinders to it, Fig. 4-6. A cylinder may also be moved by tilting it with one hand on the safety cap. The other hand is used to roll it in the direction of travel. Fig. 4-7 shows this method of moving a cylinder. Cylinders filled with gases under pressure must be

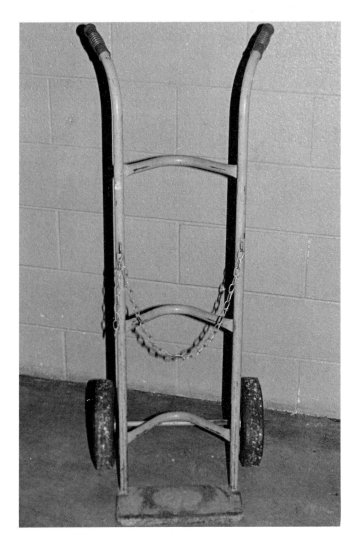

Fig. 4-6. This specially designed hand truck is used to move oxygen, acetylene, or other gas cylinders. The safety chain secures the cylinder in an upright position.

Fig. 4-7. A cylinder being moved by tilting it with one hand and rolling it into position with the other hand. Notice that the safety cap is in place.

stored in an upright position. They should be fastened to a wall, column, or hand truck with chains or steel straps. Warning: The cylinder valve may break and leak if the cylinder falls.

Fusible plugs

Fusible plugs prevent acetylene cylinders from exploding in a fire. A *fusible plug* is a steel plug filled with a metal that melts at about 212°F (100°C). The plug is externally threaded and is screwed into an opening at the top or bottom of the acetylene cylinder, Fig. 4-8. In case of a fire, the center of the fusible plug melts. This allows the acetylene to escape slowly. Although the gas escapes and burns, the tank will not explode.

Acetylene cylinder valve

An acetylene *cylinder valve* controls the flow of gas from the cylinder. A handwheel or a cylinder valve wrench is used to open and close the valve. See Fig. 4-9. Warning: The handwheel or valve wrench should always remain on the cylinder valve when the cylinder is in use. An acetylene cylinder valve is usually opened only 1/4-1/2 turn for sufficient gas flow. This allows it to be closed quickly in case of emergency.

Fig. 4-8. Fusible plugs on an acetylene cylinder. The valve is depressed and has internal threads. This type of cylinder is known as a POL (Prestolite) cylinder.

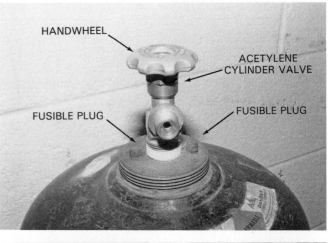

Fig. 4-9. A—Handwheel used to regulate the flow of acetylene. This type of cylinder is known as a commercial cylinder. It has an externally threaded valve and two fusible plugs. B—Cylinder valve wrench used to control the flow of acetylene. The handwheel or valve wrench should be on the valve when the cylinder is in use.

ACETYLENE SAFETY PRECAUTIONS

Acetylene can be very explosive if not handled safely. Always refer to the manufacturers' instructions when handling or using equipment. Also, you should be familiar with the following safety standards:

- American National Standards Institute (ANSI) Z49.1, "Safety in Welding and Cutting."
- National Fire Protection Association (NFPA) No. 51, "Oxygen-Fuel Gas Systems for Welding and Cutting."
- Linde Form 2035, "Precautions and Safety Practices in Welding and Cutting with Oxyacetylene Equipment."

1. Acetylene gas is unstable at pressures above 15 psig (103.42 kPa).
2. Avoid contact between acetylene and copper, silver, or mercury. Also avoid contact with their salts, compounds, or high concentrations of these metals. Under certain conditions, acetylene and these metals can form explosive compounds.
3. Fusible plugs melt at approximately 212°F (100°C).
4. Concentrations of acetylene between 2.5% and 80% by volume in air ignite easily and may cause an explosion.
5. Acetylene has a garlic-like odor. Acetylene may displace air in a poorly ventilated space. Adequate ventilation is essential. An area must contain at least 18% oxygen to prevent dizziness, unconsciousness, or possibly death.
6. Smoking, open flames, unapproved electrical equipment, or other ignition sources are not permitted in acetylene storage areas.
7. Do not place cylinders beneath overhead welding or cutting operations. Hot slag (metal) may fall on them and melt a fusible plug.
8. Keep the handwheel or valve wrench on the cylinder valve while the cylinder is in use.
9. Avoid contact between the torch flame and cylinder.
10. Secure all cylinders in an upright position.
11. Do not use safety caps for lifting cylinders.
12. Keep cylinder valves covered with safety caps when the cylinder is not in use.
13. Do not force cylinder valves open.
14. Do not use any cylinder that does not have a label or that has an illegible label. Never assume a cylinder contains a particular gas. Return the cylinder to the supplier.
15. Do not use a cylinder with a leaking valve. Return it immediately to the supplier.

OXYGEN

Oxygen is a colorless, odorless gas contained in the earth's atmosphere. The atmosphere consists of 78% nitrogen, 21% oxygen, and 1% other gases, such as argon and helium. When oxygen is added to a fuel gas

flame, an increased flame temperature and rate of combustion results.

Oxygen is produced for the welding industry using the liquefaction process. *Liquefaction* is a process that liquefies air and then separates the gases in it at their various boiling points. The process requires large, expensive equipment.

Air becomes a liquid at a very low temperature. When air is liquefied, the gases *distill* (boil away) at different temperatures. Nitrogen is distilled at −320°F (−196°C), then oxygen is distilled at −297°F (−183°C). After oxygen is distilled, it is cleaned and all traces of water are removed. After distilling, oxygen is a gas.

Oxygen is stored as a gas or a liquid. Gaseous oxygen is stored in strong cylinders at approximately 2200 psig (15.168 MPa). Liquid oxygen is stored in a Dewar flask. A *Dewar flask* is a pressurized container with an insulated double wall.

OXYGEN CYLINDER

An *oxygen cylinder* is a seamless, portable container used to store oxygen. See Fig. 4-10. Oxygen is stored in cylinders at a pressure of about 2200 psig (15.168

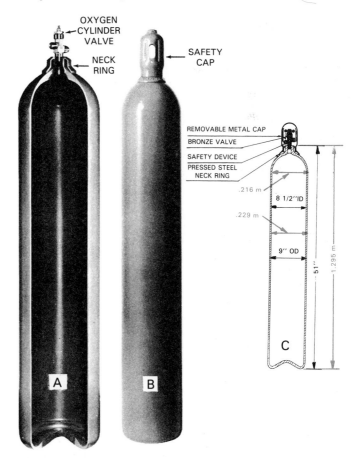

Fig. 4-10. Typical oxygen cylinder with a 244 ft.³ (6909 L) capacity. A—Internal construction of cylinder. Note the one-piece construction. B—Exterior of the cylinder. Note the safety cap over the cylinder valve. C—Dimensions of a 244 ft.³ (6909 L) cylinder. (Pressed Steel Tank Co.)

MPa). The Interstate Commerce Commission (ICC) regulates the construction of oxygen cylinders. Oxygen cylinders are made of forged steel. The minimum thickness of a cylinder is 1/4″ (6.4 mm). Fig. 4-10 shows the typical construction of an oxygen cylinder.

Oxygen cylinders are available in a variety of sizes to meet the needs of the users. The sizes differ slightly according to the manufacturer. Typical sizes are 20 ft.³ (566 L), 55 ft.³ (1557 L), 80 ft.³ (2265 L), 122 ft.³ (3455 L), 220 ft.³ (6230 L), 244 ft.³ (6909 L), and 330 ft.³ (9345 L). Fig. 4-11 shows three common sizes of oxygen cylinders.

Fig. 4-11. Three common sizes of oxygen cylinders.

An oxygen cylinder should be transported with a properly designed hand truck. A cylinder may also be moved by tilting it and rolling it along the bottom edge. An oxygen cylinder must be stored in an upright position. It should be secured to a wall, column, or hand truck with chains or straps. Warning: The cylinder valve may break and leak if the cylinder falls.

Dewar flask

Weld shops that need a large amount of oxygen commonly use liquid oxygen. Liquid oxygen is stored and shipped in a Dewar flask. The flask is designed to store liquid oxygen at −297°F (−183°C) under pressure. Oxygen is withdrawn from the flask through tubing

that passes between the inner and outer walls. This warms the liquid, causing it to vaporize (turn into a gas). See Fig. 4-12. A standard regulator is used to maintain the correct working pressure of the gas.

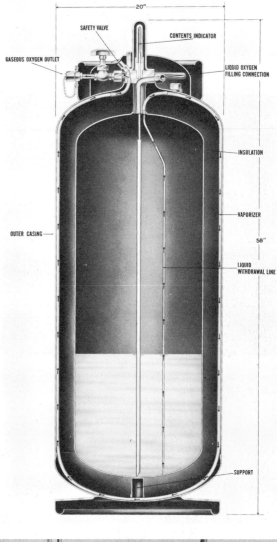

Fig. 4-12. A—Cross sectional view of a Dewar flask. Liquid oxygen vaporizes as it passes between the inner and outer walls. (Linde Div., Union Carbide Corp.) B—This Dewar flask can supply several torches at one time.

Oxygen Cylinder Valve

A forged brass cylinder valve controls the flow of oxygen from the cylinder, Fig. 4-13. The valve should be protected with a properly installed safety cap when the cylinder is being moved or is not in use.

The oxygen cylinder valve should always be completely open when in use. A backseating valve prevents oxygen from escaping around the valve stem. The backseating valve operates only when the cylinder valve is completely open. Fig. 4-14 shows a cut-away view of an oxygen cylinder.

Fig. 4-13. A forged brass oxygen cylinder valve. The white material is Teflon® tape that has been used to seal the threads between the valve and cylinder.

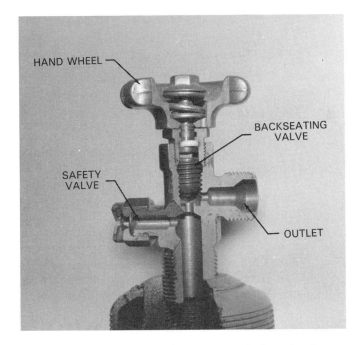

Fig. 4-14. Cut-away view of an oxygen cylinder valve. It must be opened completely for the backseating valve to function properly.

Safety Valve

A safety valve is an integral part of the oxygen cylinder valve. A *safety valve* prevents an explosion when the cylinder is exposed to high temperatures. As the temperature increases, the cylinder pressure also increases. When a predetermined pressure is reached, a safety disc within the valve ruptures, and slowly releases the oxygen in the cylinder. See Fig. 4-15.

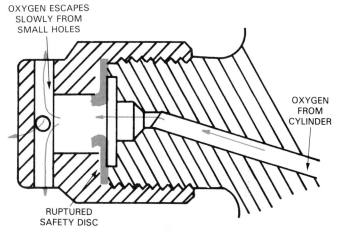

Fig. 4-15. A schematic of the safety valve on an oxygen cylinder valve. When the disc ruptures, oxygen escapes through several drilled holes.

OXYGEN SAFETY PRECAUTIONS

General safety precautions that should be followed when using oxygen include:

1. Do not place liquid oxygen equipment on asphalt or surfaces with oil or grease deposits.
2. Smoking or open flames are not permitted in areas where oxygen is stored, handled, or used.
3. Do not use a cylinder that has no label or that has an illegible label. Do not assume a cylinder contains a particular gas. Return the cylinder to the supplier.
4. Avoid contact between liquid oxygen and your skin or eyes. Liquid oxygen is at $-297°F$ ($-183°C$) and will cause freeze burns.
5. Remove all clothing that has been splashed or saturated with liquid oxygen. Oxygen-saturated clothing is highly flammable. Allow the clothing to air out at least 30 minutes so that the oxygen dissipates.
6. Avoid contact between oxygen or oxygen fittings and organic materials such as oil, grease, kerosene, cloth, wood, tar, and coal dust. These materials are highly combustible when combined with oxygen.
7. Secure all cylinders in an upright position.
8. Avoid contact between the torch flame and cylinder.
9. Do not use safety caps for lifting cylinders.

10. Always transport a cylinder by using a hand truck or by rolling it along the bottom edge.
11. Keep cylinder valves covered with safety caps when not in use.
12. Do not force cylinder valves open. If the handwheel is missing, or will not operate properly, return the cylinder to the supplier.

MANIFOLDS

Manufacturers and repair shops that need a large volume of oxygen and fuel gas for welding or cutting may use manifolds. A *manifold* is a brazed assembly of pipes that delivers gas from several cylinders into one supply pipe. The pipe distributes gas to several work stations or locations. Fig. 4-16 shows an acetylene manifold.

Fig. 4-16. An acetylene manifold. Several cylinders are connected to the manifold to supply gas for a number of users. Notice the safety sign.

Oxygen and acetylene manifolds must be kept separate to comply with local fire codes. Their construction is also governed by fire codes. Acetylene may form explosive compounds when combined with copper, mercury, or silver. For this reason, copper tubing or pipe must not be used in an acetylene manifold system. Manifolds also must be enclosed, ventilated, and located outside of the welding or cutting area.

Cylinders of oxygen and acetylene must never be connected to the wrong manifold. Intermixing gases in the delivery pipe could result in an explosive condition. Connecting oxygen and acetylene cylinders to the wrong manifold is virtually impossible due to the fittings. Acetylene fittings have left-hand threads; oxygen, right-hand.

Check the following safety standards for more information on the use of oxygen and fuel gas cylinders and manifolds:

• ANSI Z49.1 "Safety in Welding and Cutting."

• NFPA No. 50, "Bulk Oxygen Systems at Consumer Sites."
• NFPA No. 51, "Oxygen-Fuel Gas Systems for Cutting and Welding."
• Linde Form 9888, "Precautions and Safe Practices—Liquid Atmospheric Gases."
• Linde Form 2035, "Precautions and Safety Practices in Welding and Cutting with Oxygen-Fuel Gas Equipment."
• Local and national safety codes.

PRESSURE REGULATORS

A *pressure regulator* is a device used to reduce the pressure at which oxygen or fuel gas is delivered. Gas pressure in acetylene and oxygen cylinders is high. The working pressure for welding, however, may be as low as 1 psig (6.9 kPa). A pressure regulator is used to reduce the pressure of the gas.

A *single-stage regulator* reduces cylinder pressure to working pressure in one stage (step). The *regulator adjusting screw* controls the working pressure of gas delivered by the regulator. See Fig. 4-17.

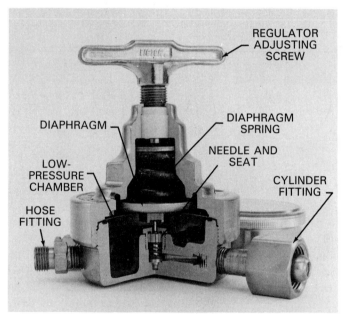

Fig. 4-17. Cut-away view of a single-stage regulator. (Victor Equipment Co.)

A *two-stage regulator* reduces cylinder pressure to working pressure in two stages. Cylinder pressure is first reduced to an intermediate pressure. The intermediate pressure is then reduced to the working pressure in the second stage. See Fig. 4-18. Two-stage regulators are more expensive than the single-stage regulators, but provide more precise control.

Pressure adjustments are made by using the regulator adjusting screw. When the screw is turned clockwise (in), the working pressure increases. Turning the screw

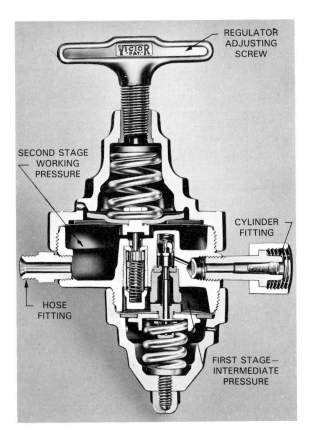

Fig. 4-18. Cross-sectional view of a two-stage regulator. This regulator uses stem-type valves in both stages. (Victor Equipment Co.)

Fig. 4-19. An acetylene regulator and gauges. The working pressure gauge is shaded in red above 15 psi, because acetylene is dangerous above this point.

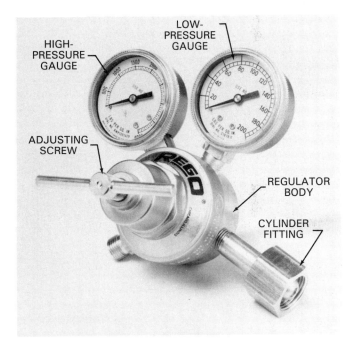

Fig. 4-20. Typical oxygen regulator. The pressure regulating mechanism is built into the regulator body. (Rego Co.)

counterclockwise (out) decreases the working pressure. The regulator is turned completely off when the screw is turned counterclockwise until it feels loose.

Pressure gauges

Pressure regulators usually have two gauges. The *cylinder pressure gauge* indicates cylinder pressure. The *working pressure gauge* shows working pressure. Gauges are marked with pressure readings at least 50% higher than the highest pressure that is expected.

An acetylene cylinder pressure gauge is often marked to indicate pressures of 400 or 500 psig (2.76 or 3.45 MPa). Acetylene working pressure gauges may indicate pressure up to 30 psig (207 kPa). Several models of acetylene working pressure gauges indicate pressure up to 15 psig (103.42 kPa) and identify dangerous pressure levels with a red background. See Fig. 4-19. For safe use, acetylene working pressure must be kept below 15 psig (103.42 kPa).

An oxygen cylinder pressure gauge is commonly marked to indicate pressures up to 3000 or 4000 psig (20.7 or 27.6 MPa). Oxygen working pressure gauges usually show pressure up to 50 psig (345 kPa). See Fig. 4-20. Special working pressure gauges, with markings as high as 1000 psig (6.89 MPa), may be needed for heavy cutting operations.

HOSES AND FITTINGS

A *hose* is a flexible rubber tube used to convey gases from a pressure regulator to a welding torch. Hoses are designed to withstand high pressure. The oxygen hose is green. The fuel gas hose is red. Both single and dual (Siamese) hoses are available. See Fig. 4-21.

A *fitting* is used to connect the hose to the regulator or torch. Each end of the hose has a fitting consisting of a brass nut and gland. Oxygen hose nuts have right-hand threads. Fuel gas hose nuts have left-hand threads, and a groove machined around the nut for identification. See Fig. 4-22. The left- and right-hand threads and different color hoses are safety precautions.

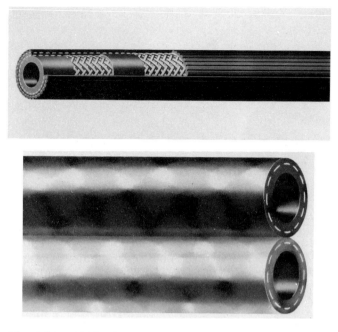

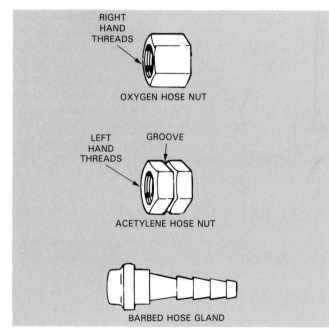

Fig. 4-21. Welding hoses. A—Single welding hose. Notice the inner hose, the two layers of fabric, and the ribbed rubber covering. B—Dual or Siamese welding hose. One hose carries the oxygen and the other hose carries the fuel gas. (Anchor Swan Corp.)

RIGHT HAND THREADS

OXYGEN HOSE NUT

LEFT HAND THREADS   GROOVE

ACETYLENE HOSE NUT

BARBED HOSE GLAND

Fig. 4-22. Hose fittings. The fuel gas hose nut has left-hand threads and a groove cut around it. (Airco Welding Products, Div. of Airco, Inc.)

They prevent the hoses from being connected incorrectly. Oxygen flowing through an acetylene hose could cause an explosion.

Hoses should not be laid across an area where they may be damaged or run over by a vehicle. When hoses must be temporarily run across a traffic path they can be protected by placing them under a section of inverted channel iron or steel angle.

BACKFIRES AND FLASHBACKS

A *backfire* is a small explosion that produces a sharp popping sound. After a backfire, the flame may be extinguished or continue to burn. A backfire usually remains in the torch head. It occurs when the torch head or tip overheats, when the tip is held too close to the work, or when the tip is dirty.

One of the greatest dangers in oxyfuel gas welding is a flashback. A *flashback* occurs when the flame moves into or beyond the *mixing chamber* of the torch. A flashback may move into the hoses, regulator, and possibly into the cylinder. The flame burning back into the torch can result in a damaged torch or hose. It can also cause a violent explosion which may destroy the regulator or cylinder. The explosion could also cause serious personal injury and fire.

Flashbacks are controlled using check valves and flashback arrestors. Flashbacks may be caused by a reverse flow of gases or by loose and leaking connections. Leaks reduce the gas flow. The lower flow may not be able to support the flame at the torch tip. The flame may then burn back into the torch or even farther.

Check valves and flashback arrestors

A *check valve* is a valve that allows the flow of gas in only one direction. Such valves are used to prevent the reverse flow of gases through the torch, hoses, and/or regulators. Gas pressure opens the valve to allow gas to flow. The valve closes when the flow stops or when gas tries to flow in a reverse direction. See Fig. 4-23. Check valves are most often installed at the torch inlet. They sometimes are placed at the regulator outlet, or at both the torch and regulator.

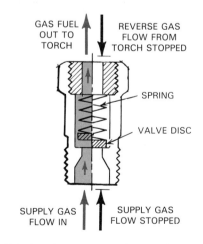

Fig. 4-23. Check valve. In normal operation (left side), gas from the regulator is at a higher pressure than gases in the torch. Pressure from the regulator lifts the valve disc and allows gas to flow into the torch. If pressure within the torch, plus force of the spring on the valve disc, exceeds pressure from the regulator (right side), the disc seats and does not allow gas to flow from the torch into the hose. A check valve is not reusable after a flashback. (Welding Design & Fabrication)

44

A *flashback arrestor* is a device used to prevent the flow of a burning fuel gas and oxygen mixture from the torch back into the hoses, regulators, and cylinders. It is designed to eliminate the possibility of an explosion in the regulator or cylinder. See Fig. 4-24. Flashback arrestors are installed between the torch and hose. An arrestor consists of the following safety valves:

1. Reverse-flow check valve. Prevents the flow of gas in the wrong direction.
2. Pressure-sensitive cut-off valve. Stops gas flow in case of an explosion.
3. Stainless steel filter. Prevents the flame from entering the hose.
4. Heat-sensitive check valve. Stops gas flow if the arrestor reaches 220°F (104°C).

TORCHES

An oxygen gas welding *torch* controls and mixes the fuel gas and oxygen. It is also used to direct the gas flame to the welding, brazing, or soldering work area. A complete welding torch has several parts. These include the torch valves, torch body, mixer or injector, torch tube, and torch tip. See Fig. 4-25.

Either an injector-type torch or a positive pressure torch may be used for oxyfuel gas welding. An *injector-type torch* is used with a low-pressure acetylene generator. See Fig. 4-26. Acetylene pressures as low as .25 psig (1.7 kPa) are used with an injector-type torch. When using the injector-type torch, acetylene is drawn into the injector by oxygen traveling through the injector. The oxygen and acetylene are mixed in the mixing chamber. The mixed gases flow to the tip of the torch where they are burned.

A positive pressure torch is used with acetylene pressures above .25 psig (1.7 kPa). The oxygen and

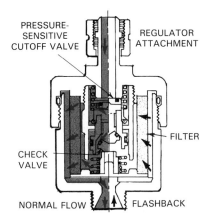

Fig. 4-24. Flashback arrestor. In normal operation (left side), gas flows through the open cutoff valves, check valve, and flame arrestor filter and into the hose. In case of a flashback (right side), the stainless steel filter stops the flame and the pressure fluctuations activate the cutoff valve. The flow of gas is halted, extinguishing the flame. The check valve operates when gas flows toward the cylinder. If the arrestor is exposed to fire, the thermal cutoff valve shuts off the gas supply. A flashback arrestor is reusable after a flashback.
(Welding Design & Fabrication)

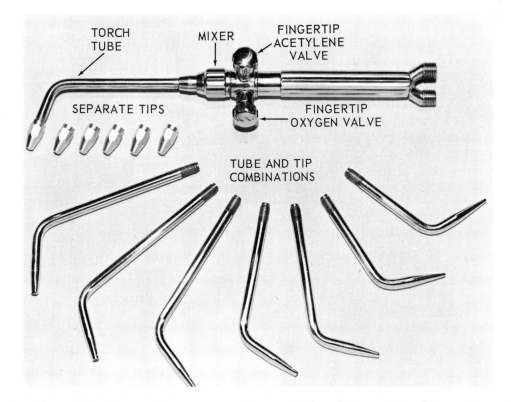

Fig. 4-25. A light-duty positive pressure welding torch. Two different styles of tips can be used with the same mixer. Note the location of the valves. (Air Products and Chemicals, Inc.)

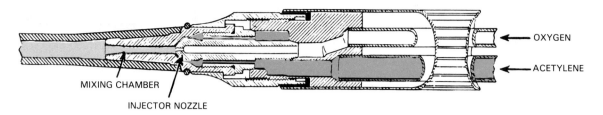

MIXING CHAMBER

INJECTOR NOZZLE

OXYGEN

ACETYLENE

Fig. 4-26. Cross-sectional view of an injector-type welding torch. The acetylene is injected (drawn) into the mixing chamber by suction created by the flow of oxygen.

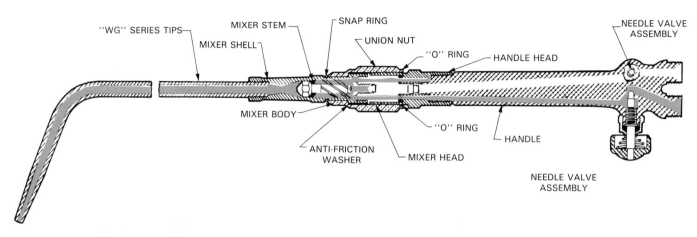

"WG" SERIES TIPS — MIXER STEM — SNAP RING — NEEDLE VALVE ASSEMBLY

MIXER SHELL — UNION NUT — "O" RING — HANDLE HEAD

MIXER BODY — ANTI-FRICTION WASHER — MIXER HEAD — "O" RING — HANDLE

NEEDLE VALVE ASSEMBLY

Fig. 4-27. A cross-sectional view of a positive pressure torch. The working pressure of the oxygen and acetylene is high enough to force the gases into the mixing chamber.   (Modern Engineering Co., Inc.)

acetylene pressures are high enough to force gas through the torch into the mixing chamber. The mixing chamber may be in the torch body or at the torch end of the tip. See Fig. 4-27. The gases are mixed, then flow to the tip where they are burned.

Torch valves

Torch valves control the flow of oxygen and fuel gas into the torch. The valves of light-duty positive pressure torches are usually found near the discharge end of the torch. The valves of medium- and heavy-duty torches are found near the hose end of the torch body. A mark on the torch body near each valve indicates whether it is for fuel gas or for oxygen.

Needle-and-seat or ball-and-seat valve construction is used to control the gas flowing to the mixing chamber. In a needle-and-seat valve, a conical stem and beveled seat are used to obstruct the passageway for gases. See Fig. 4-28. A hardened steel ball and hemispherical seat are used to control gas flow in a ball-and-seat valve.

Torch valves are opened and closed by hand. Only finger force should be used. Do not use a wrench to open and close the valves. One-half to one full turn counterclockwise generally will open the valves completely. The valves are turned clockwise to close them.

Torch tips

A *torch tip* is the part of the end of the torch where the fuel gas and oxygen are ignited. Two types of torch

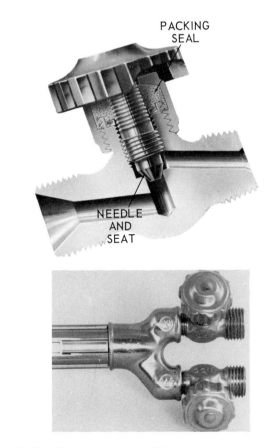

PACKING SEAL

NEEDLE AND SEAT

A

B

Fig. 4-28. Torch valves. A—The valve shown uses needle-and-seat construction. B—Note that the valves are marked (OXY) and (GAS).   (Veriflow Corp.)

tips are available. A welder may choose a tube and tip combination or a separate tube and tip.

An *orifice* (precise hole) is bored into the end of the torch tip when it is made. The orifice size number is stamped on the tip by the manufacturer. Unfortunately, tip identification numbering systems differ among tip manufacturers. Generally, a small orifice number indicates a small hole. A small orifice delivers a small amount of heat to the base metal. A large orifice delivers a greater amount of heat to the base metal.

The manufacturer's recommendation should be followed when selecting the correct tip size for a welding job. Welding supply companies supply tip size recommendation charts to their customers. Fig. 4-29 is a chart of the approximate tip size to use with various thicknesses of metal. The drill size of an orifice may be found using a set of numbered drill bits. Care should be taken not to scratch or distort the orifice when using the bits.

Molten metal or dirt may collect on the end of the tip. This may change the size or shape of the orifice. A drill bit or broaching wire is used to clean the orifice. See Fig. 4-30A. The end of the tip may need to be cleaned or filed flat. Dirt, filler metal, weld metal, or other materials may build up on the tip. A mill file may be used as shown in Fig. 4-30C. After the tip is filed flat, the orifice may need to be cleaned again.

WELDING RODS

The edges of thick metal are commonly ground, flame cut, or machined before welding. Edge preparation removes a portion of the base metal and thins it.

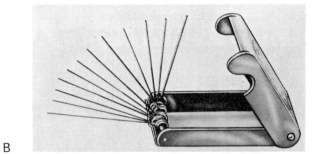

A

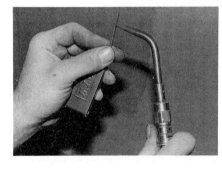

B

C

Fig. 4-30. A—Cleaning an orifice using a broaching wire. B—A broaching tool has many sizes of broaching wires. (Thermacote—Welco Co.) C—A small file in the broaching tool can be used to clean the end of the tip.

METAL THICKNESS	WELDING TIP ORIFICE SIZE*	WELDING ROD DIAMETER	OXYGEN		ACETYLENE		WELDING SPEED FT./HR
			PRESSURE (PSIG)	CFH	PRESSURE (PSIG)	CFH	
1/32	74	1/16 in.	1	1.1	1	1	
1/16	69	1/16 in.	1	2.2	1	2	
3/32	64	1/16 in. or 3/32 in.	2	5.5	2	5	20
1/8	57	3/32 in. or 1/8 in.	3	9.9	3	9	16
3/16	55	1/8 in.	4	17.6	4	16	14
1/4	52	1/8 in. or 3/16 in.	5	27.5	5	25	12
5/16	49	1/8 in. or 3/16 in.	6	33.	6	30	10
3/8	45	3/16 in.	7	44.	7	40	9
1/2	42	3/16 in.	7	66.	7	60	8

*Note the tip orifice size as shown is the number drill size. These recommendations are approximate. The torch manufacturers' recommendations should be carefully followed.

Fig. 4-29. Table used for oxyacetylene welding with a positive pressure torch. Notice the relationships between tip orifice size, gas pressures, welding rod diameter, and metal thickness.

Even though metal is removed from the edges, a completed weld must still be at least as strong as the base metal. Filler metal is generally added to the weld bead to increase the thickness of the weld.

A *welding rod* is a long thin rod of a similar type of metal to that being welded. As the rod is melted, it is added to the weld bead as filler metal.

Welding rod is available in 36″ (914 mm) lengths. Diameters range from 1/16″ (1.6 mm) to 3/8″ (9.5 mm). See Fig. 4-31. The rods are usually packaged in 50 pound (22.7 kg) bundles. Welding rod is made from various metals, such as carbon steel, stainless steel, bronze, aluminum, and magnesium. Carbon steel welding rods are usually copper-coated to protect them from rusting.

Fig. 4-32. A flint-and-steel sparklighter being used to light an oxyfuel gas welding torch.

Fig. 4-31. Welding rods are available in several diameters. Five are shown: 1/16″ (1.6 mm), 3/32″ (2.4 mm), 1/8″ (3.2 mm), 3/16″ (4.8 mm), and 1/4″ (6.4 mm).

TORCH LIGHTERS

A torch lighter provides safe ignition for oxyfuel gas welding torches. Never use a match or butane lighter to light a torch. Severe burns may occur.

The flint-and-steel sparklighter is the most common torch lighter. A spark is produced by a sparklighter by squeezing the handles together. As the handles are squeezed together, a replaceable flint rubs across a file segment to create a spark. See Fig. 4-32.

In large production shops, an economizer and pilot light may be used to light a torch. The pilot light is fueled by acetylene or city gas. It remains burning during welding or cutting operations. See Fig. 4-33. When the torch is not being used, it is hung on the economizer. That device pivots to shut off the supplies of acetylene and oxygen gas. When the torch is lifted from the economizer, the gases flow again.

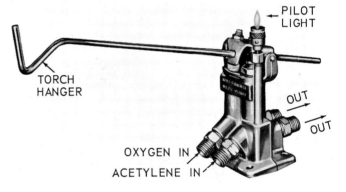

Fig. 4-33. Combination economizer and pilot light.

PROTECTIVE CLOTHING

A welder must be properly dressed to avoid injury due to flying sparks and metal. She or he should wear coveralls or a shirt or jacket that buttons at the collar, Fig. 4-34. Dark-colored clothing is preferred because it reflects less light. Clothing must be flame-resistant and have covered pockets. Pant legs should be cut to length. Cuffs or folds at the bottom of the leg are dangerous. They may catch sparks or hot metal.

Flammable objects, such as matches and butane lighters, should not be carried in your pockets. They could be ignited by a spark while welding.

A cap should be worn to keep your hair clean and free from sparks and hot metal.

Gloves must be worn while welding. Leather gloves or those with leather palms are preferred. Gloves with tight-fitting cuffs can be used for light-duty welding. Gauntlet gloves should be worn for heavy duty or out-of-position welding.

Leather safety (hard toe) shoes are recommended to protect the welder's feet from falling objects. The shoe tops should be high enough to fit under the pant leg.

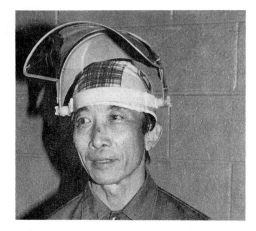

Fig. 4-34. This welder is wearing a dark-colored shirt that is buttoned at the collar. Notice the filter-quality face shield.

Leathers should be worn when overhead welding. The term *leathers* refers to the leather coat, hood, cape, leggings, and chaps worn by welders. See Fig. 4-35.

Goggles

Ultraviolet and *infrared rays* are created by most welding processes. These rays are harmful to the eyes, so direct exposure should be avoided.

Welding goggles with protective filter lenses must be worn. Welding goggles have two round filter lenses or one rectangular filter lens. The rectangular type can be worn over prescription glasses. Filter-quality face shields are also available.

Filter lenses range in shade from a #1 to #14. A higher number indicates a darker lens. A #4-#6 lens is recommended for oxyfuel gas welding.

Clear glass or plastic cover plates are installed outside the filter lenses. *Cover plates* protect the costly filter lenses from damage. Cover plates are inexpensive. They should be replaced when badly spattered or scratched. See Fig. 4-36.

Safety glasses

Shatter-resistant safety glasses are designed to protect your eyes when grinding, chipping, or performing other activities that produce flying particles. The glasses should be worn under a welding hood or welding goggles. Wear safety glasses whenever you work in an area where eye hazards are present. These include welding, cutting, and grinding areas, as well as areas used for machining, turning, stamping, or similar operations. Safety glasses also may be tinted to help protect your eyes from ultraviolet and infrared rays.

Fig. 4-35. Gauntlet gloves and a leather jacket are worn by this welder.

A

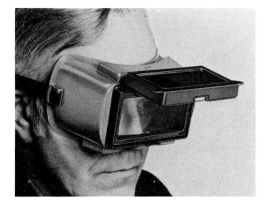

B

Fig. 4-36. Welding goggles. A—With 50 mm round filter lenses. B—Rectangular filter lens measures 2″ x 4 1/4″ (50.8 mm x 108 mm). Rectangular goggles are easily worn over prescription glasses. (Dockson Corp.)

REVIEW QUESTIONS

Write your answers on a separate sheet of paper.
1. List five fuel gases used in oxyfuel gas welding or cutting.
2. A low-pressure acetylene generator produces acetylene gas at _____ psig (_____ kPa).
3. At which end of the torch body are valves generally located on a light-duty torch?
4. Will oxygen or nitrogen boil away first when air is liquefied?
5. Acetylene becomes unstable and explosive at _____ psig (_____ kPa) if it is allowed to collect in large volume as a gas.
6. Which type of gas cylinder uses a fusible plug?
7. Why should a safety cap be used on a cylinder when it is moved or stored?
8. What type of safety device is used in an oxygen cylinder to prevent an explosion during a fire?
9. Liquid oxygen is stored in a _____ flask.
10. A _____ is used to connect several cylinders to one delivery pipe.
11. What metals should not be used for tubing or pipes that carry acetylene.
12. A _____ is used to reduce and control the pressure of the welding gases.
13. When the regulator adjusting screw feels loose in its threads, is the regulator opened or closed?
14. A high-pressure oxygen gauge is usually marked to indicate pressures up to _____ psig (_____ MPa).
15. Low-pressure acetylene gauges are often red lined above _____ psig (_____ kPa).
16. What is a Siamese welding hose?
17. Oxygen flowing through an acetylene hose may cause an explosion. True or False?
18. A _____ welding torch is used with acetylene and oxygen that is stored in cylinders.
19. When opening or closing torch valves, how much force should be applied?
20. A _____ or _____ should be used to clean the torch tip orifice.
21. The recommended filter lens for oxyfuel gas welding is #_____-#_____.
22. A _____ is a one-way valve that prevents gases from burning back to the regulator.
23. Explain the difference between acetylene and oxygen hose nuts.
24. Pant leg cuffs and pockets are not recommended on welding clothing. Why?
25. Oil, grease, kerosene, cloth, tar, wood, and coal dust must be kept away from all _____ fittings to prevent fires or explosions.

Chapter 5

Oxyfuel Gas Welding Equipment: Assembly and Adjustment

After studying this chapter, you will be able to:
☐ List the steps required to assemble an oxyfuel gas welding outfit.
☐ List the steps required to turn on an oxyacetylene welding outfit.
☐ Describe the procedure used to check for leaks in an oxyacetylene welding system.
☐ List the steps required to light and adjust the flame on an oxyacetylene torch.
☐ Identify the three types of flames.
☐ Describe the procedure for shutting off an oxyacetylene welding outfit.

TECHNICAL TERMS

Carburizing flame, neutral flame, non-petroleum-based, oxidizing flame, petroleum-based, purge.

ASSEMBLING THE WELDING OUTFIT

Proper assembly, care, and security of the oxyfuel gas welding outfit is necessary for its safe and effective use. Care must be taken when assembling the various threaded fittings. When tightened, the fittings must not leak. The fittings are generally made from soft metal like brass. Do not overtighten fittings, or the threads may be damaged.
Petroleum-based oil, grease, or soap must not be used to lubricate any part of a welding or cutting outfit. These materials could ignite and cause a fire.
Oxygen and acetylene cylinders should be secured vertically to a wall, column, or hand truck, Fig. 5-1.

Chains or steel straps are commonly used. Safety caps may be removed after the cylinders are secured.

Fig. 5-1. Oxygen and acetylene cylinders secured to a hand truck with a steel band. The band is adjusted by turning the threaded handwheel.

The oxygen and fuel gas cylinder outlets should be cleaned before attaching the regulators. This is done by briefly opening and closing the cylinder valves. Fig. 5-2 shows a cylinder valve being opened. The escaping gas cleans any dust or dirt from the outlet. Do not point the cylinder outlet toward workers in the area. Flames or sparks must not be present when opening the fuel gas cylinder valve.

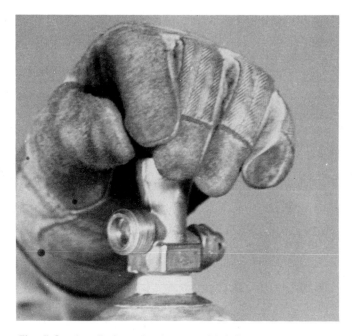

Fig. 5-2. A cylinder valve is opened briefly to remove dirt or foreign debris from the outlet. The cylinder outlet should be pointed away from nearby workers.

If cylinder valves are damaged or leaking, the cylinders must not be used. Return defective cylinders to the supplier.

Regulators are attached to the cylinder outlets and tightened so they are snug and leakproof. The brass nuts used on regulators and hoses should not be over-tightened, or they may be damaged. A proper size open-end wrench should be used to tighten the regulator nut, as shown in Fig. 5-3. The oxygen

Fig. 5-3. The regulator nut is tightened on the cylinder valve with a proper size wrench.

regulator nut has right-hand threads. Large commercial acetylene cylinders may have left- or right-hand cylinder outlet threads. Small noncommercial acetylene cylinders have right-hand cylinder outlet threads.

An acetylene hose is red. A groove is machined around the nuts used on acetylene hoses. One end of the hose is connected to the acetylene regulator outlet fitting, Fig. 5-4A. The other end is connected to the torch inlet fitting, Fig. 5-4B.

An oxygen hose is green, and the nuts do not have a groove. One end of the oxygen hose is connected to the oxygen regulator outlet fitting. The other end is connected to the torch inlet fitting.

A torch tip is selected and threaded into the torch or into the torch tube. The combination tube and tip is tightened by hand. When a separate tube and tip are used, select a proper size box or open-end wrench to tighten the tip.

TURNING ON AN OXYACETYLENE WELDING OUTFIT

The cylinder, regulator, and torch valves must be turned on in a given order. Follow these steps when

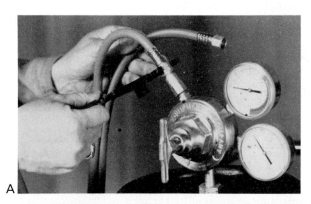

Fig. 5-4. A—One end of a hose is connected to the regulator outlet using a proper size wrench. Notice the check valve between the hose and regulator. B—The other end of the hose is connected to the torch. Be careful not to overtighten brass fittings.

turning on an oxyacetylene welding or cutting outfit:

1. Visually check the torch, valves, hoses, fittings, regulators, gauges, and cylinders for damage.
2. Make certain the regulators are closed before opening the cylinder valves. This will prevent damage to the regulators and gauges. See Fig. 5-5. Turn the regulator adjusting screws on the oxygen and acetylene regulators *counterclockwise* (to the left). Continue to turn the screws counterclockwise until they feel loose.
3. Stand to one side of the regulators while opening the cylinder valves, as shown in Fig. 5-6. A regulator or gauge could burst, causing severe injury.

4. Slowly turn the acetylene cylinder valve counterclockwise until it is open (about 1/4 to 1/2 turn). See Fig. 5-7. This provides enough acetylene flow for most purposes. Use the proper size wrench for the cylinder valve. Leave the wrench in place so that the valve can be closed quickly in case of emergency.
5. Slowly turn the oxygen cylinder valve counterclockwise until it is fully open (the valve may leak if it is not fully opened). Remember that cylinder pressure may be 2200 psig (15.17 MPa). A rapid flow of high-pressure gas could rupture the regulator diaphragm or gauges.
6. Turn the acetylene torch valve one complete turn counterclockwise.

Fig. 5-5. Closing the regulator. Turn the regulator adjusting screw counterclockwise until it feels loose in the threads. Note that both pressure gauges read zero.

Fig. 5-7. Opening the acetylene cylinder valve. A special wrench may be required. The wrench should remain on the valve so that it can be turned off quickly in an emergency.

7. Turn the acetylene regulator screw *clockwise* (to the right) until the low-pressure gauge shows the correct working pressure. See Fig. 5-8. The correct working pressure depends on the tip size used. Refer to Fig. 4-29 for a table showing the tip manufacturer's size recommendations.

Fig. 5-6. Opening the oxygen cylinder valve. Always stand to one side when opening any cylinder valve.

Fig. 5-8. Adjusting the acetylene working pressure. The torch valve must be open to properly adjust the working pressure. The torch valve is closed after the pressure is set.

8. Close the acetylene torch valve. Check the acetylene regulator for possible leaks. Checking for leaks is discussed in the next section.
9. Open the oxygen torch valve about one turn counterclockwise.
10. Turn the oxygen regulator adjusting screw clockwise until the desired working pressure is obtained.
11. Close the oxygen torch valve. Check the oxygen regulator for leaks. Checking for leaks is discussed in the next section.

The outfit is now ready for lighting.

CHECKING FOR LEAKS

A welding outfit should be checked for leaks each time it is turned on. External leaks may be found using soap suds. Both internal and external leaks may be located by watching the low-pressure gauge. A *non-petroleum-based* soap must be used to check for leaks. *Petroleum-based* products may cause fires when in contact with oxygen. Saturate fittings with soap suds when locating leaks. Bubbles will occur in the soap suds when a leak is present.

REGULATORS

The high-pressure valve under the regulator diaphragm opens and closes to control the working pressure. A leaking regulator valve allows pressure to rise uncontrollably inside the regulator. The regulator or low-pressure gauge may burst if there is excessive pressure below the diaphragm. Check regulators for leaks immediately after setting the working pressure. The regulator may be checked for leaks as follows:
1. Turn on the welding or cutting outfit.
2. Close both torch valves.
3. Carefully watch the acetylene and oxygen low-pressure gauges. The pressure readings should remain constant. A leaky regulator valve is indicated if the gauge pressure continues to rise. Immediately close the cylinder valves and shut off the outfit.
4. Send the bad regulator out for repair.

CYLINDER-TO-REGULATOR CONNECTIONS

Cylinder-to-regulator fittings may be checked for leaks by intentionally trapping pressurized gas between the regulator and cylinder. A leak is indicated by a decrease of pressure on the high-pressure gauge. This test must be made on both the oxygen and acetylene systems.
1. Turn on the outfit. Note the pressure on the high-pressure gauge.
2. Close the regulator by turning the adjusting screw counterclockwise. Close the cylinder valve. High pressure gas is now trapped between the

regular and cylinder valve.
3. The high-pressure gauge reading should remain constant. If it drops, there is a leaking connection between the regulator and the cylinder. Check the fitting for tightness and repeat steps 1 and 2. If a leak is still indicated, the cylinder outlet fitting may be bad. If so, return the cylinder to the vendor.
4. Try the regulator on another cylinder and repeat steps 1, 2, and 3. The regulator fitting itself may be bad if there is still a leak.

HOSES AND HOSE FITTINGS

Gas leaks in the hoses or fittings between the regulators and the torch valves will show as a pressure drop on the low-pressure gauge. Gas must be trapped between the regulator and the torch valves to perform this test. This procedure must be used to test the oxygen and acetylene hoses and fittings:
1. Turn on the outfit.
2. Close the torch valves.
3. Note the reading on the low-pressure gauge.
4. Close the regulator with the adjusting screw.
5. The low-pressure gauge reading should remain constant. If the pressure drops, there is a leak in the hose or fittings. Locate it using non-petroleum-based soap suds.

FLAMES

Three types of flames can be produced with oxygen and acetylene. They are the carburizing, neutral, and oxidizing flames.

CARBURIZING FLAME

A *carburizing flame* is produced when too little oxygen is present. Three flame areas are visible, Fig. 5-9A. The end of the inner flame is ragged or rough. A carburizing flame may add carbon to the weld area. This may cause undesirable hardening of the weld. A carburizing flame is used whenever it is necessary to avoid adding oxygen to weld or braze joints. Slightly carburizing flames are used for brazing, braze welding, soldering, and welding certain metal alloys.

NEUTRAL FLAME

A *neutral flame* results from the correct balance of oxygen and acetylene. It is used for most welding and cutting operations. There are two flame areas. The end of the inner cone is smooth and shaped like a bullet. See Fig. 5-9B. A neutral flame will not burn the weld or base metal because it does not have excessive oxygen. It will not add carbon to the weld because all the carbon is burnt in the flame.

OXIDIZING FLAME

An *oxidizing flame* is produced when too much oxygen is present. Two flame areas are seen. The end of the inner flame is smooth and pointed, Fig. 5-9C. The flame is loud and makes a hissing noise. An oxidizing flame is usually not desired because it burns the surface of the base metal and weld.

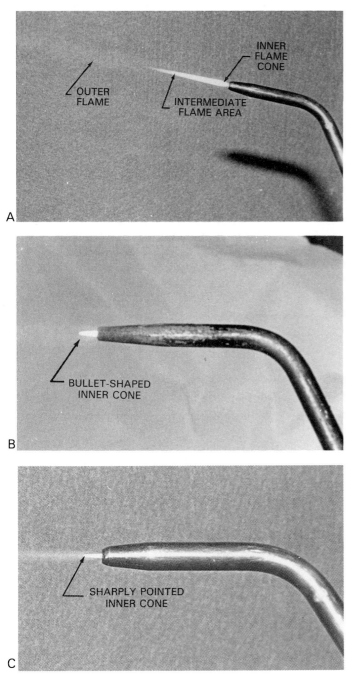

LIGHTING AND ADJUSTING THE FLAME— RECOMMENDED METHOD

A positive pressure torch is used for many oxyacetylene welding operations. The procedures used when lighting a positive pressure torch are as follows:

The oxygen and acetylene systems must be purged. *Purging* is the process of passing the correct gas through the entire system to remove air or undesirable gases. Purging ensures that the correct gas is in the appropriate regulator, hose, and torch passage. This is done by allowing acetylene to flow through the acetylene hose and oxygen to flow through the oxygen hose for a short period of time.

A sparklighter should be used to light an oxyacetylene welding or cutting torch. The flint and steel in the sparklighter create a spark that will safely light the acetylene at the torch tip. A sparklighter should be long enough to keep your hand safely away from the acetylene flame. Never use a match or butane lighter to light an oxyacetylene torch. Severe burns could result.

Use the following procedure to light and adjust the flame.

1. Open the acetylene torch valve slightly (approximately 1/16 turn).
2. Light the acetylene at the torch tip, using a sparklighter.
3. Open the acetylene torch valve until flame becomes turbulent (rough) about 3/4" to 1" (19-25 mm) from the end of the tip. Adjust the acetylene so the flame no longer smokes or releases soot. See Fig. 5-10.
4. Turn on the oxygen torch valve slowly after the acetylene is regulated. Adjust the oxygen torch valve until a neutral flame is obtained.

Fig. 5-9. A—Carburizing flame. The flame begins to burn cleanly as oxygen is turned on. Three distinct areas of the flame can be seen. B—Neutral flame. When the oxygen and acetylene are properly adjusted the inner cone is bullet-shaped or rounded on the end. C—Oxidizing flame. When too much oxygen is present, the inner cone becomes pointed. The flame is noisy and gives off a hissing sound. This flame is undesirable because it causes the weld area to oxidize (rust).

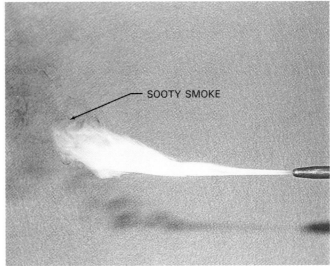

Fig. 5-10. Acetylene burning at the torch tip. The flame is long and gives off large amounts of black soot when oxygen is not present.

LIGHTING AND ADJUSTING THE FLAME—ALTERNATE METHOD

Another method of lighting and adjusting the flame may be used when the actual orifice size is unknown. This method is also used to compensate for the pressure drop in a long hose.

1. Visually check the torch, valves, hoses, fittings, regulators, gauges, and cylinders for damage.
2. Make certain the regulators are closed before opening the cylinder valves. This will prevent damage to the regulators and gauges. See Fig. 5-5. Turn the regulator adjusting screws on the oxygen and acetylene regulators counterclockwise (to the left). Continue to turn the screws counterclockwise until they feel loose.
3. Stand to one side of the regulators while opening the cylinder valves, as shown in Fig. 5-6. A regulator or gauge could burst causing severe injury.
4. Slowly turn the acetylene cylinder valve counterclockwise until it is open (about 1/4 to 1/2 turn). See Fig. 5-7. This provides enough acetylene flow for most purposes. Use the proper size wrench for the cylinder valve. Leave the wrench in place so that the valve can be closed quickly in case of emergency.
5. Slowly turn the oxygen cylinder valve counterclockwise until it is fully open (the cylinder valve may leak if it is not fully opened). Remember that cylinder pressure may be 2200 psig (15.17 MPa). A rapid flow of high-pressure gas could rupture the regulator diaphragm or gauges.
6. Open the acetylene torch valve one complete turn. Slowly turn the acetylene regulator adjusting screw clockwise until the gas begins to flow.
7. Light the acetylene at the torch tip with a sparklighter.
8. Keep turning the acetylene regulator adjusting screw clockwise until the flame stops smoking or releasing soot. The flame must remain in contact with the end of the tip.
9. Open the oxygen torch valve one turn.
10. Turn the oxygen regulator adjusting screw until

the correct flame is obtained. A neutral flame is usually preferred.

SHUTTING DOWN AN OXYACETYLENE WELDING OUTFIT

The flame must be turned off whenever the oxyacetylene torch is not in your hand. The flame is turned off by closing the acetylene torch valve and then the oxygen torch valve.

The welding outfit should be shut down when not in use, following this procedure:

1. Turn off the flame. First, turn off the acetylene torch valve. Then, turn off the oxygen torch valve.
2. Completely close the acetylene and oxygen cylinder valves.
3. Open the oxygen and acetylene torch valves. This allows all the gases in the system to escape.
4. Close the torch valves after the high- and low-pressure gauges read zero.
5. Turn the adjusting screws on both regulators counterclockwise until they feel loose.

REVIEW QUESTIONS

Write your answers on a separate sheet of paper.

1. How are the cylinder valve outlets cleaned before installing the regulators?
2. The _____ must be closed before the cylinder valves are opened.
3. The regulator is closed when the adjusting screw feels _____.
4. Why must you stand to the side of the regulators while opening the cylinder valves?
5. The torch valve must be _____ before adjusting the working pressure on the regulator.
6. The _____ and _____ are checked for leaks when the welding outfit is turned on and the regulator and torch valves are closed.
7. What type of soap must never be used when checking oxygen fittings?
8. How far should you open the acetylene torch valve when lighting the torch?
9. Describe the appearance of a neutral flame.
10. Describe the characteristics of an oxidizing flame.

Chapter 6

OFW: Flat Welding Position

After studying this chapter, you will be able to:
☐ Identify the proper protective clothing for welding in the flat welding position.
☐ Explain how to hold a torch when forehand welding.
☐ Explain how to hold a torch when backhand welding.
☐ Carry a weld pool along a weld joint.
☐ Weld an edge joint without a welding rod.
☐ Weld a corner joint without a welding rod.
☐ Weld a butt joint without a welding rod.
☐ Select a welding rod.
☐ Lay a bead along a weld joint.
☐ Fillet weld a lap joint in the flat position.
☐ Fillet weld a T-joint in the flat position.
☐ Weld a butt joint in the flat position.
☐ Identify weld defects.

TECHNICAL TERMS

Backhand welding, carrying a weld pool, concave bead, convex bead, creating a continuous weld pool, C-shaped weld pool, downhand welding, flat bead, forehand welding, freeze, inclusions, keyhole, keyhole welding, laying a bead, leading edge, overlap, pass, root of the weld, tack weld, undercutting, weld pool.

PREPARING TO WELD

Flat position welding is also known as *downhand welding.* Downhand welding is easier than welding in any other position.

Proper protective clothing must be worn for oxyfuel gas welding. A welder must wear coveralls, or a shirt or jacket that buttons at the collar. All clothing must be flame resistant. Pant legs should have no cuffs or folds. A cap should be worn to protect your hair from sparks and hot metal. Gloves must be worn to protect your hands. Welding goggles with a #4-#6 filter lens should be worn.

Review the safety precautions for oxyfuel gas welding listed in Chapter 4. Check the oxyfuel gas welding station to be sure it is assembled correctly. All gas connections must be tight and leakproof.

Select a welding tip with the proper orifice size. Refer to Fig. 4-29 to select a suitable tip for the thickness of the metal being welded.

Turn on the oxyfuel gas welding outfit. Adjust the oxygen and fuel gas pressures for the orifice size being used.

HOLDING THE TORCH

The welding torch may be held like a pencil or a hammer. See Fig. 6-1. With lightweight torches, the pencil-like grip is comfortable and effective. This method is also effective when welding vertically, horizontally, or overhead. To reduce fatigue and make the torch feel lighter, the hoses may be hung over the welder's arm.

Two methods may be used to direct the flame to the weld area. In the *forehand welding* method, the torch tip is normally held so that the flame is pointing in the direction of travel, Fig. 6-2A. The flame melts the base metal and preheats the welding rod. Excess heat is reflected off the base metal.

The backhand welding method may be used when welding thick metal. In the *backhand welding* method, the flame is pointed opposite to the direction of travel, Fig. 6-2B. For a right-handed welder, backhand welding is begun at the left side. The flame is pointed toward the beginning of the weld. It is used to create the weld pool, sometimes simply called the "pool." The flame also warms the welding rod. Only a small amount of heat is reflected off the metal and lost. Therefore, more heat is directed at the base metal.

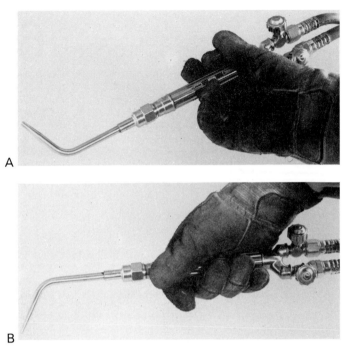

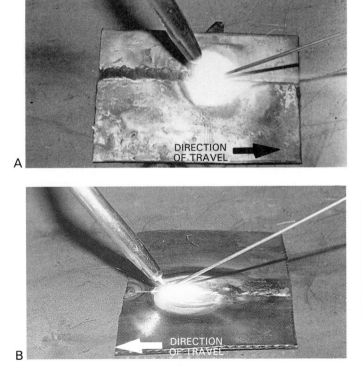

Fig. 6-1. A—The torch may be held like a pencil for vertical, horizontal, or overhead welding. B—Holding the torch like a hammer.

Fig. 6-2. A—Forehand welding. The torch and flame are pointed in the direction of travel. B—Backhand welding. The torch and flame are pointed away from the direction of travel. This method is often used on thick metal.

CREATING A CONTINUOUS WELD POOL

A weld *pool* (puddle) is the small pool of molten metal that is formed directly below the tip of the welding flame. See Fig. 6-3. *Creating a continuous weld pool* is the process of creating a pool and moving it along a line for several inches. It provides practice in

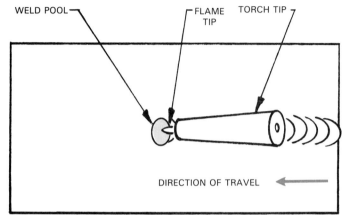

Fig. 6-3. The weld pool. The pool is shown from above.

torch and flame control. Good penetration must be achieved in all welding processes. The amount of penetration cannot always be seen, especially when welding thick metals. Creating a continuous weld pool on thin metal allows the amount of penetration to be seen.

Penetration is controlled by the width of the weld pool. The width of the pool should increase as the thickness of the metal increases. A wider pool results in deeper penetration on the same thickness of metal.

The weld pool must be wide enough to permit the molten metal to sag slightly. The deformation caused by this sag should be visible on the underside of the metal after the pass is completed. This deformation shows that total penetration has occurred. The deformation also will indicate whether the flame and weld pool have been properly controlled.

EXERCISE 6-1 — CREATING A CONTINUOUS WELD POOL

1. Obtain a piece of mild steel measuring 1/16″ x 3″ x 6″ (1.6 mm x 76 mm x 152 mm).
2. Clean both surfaces of the metal with steel wool or abrasive paper.
3. Lay out five straight lines on the metal using a ruler and soapstone.
4. Select the correct tip size and install the tip in the torch. See Fig. 4-29.
5. Set the desired welding pressures.
6. Light the torch and adjust to a neutral flame.
7. Begin the pool 1/2″ (12.7 mm) from the edge of

the metal. Hold the flame about 1/16″ — 1/8″ (1.6 mm — 3.2 mm) from the metal. See Fig. 6-4. Tilt the torch tip to the appropriate angle.

8. Watch the metal melt as the weld pool forms. When the metal sags, quickly note the pool width. Begin to move the flame ahead slowly and carry the weld pool forward. Maintain a steady forward speed and a consistent side-to-side motion. This keeps the pool the same width as it is carried along. Continue the pool to the end of the line. Inspect your first pass. Try to prevent any defects in your next pass.

9. Carry a weld pool along the length of the other four lines. Inspect each pass as it is completed.

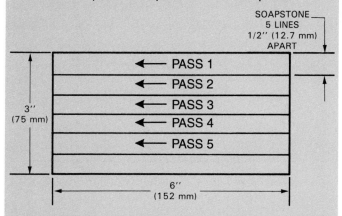

Inspection:

Turn the metal over, using pliers. A small continuous deformation (bump) should appear on the underside. The weld pool width should be uniform. This indicates continuous penetration and good torch and flame control. Holes should not be present.

All metal in a weld shop should be considered hot. Use pliers to pick up metal if you are not sure of its temperature. If a piece of hot metal must be left unattended, mark it ''HOT'' with chalk or soapstone.

FORMING AND CARRYING A CONTINUOUS WELD POOL

A welding rod is not needed to carry a weld pool. *Carrying a weld pool* refers to creating a pool and moving it along a line. The pool is created with molten base metal. The flame should be held about 1/16″ — 1/8″ (1.6 mm — 4.2 mm) above the surface of the base metal. The torch should be held at a 30° — 45° angle, with the flame pointed in the direction of travel. See Fig. 6-4. The base metal will begin to melt and form a weld pool. The metal begins to sag as the pool increases in diameter. Notice the width of the weld pool when the metal sags. Move the torch flame ahead slowly and the pool will follow. Continue moving the torch ahead slowly, carrying a consistent-width weld pool to the end of the joint.

The end of the metal piece may become overheated as the weld pool approaches the end of the joint. The flame should be raised about 1″ (25 mm) from the surface. This allows the metal to cool. Return the flame

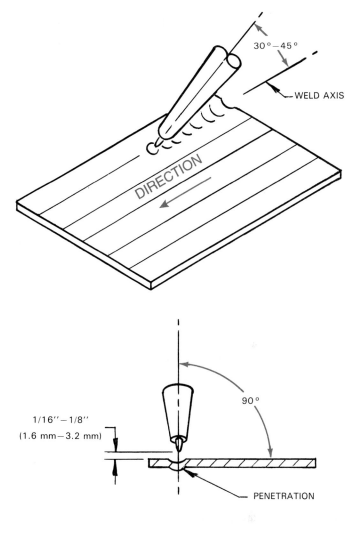

Fig. 6.4 Creating a continuous weld pool. The 30° — 45° angle may vary, depending on the amount of heat required. The 90° angle keeps the heat equal on both sides of the centerline. Notice how the weld pool sags under the flame, creating penetration.

to within 1/16″ — 1/8″ (1.6 mm — 3.2 mm) of the metal to complete the pass.

A side-to-side torch motion may be required to widen the weld pool. Use a crescent-shaped motion or one similar to those shown in Fig. 6-5 to make a wide bead. Some type of flame motion is required to keep the surface of the pool agitated. This prevents the surface from cooling too rapidly.

A weld pool should remain the same width its entire length. This ensures the same amount of deformation on the underside of the metal. See Fig. 6-6.

If the pool becomes too large, the metal begins to sag excessively. The molten metal may drop through, creating a hole. A hole in the metal indicates that too much time was spent welding in one spot. A hole also indicates that the weld pool was allowed to become too large. If a weld pool becomes too large or too deep, withdraw the flame about 1″ (25 mm). When the metal cools slightly, return the flame to within 1/16″ − 1/8″ (1.6 mm − 3.2 mm) of the metal. The tip size may be too large if the flame must be withdrawn often or if the weld pool must be moved too rapidly.

A lack of penetration or sag indicates that not enough time was spent welding in an area. A narrow weld pool is generally the cause of little or no penetration. If the weld pool cannot be made wide enough, the tip size may be too small. Select a tip with a larger orifice to produce more heat and create a wider pool.

WELDING WITHOUT A WELDING ROD

Several types of welds can be made without using a welding rod. The base metal is used as the filler metal in these welds. When made properly, these welds are as strong as any made using a welding rod.

The edge joint, outside corner joint, and an edge-flanged weld in a butt joint can be made without welding rods. See Fig. 6-7. The flame is used to melt the edges of the metal.

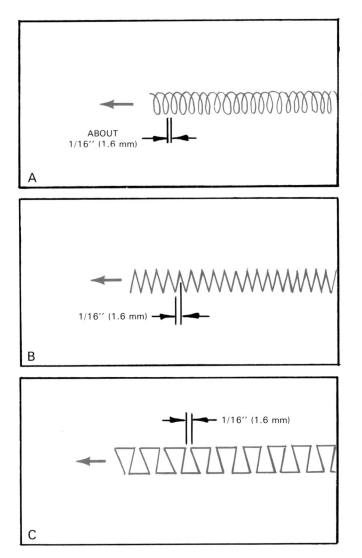

Fig. 6-5. *Three suggested flame motions. In each case, the torch should move forward about 1/16″ (1.6 mm).*

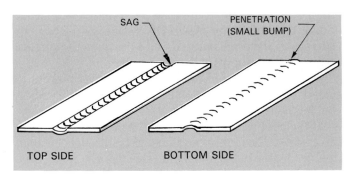

Fig. 6-6. *The appearance of a good weld pool pass and the penetration on the underside.*

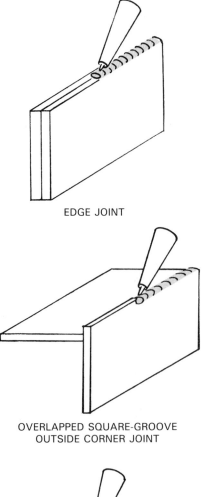

EDGE JOINT

OVERLAPPED SQUARE-GROOVE
OUTSIDE CORNER JOINT

EDGE WELD IN A
FLANGE BUTT JOINT

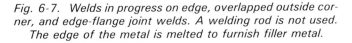

Fig. 6-7. *Welds in progress on edge, overlapped outside corner, and edge-flange joint welds. A welding rod is not used. The edge of the metal is melted to furnish filler metal.*

EDGE JOINT

Welds on edge joints of thin metal assemblies are often made without filler metal. The flame is applied to the edges to melt them and form a weld bead. The resulting weld can have deep penetration and a very uniform bead. The joint must be tightly fitted, run from edge to edge, and not have an overlap.

EXERCISE 6-2 — WELDING AN EDGE JOINT

1. Obtain two pieces of low carbon steel, each measuring 1/16″ x 1 1/2″ x 6″ (1.6 mm x 38 mm x 152 mm).
2. Clean the surface at least 1/2″ (12.7 mm) from the edges. Place the pieces in the desired position.
3. Select the correct torch tip and install it in the torch. See Fig. 4-29.
4. Start the welding outfit. Adjust the outfit to obtain the correct pressures.
5. Light the torch and adjust to a neutral flame.
6. Tack weld the pieces in three places.
7. Carry a weld pool along the edges for the length of the metal. See Fig. 6-7.

Inspection:
The bead should be smooth. It should fully extend across the joint from one edge to the other.

CORNER JOINT

Inside corner joints are generally not welded without a welding rod. The rod furnishes the additional metal normally needed to reinforce the weld.

Several outside corner joints are designed to provide filler metal to reinforce the weld. The pieces used to form an outside corner joint may be overlapped as shown in Fig. 6-7. They may also be bent in a brake to form a corner-flange or flare-bevel joint. The overlapped or bent metal edge is melted with the flame to form a uniform weld bead. The bead width should be twice the base metal thickness.

EXERCISE 6-3 — WELDING AN OUTSIDE CORNER JOINT

1. Obtain two pieces of low carbon steel, each measuring 1/16″ x 1 1/2″ x 6″ (1.6 mm x 38 mm x 152 mm).
2. Clean the surfaces at least 1/2″ (12.7 mm) from the edges. Align the pieces as shown in Fig. 6-7.
3. Select the correct torch tip and install it in the torch. See Fig. 4-29.
4. Start the welding outfit. Adjust the outfit to obtain the correct pressures.
5. Light the torch and adjust to a neutral flame.

6. Tack weld the pieces in three places.
7. Melt the overlapped edge until it is flush with the horizontal surface. See Fig. 6-7.

Inspection:
The completed weld should have a smooth bead. The overlapped edge should be flush with the surface of the horizontal piece. The bead should be about 1/8″ (3.2 mm) wide.

EDGE WELD IN A FLANGED BUTT JOINT

The flanged butt joint is prepared by bending the metal edge in a brake. These bent edges furnish the filler metal. The edges must fit tightly together, since they are melted to form a weld that runs from outside edge to outside edge. The finished weld has an unusually uniform bead. See Fig. 6-7.

EXERCISE 6-4 — WELDING AN EDGE WELD IN A FLANGED BUTT JOINT

1. Obtain two pieces of low carbon steel, each measuring 1/16″ x 1 1/2 ″ x 6″ (1.6 mm x 38 mm x 152 mm).
2. Create a 1/4″ (6.4 mm) flange along one edge of each of the pieces. Use a brake press to bend the metal.
3. Clean the surfaces at least 1/2″ (12.7 mm) from the edges. Align the pieces as shown in Fig. 6-7.
4. Select the correct torch tip and install it in the torch. See Fig. 4-29.
5. Start the welding outfit. Adjust the outfit to obtain the correct pressures.
6. Light the torch and adjust to a neutral flame.
7. Tack weld the pieces in three places.
8. Run a weld pool along the edges to form a smooth bead. See Fig. 6-7.

Inspection:
The completed bead should be straight, even in width, and free of holes.

SELECTING A WELDING ROD

A welding rod is added to a weld pool to:
1. Fill a groove weld.
2. Form a fillet weld.
3. Fill a weld pool that has a depression in it.
4. Make a completed weld as strong as the base metal.

Welding rods are available in the following diameters: 1/16″ (1.6 mm), 3/32″ (2.4 mm), 1/8″ (3.2 mm), 5/32″ (4.0 mm), 3/16″ (4.8 mm), 1/4″ (6.4 mm), 5/16″ (7.9 mm), 3/8″ (9.5 mm). They are available in standard 36″ (.91 m) lengths.

The required amount of filler metal should be added to the weld pool each time that the welding rod is dipped into the weld pool.

A welding rod of the correct diameter forms a good convex bead. The correct size rod also permits the weld pool to remain fluid as the rod is added. See Fig. 4-29 for a table of suggested welding rod sizes.

A rod that is too small does not add enough filler metal to the pool. The rod must be dipped into the weld pool repeatedly to form a good bead. A welding rod that is too large in diameter can cause the pool to *freeze.* This occurs when the welding rod is too large and the attempt to melt it draws too much heat from the pool. The weld pool cools and traps the welding rod. The pool must be reheated to release the rod. A torch tip orifice that is too small in diameter may also cause a welding rod to freeze in the pool.

Your hand position on the welding rod must change as it melts and gets shorter, Fig. 6-8. To change hand position, place the rod on the work table or the weldment, with the hot end away from your body. Move your hand to a new position and pick up the rod. Do not place the rod against your body to change hand position.

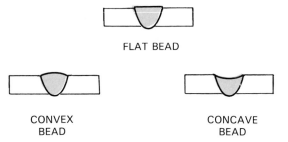

FLAT BEAD

CONVEX BEAD

CONCAVE BEAD

Fig. 6-9. Flat, convex, and concave beads. The convex bead is generally preferred. A flat bead is used when it is to be ground down after welding. A concave bead is weaker because it is thinner.

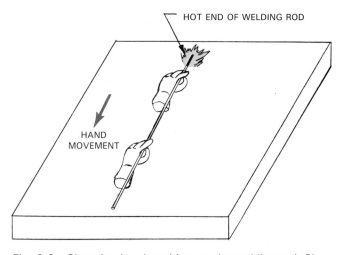

HOT END OF WELDING ROD

HAND MOVEMENT

Fig. 6-8. Changing hand position on the welding rod. Place the rod on a table or the weldment with the hot end facing away from you. Then, move your hand to a new position.

LAYING A BEAD

A weld bead results from adding filler metal to a weld pool. In this case, the filler metal is a welding rod. The welding rod is added to the weld pool until the pool is flush with the surface or slightly convex. See Fig. 6-9. A *convex bead,* one that is raised slightly above the surface, is commonly used. The filler metal increases the strength of the metal in the area of the joint.

A *flat bead* is used whenever the bead is to be ground down after welding. A *concave bead* is sunk below the surface. It is the weakest type of bead because it is thinner. It is used in fillet welds when a blended appearance is more important than strength.

Laying a bead requires the use of both hands. A right-handed welder holds the torch in the right hand and the welding rod in the left hand. Fig. 6-10 shows the positions of the welding rod and torch.

The torch tip is generally inclined at 30°−45° angle from the base metal. To alter the amount of heat, the torch angle can be changed. Lower angles cause more heat to be reflected from the base metal. Higher angles concentrate more heat on the metal. The torch should be held at 90° (perpendicular) to the surface of the base metal, as shown in the end view, Fig. 6-10. At this

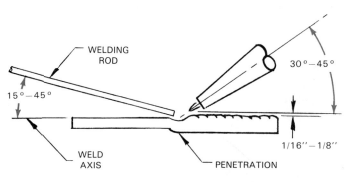

WELDING ROD

30°−45°

15°−45°

WELD AXIS

PENETRATION

1/16″−1/8″

SIDE VIEW

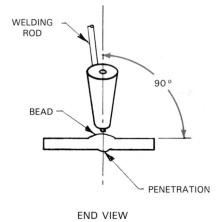

WELDING ROD

90°

BEAD

PENETRATION

END VIEW

Fig. 6-10. Torch tip and rod positions for welding a butt joint.

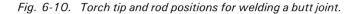

angle, the metal is heated equally on each side of the weld axis. If metals of different thicknesses are being welded, incline the torch toward the thicker piece.

The flame is hottest at a distance of 1/16″ − 1/8″ (1.6 mm − 3.2 mm) from the tip of the flame. Therefore, the tip of the flame should be held at that distance from the base metal. The flame may be drawn away to reduce the heat input to the metal.

The welding rod is normally inclined at 15° − 45° to the weld surface. More of the rod is heated by the reflected heat from the torch flame when the rod is held at a low angle.

The end of the welding rod must be held near the weld pool to keep it hot. A welding rod will cool if it is held too far from the pool. The pool may freeze if a cool welding rod is placed in it.

A weld pool must be formed first when laying a bead. The welding rod is added when the pool sags. The rod is added by dipping it into the *leading* (forward) *edge* of the weld pool. The heat melts the end of the welding rod, filling the pool. A welding rod that is the correct diameter will only need to be dipped into the weld pool every few seconds to form a good bead.

The flame movement and the use of the welding rod must be coordinated to produce a good bead. Move the flame and weld pool along the weld line at a constant rate. Control the width and depth of the pool, and move the welding rod in and out of the pool consistently. This will produce a uniform ripple in the bead. Fig. 6-11 shows a well-formed bead with adequate penetration.

Raise the flame slightly toward the end of the bead. This allows the weld pool to cool as the welding rod is added to fill the end of the bead.

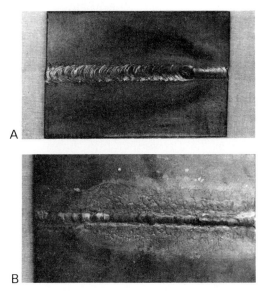

Fig. 6-11. A—A well-formed cover pass bead in a butt weld on 1/4″ mild steel. The root pass bead is visible at the right side of the incomplete cover pass bead. B—The underside of the same weld. Note that 100% penetration is indicated by the continuous sag of the metal at the root of the weld.

EXERCISE 6-5 — LAYING A BEAD

1. Obtain one piece of mild steel measuring 1/16″ x 3″ x 6″ (1.6 mm x 76 mm x 152 mm).
2. Clean the surfaces.
3. Lay out four lines on the surface 3/4″ (19.1 mm) apart with chalk or soapstone.
4. Select the proper size welding tip and welding rod size. See Fig. 4-29.
5. Start the welding outfit and set the correct pressures. Light the torch and adjust for a neutral flame.
6. Form a weld pool on the first line. Add the welding rod when the pool is the correct width and after it sags.
7. Add enough filler metal to form a convex bead.
8. Carry the weld pool, add filler metal, and form a bead in a smooth, continuous motion.
9. Continue the bead to the end of the metal. See Fig. 6-11. Repeat the procedure on the other three lines.

Inspection:

The completed bead should be straight, consistent in width, and free of holes. It should have evenly spaced ripples and 100% penetration. On thin metal, penetration is indicated as a continuous sag on the reverse side of the metal.

LAP JOINT

A machined or a flame cut edge is not required for a lap joint. The two pieces must be in good contact with one another. They should be held tightly together with clamps and then tack welded. A *tack weld* is a small weld used to hold pieces in alignment.

A lap joint is welded along the edge of one piece and the surface of the other piece. The flame should be mainly directed toward the surface. See Fig. 6-12. The surface needs more heat to melt than does the edge.

A *C-shaped weld pool* should be formed when welding a lap joint. See Fig. 6-13. The C-shaped weld pool is the indicator that the surface has melted enough to create a good weld. It also indicates good fusion on the surface of the piece. The welding rod is added to the center of the C-shaped weld pool when it forms. Continue to dip the rod into the pool until the correct bead shape is formed.

After adding the welding rod, move the flame ahead about 1/16″ (1.6 mm). Wait for the C-shaped weld pool to form again. Add the welding rod and move on. Move ahead at a steady rate, adding the welding rod each time the C-shaped pool is well-formed. Continue this procedure to the end of the weld.

The edge of the lap joint will sometimes melt away rapidly. This is commonly caused when the two pieces of metal are not in good contact. The flame travels between the pieces, melting the edge rapidly and uncontrollably. To slow melting of the edge, rest the welding rod on it. The welding rod will draw some heat away from the edge.

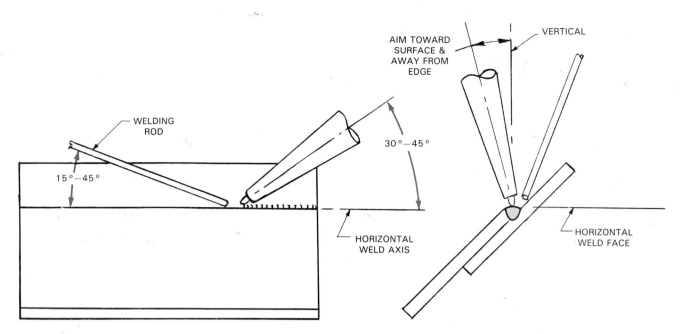

Fig. 6-12. A fillet weld on a lap joint. The edge melts before the surface. Aim the flame slightly more toward the surface to apply additional heat there.

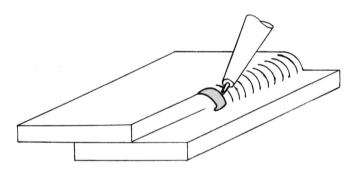

Fig. 6-13. A C-shaped weld pool indicates that the surface and edge of the base metal have melted. When making a fillet weld on a lap or corner joint, add the welding rod after the C-shaped weld pool forms.

EXERCISE 6-6 — FILLET WELD ON A LAP JOINT

1. Obtain two pieces of mild steel, each measuring 1/16'' x 1 1/2'' x 6'' (1.6 mm x 38 mm x 152 mm).
2. Clean the surfaces and edges of both pieces.
3. Select the correct torch tip and welding rod size. See Fig. 4-29.
4. Start the welding outfit and set the working pressures. Light the torch and adjust to a neutral flame.
5. Position the pieces so that they overlap about 3/4'' (19 mm).
6. Clamp the pieces tightly, using C-clamps.
7. Tack weld the joint three times on the top and bottom sides.
8. Weld the fillet weld to create a convex bead. See Fig. 6-12. Wait for a C-shaped weld pool to form before adding the welding rod.

9. Make a second fillet weld on the other side.
Inspection:
The completed welds must be straight and have evenly spaced ripples. The bead should be convex and have consistent width.

INSIDE CORNER AND T-JOINTS

Inside corner and T-joints may be welded with or without edge preparation. On thin metal, a square-groove weld is used. A bevel-groove weld may have to be used on metal over 3/16'' (4.8 mm) thick to ensure good penetration. On thicker metal, a bevel-groove weld is followed by a fillet weld.

A fillet weld alone is used on square-groove inside corner and T-joints. The inside corner joint is welded from one side only; T-joints may be welded from both sides.

The leg size of a fillet weld is determined by the weld joint designer. As a rule of thumb, however, the leg size should be equal to, at least, the metal thickness. For example, 1/8'' (3.2 mm) thick metal requires 1/8'' (3.2 mm) leg sizes. The leg size should be 1/4'' (6.4 mm) for 1/4'' (6.4 mm) thick metal.

The weld for an inside corner or a T-joint is made on two surfaces. The torch flame must heat each piece equally. Fig. 6-14 shows the position of the pieces of metal when welding an inside corner or T-joint in the flat position. The correct angles for the welding tip and welding rod are also shown.

A C-shaped weld pool must be formed when fillet welding an inside corner or T-joint. The weld pool shows that the metal surfaces have melted properly. The welding

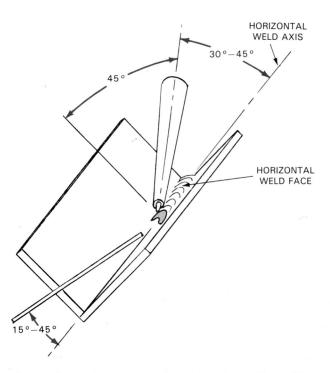

Fig. 6-14. Torch, flame, and rod positions for making a fillet weld on an inside corner joint in the flat welding position. The weld face is horizontal. The torch tip is at a 45° angle to the metal and 30°—45° to the weld axis. Notice the C-shaped weld pool.

rod is dipped into the upper edge of the pool after it is formed. Continue dipping the rod into the weld pool until the desired bead shape is formed. If the shape is not specified, use a convex bead for greatest strength. Move the torch ahead smoothly and dip the welding rod into the weld pool after the C-shape re-forms. Continue the procedure to the end of the joint.

EXERCISE 6-7 — WELDING A T-JOINT WITH A FILLET WELD

1. Obtain two pieces of mild steel, each measuring 1/16″ x 1 1/2″ x 6″ (1.6 mm x 38 mm x 152 mm).
2. On one piece, clean both sides along one 6″ edge. Clean one surface of the second piece.
3. Select the correct torch tip and welding rod. See Fig. 4-29.
4. Start the welding outfit and adjust the pressures. Light the torch and adjust to a neutral flame.
5. Place the clean edge in the middle of the clean surface. Use a brick to prop up the vertical piece.
6. Tack weld the joint three times on each side. Alternate the tack welds from side-to-side and end-to-end. This prevents the vertical piece from pulling in one direction.
7. Weld a fillet weld on one side using the correct torch tip and welding rod positions. Do not dip the welding rod until a good C-shaped pool is formed. See Fig. 6-14.

8. Weld a fillet weld on the other side.
Inspection:
The completed welds should be convex and even in width. The bead should be straight and have evenly spaced ripples.

BUTT JOINT

The edges of metal used to form a butt joint are generally machined or flame cut. Metal thicker than 3/16″ (4.8 mm) must have its edges prepared as a bevel-groove, V-groove, J-groove, or U-groove. Thicker metal may be welded from one or both sides.

Weld beads seldom should be thicker than 1/4″ (6.4 mm). Thicker beads will cool before gases and impurities in the weld rise to the surface. Gases and impurities trapped within a weld weaken it.

Welds made on thick metal may require more than one bead. Each completed bead requires one welding *pass.* A weld that requires three beads needs three passes to complete it. Complete penetration is possible on metal 3/16″ (4.8 mm) or less without flame cutting or machining the edges.

Metal in a butt joint will pull together due to expansion and contraction that takes place during welding. The expansion may be so great that the adjoining edges overlap at the end of a long butt joint. To prevent overlapping or reducing the weld root, the pieces must be clamped together or tack welded. Tack welds are often placed 3″ (76 mm) apart. See Fig. 6-15. The tack weld melts and becomes part of the main weld as welding proceeds. A tack weld must be well-made so it will not break as the weld is made.

Expansion may be compensated for by yet another method. In this method, the root opening tapers from beginning to end. See Fig. 6-16. The amount of taper

Fig. 6-15. Two tack welds being made to hold a butt joint in alignment for welding.

Fig. 6-16. *The root opening is tapered prior to welding this butt joint. As the weld progresses from the narrow end, the metal pulls together, closing the gap.*

may be difficult to estimate. Therefore, this method should be used only when identical welds are to be made repeatedly. Several factors must be considered in determining the amount of taper needed to avoid overlap. These include metal thickness, type of metal, tip size, and the length of the weld.

Welds made on butt joints must totally penetrate the base metal. Thick metal may be welded from both sides to ensure 100% penetration. Multiple passes may also be required on thick metal.

The edges of a piece of metal that have been prepared by machining or flame cutting are thinner at the root. This causes the *root of the weld* to melt more rapidly. The root opening enlarges due to this rapid melting. The enlarged root opening looks like an old-fashioned keyhole. See Fig. 6-17. The *keyhole* indicates that the metal has been melted through to the reverse side. Total

penetration is ensured if the keyhole develops before adding the welding rod. This procedure is known as *keyhole welding* or keyholing.

EXERCISE 6-8 — WELDING A SINGLE SQUARE-GROOVE BUTT JOINT

1. Obtain two pieces of mild steel, each measuring 1/16″ x 1 1/2″ x 6″ (1.6 mm x 38 mm x 152 mm).
2. Clean the edges. Clean the surface of the pieces 1/2″ (12.7 mm) from the edge.
3. Select the correct torch tip and welding rod. See Fig. 4-29.
4. Start the welding outfit and set the pressures. Light the torch and adjust to a neutral flame.
5. Tack weld the pieces near the beginning, middle, and end of the joint. Be sure to maintain a 1/16″—3/32″ (1.6 mm—2.4 mm) root opening.
6. Melt the surfaces of the pieces evenly. Side-to-side torch motion may be needed. The molten metal fills the root opening. It also forms a weld pool and sags. Notice the width of the pool when this occurs. Add the welding rod to the leading edge of the pool as needed to fill the pool and form a bead. Move the torch flame slowly and continuously ahead. Add the welding rod as the correct width weld pool re-forms. Continue the procedure to the end of the joint.

Inspection:

The completed weld should have a uniform width and consistent ripple pattern. Total penetration should be indicated on the underside.

WELD DEFECTS

Welders should be able to identify a good weld. They also must be able to recognize defects and know how

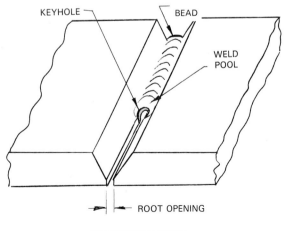

PERSPECTIVE VIEW

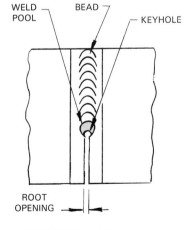

VIEW FROM ABOVE

Fig. 6-17. *Two views of a keyhole. The keyhole indicates that the weld pool just ahead of the bead has penetrated through the metal. A keyhole should also be present in SMAW, GMAW, and GTAW.*

to prevent them.

Most often welds are visually inspected. Fig. 6-18 shows fillet welds with different bead contours and some common weld defects. Fig. 6-18A—C shows well-formed fillet welds with different bead contours. Each weld has good penetration. In Fig. 6-18D, the vertical piece of metal has been thinned and weakened by an undercut weld. A weld bead with overlap is shown in Fig. 6-18E. The overlapped bead does not add strength to the weld. The actual size of the completed weld as shown by the dashed line, is smaller than desired. The weld in Fig. 6-18F will be weak because of poor penetration.

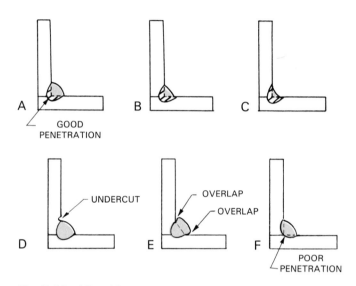

Fig. 6-18. Visual inspection of fillet welds on inside corner joints. The penetration at A, B, and C is good. A convex bead, as shown at A, is generally preferred. At D, E, and F the defects of undercut, overlap, and poor penetration are shown.

Fig. 6-19 shows one well-formed and two poorly made butt welds. A narrow bead, poor penetration, and overlap are caused by insufficient heat. Insufficient heat is the result of using too small a welding tip or a welding speed that is too fast. If the weld pool is not large enough, the weld will not penetrate, the bead will be narrow, and overlap may occur. If the torch tip is too large, a wide bead and excessive penetration may result.

Undercutting is the thinning of the base metal along the toe of the weld. Undercutting can occur in fillet welds and in welds made on groove joints. It is eliminated by adding more filler metal to the weld pool. Undercutting occurs when the incorrect torch flame angle is used. On a horizontal fillet weld, the filler metal sags from the vertical piece causing an undercut condition. The problem is corrected by adding filler metal to the upper edge of the weld pool.

Overlap is a condition in which the bead is not fused into the base metal. Overlap occurs when the base metal is not molten at the time the welding rod is added.

Incomplete penetration results in a weak weld. When welding thick metal, use a groove-type weld. A larger tip size and carrying a deeper weld pool also will eliminate poor penetration.

Inclusions are foreign materials in the weld pool. Cleaning the metal before welding, and keeping a clean welding area, will help to eliminate inclusions. Air pockets also may occur. The welding flame should be constantly kept in motion to stir the weld pool and eliminate this problem.

Flat or concave beads may be required on a weld. A convex bead is generally preferred, however, since it is stronger. A convex bead is obtained by adding more filler metal to the weld pool.

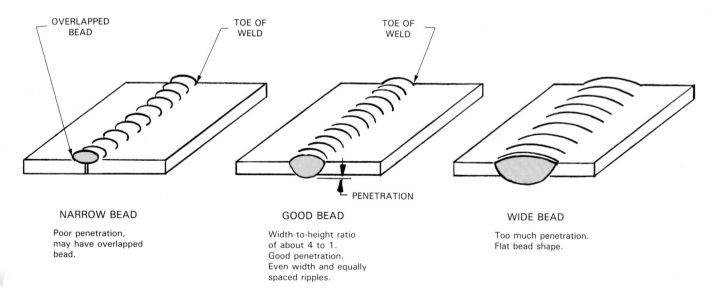

Fig. 6-19. Visual inspection of welds on a butt joint. A good weld bead is convex unless a different shape is specified. The weld must penetrate through the entire thickness. A narrow bead will not penetrate deeply and may show overlap. Too much heat produces a bead that is too wide and has too much penetration.

REVIEW QUESTIONS

Write your answers on a separate sheet of paper.

1. Refer to Fig. 4-29. What welding rod and torch tip size should be used for welding 1/8″ (3.2 mm) thick mild steel?
2. When creating a continuous weld pool, how is penetration indicated?
3. If the weld pool must be moved quickly or if holes occur in the pool, the tip size may be too _____.
4. How does a welder know when the welding rod is too large?
5. List three types of joints that can be welded without using welding rod.
6. Metal that is _____″ (_____ mm) and thicker should have its edges prepared by machining or flame cutting.
7. Why should pieces be tack welded before they are welded?
8. Which of the following will help ensure 100% penetration of a butt joint on thick metal?
   a. Form a bevel or groove.
   b. Weld from both sides.
   c. Use multiple passes.
   d. Use bigger tip size.
   e. All of the above.

9. How do you know when to dip the rod in the weld pool while making a fillet weld on a T-, lap, or inside corner joint?
10. Complete the sketch below to show what a "keyhole" looks like when viewed from above.

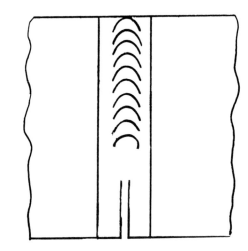

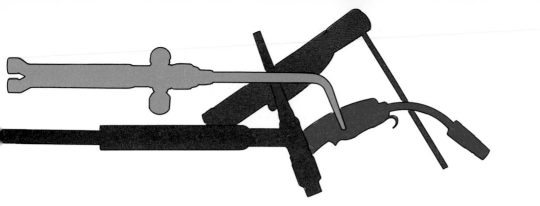

Chapter 7

OFW: Horizonal, Vertical, and Overhead Welding Positions

After studying this chapter, you will be able to:
☐ Define out-of-position welding.
☐ Identify safety measures to be taken when welding out of position.
☐ Describe methods used to perform welds in the horizontal welding position.
☐ Weld in the horizontal welding position.
☐ Describe methods used to perform welds in the vertical welding position.
☐ Weld in the vertical welding position.
☐ Describe methods used to perform welds in the overhead welding position.
☐ Weld in the overhead welding position.

TECHNICAL TERMS

Crescent-shaped motion, out-of-position welding, positioner, sagging.

OUT-OF-POSITION WELDING

Welding is easiest when done in the flat welding position. However, parts that are fixed in position or that are too heavy to move must be welded in another position. Welding in a position other than the flat position, is commonly referred to as *out-of-position welding.*

Welds made out-of-position must be as strong as welds made in the flat welding position. An out-of-position weld should look as good as a weld made in the flat welding position. More training and talent is required to weld out-of-position. Those who can weld out-of-position have a chance to earn more money.

Some companies use positioners to rotate large parts. *Positioners* are large machines that rotate parts so that all welds are made in the flat welding position. See Fig. 7-1.

PREPARING TO WELD

Many out-of-position welds are performed at or above eye level. Similar protective clothing is worn for

Fig. 7-1. A large rotary welding positioner. (Kioke Aronson, Inc.)

out-of-position welding and flat position welding. Coveralls or jackets should be buttoned at the collar to prevent hot metal from getting inside clothing. Pockets must be buttoned. Flammable items should not be carried in pockets. A cap should be worn when performing overhead welds. Flame-resistant leather clothing may also be worn for additional protection. Welding goggles with #4 — #6 lenses are recommended.

The torch tip and welding rod sizes are selected based on the metal thickness and/or the manufacturer's recommendations. Fig. 4-29 shows a table of recommended tip orifice and welding rod sizes. The welding outfit is turned on and the flame is lit and adjusted. A neutral flame should be obtained.

WELDING IN THE HORIZONTAL WELDING POSITION

Welds made in the horizontal welding position must meet the following conditions:
1. The weld axis is within 15° of horizontal.
2. The weld face is between 80° — 150° or 210° — 280°.

Fig. 3-21 shows the position of a horizontal weld. For practice welding, the weld face is usually vertical with a horizontal weld axis. Parts for practice welds may be held in a fabricated welding positioner. See Fig. 7-2.

LAP JOINT

The pieces of metal for a lap joint in the horizontal position should be arranged as shown in Fig. 7-3. The weld face for practice pieces should be at an angle of 135° or 225°

EXERCISE 7-1 — FILLET WELD ON A LAP JOINT IN THE HORIZONTAL WELDING POSITION

1. Obtain two pieces of mild steel that measure 1/16″ x 1 1/2″ x 6″ (1.6 mm x 38 mm x 152 mm).
2. Clean the surfaces of both pieces.

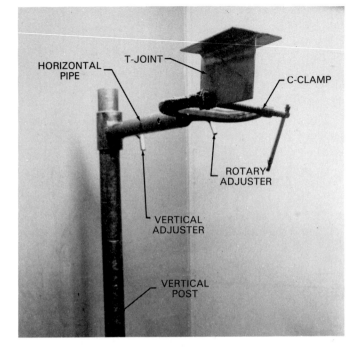

Fig. 7-2. A fixture used to hold practice welds while welding out of position. This fixture is adjustable in three ways. A C-clamp is used to hold the weldment.

3. Place one piece on the other with a 3/4″ (19.1 mm) overlap.
4. Clamp the parts tightly together.
5. Tack weld in three places on both sides.
6. Remove the clamps.
7. Form a C-shaped weld pool before adding the welding rod to the pool. Weld both sides.

Inspection:
The completed welds should have convex beads with a consistent width and even ripples.

Torch and flame position

When welding a lap joint, the edge of one piece melts more rapidly than the surface of the other. Therefore,

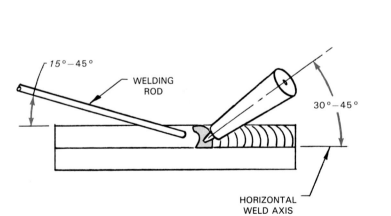

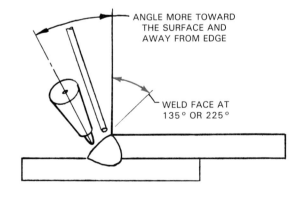

Fig. 7-3. Torch, rod, and flame positions for lap welding. Note: These angles are typical for lap welding in any position.

the center line of the tip and flame should point more at the surface than at the edge. This lessens the amount of heat applied to the edge. The torch tip and flame should be held at 30°−45° to the weld axis.

Welding rod position

The edge of thin metal tends to melt quickly when welded. To prevent this, hold the welding rod on the edge of the metal before inserting it in the weld pool. The rod draws heat away from the pool and decreases the amount of heat applied to the edge. This helps slow the melting of the edge. The rod is held at a 15°−45° angle from the metal surface, Fig. 7-4. It is usually aligned with the weld axis.

Fig. 7-4. A lap joint being welded in the horizontal welding position. The torch and flame are held at an angle so that more heat is applied to the surface than to the edge.

Weld pool

A C-shaped weld pool should form before adding the welding rod. Insert the welding rod at the upper edge of the pool. This helps to overcome gravity and improve the shape of the bead. A convex bead should normally be formed. A flat or concave bead may be specified in the working drawing. A concave bead is not as strong as a convex bead. The concave bead may be required when appearance is important such as on the outside of an appliance.

CORNER AND T-JOINTS

A procedure similar to the one used for a fillet weld on a lap joint is used to place a fillet weld in an inside corner or T-joint. The procedure for welding an out-side corner joint is similar to welding a butt joint. Fig. 7-5 shows a fillet weld being made on an inside corner joint. The weld face for most horizontal practice welds should be at about 135°.

Fig. 7-5. A fillet weld being made on an inside corner or T-joint. Note that the tip is held at 45°, so that the heat is applied evenly to both surfaces. If undercutting occurs, the flame is aimed more toward the vertical piece.

EXERCISE 7-2 — INSIDE CORNER JOINT IN THE HORIZONTAL WELDING POSITION

1. Obtain two pieces of mild steel that measure 1/16″ x 1 1/2″ x 6″ (1.6 mm x 38 mm x 152 mm).
2. Clean the surfaces of both pieces.
3. Place the metal together to form a corner joint. Use a welding positioner or metal block to support the vertical piece.
4. Tack weld the pieces in three places.
5. Hold the torch tip, flame, and welding rod at the suggested angles. Refer to Fig. 7-6 for angles.
6. Melt the two pieces evenly and watch for a C-shaped weld pool to form. When the pool forms, add the welding rod to the upper edge of the pool.
7. Continue this procedure to the end of the joint.

Inspection:

The finished weld should not penetrate through either piece. The bead should be convex, even in width, and have evenly spaced ripples.

Torch and flame position

The center line of the torch tip and flame should be centered between the surfaces of the two pieces of metal. The torch tip should also be at a 30°−45° angle to the weld axis. See Fig. 7-6. Both surfaces must be heated evenly. If you notice one piece heating more than the other, move the tip slightly to compensate.

Welding rod position

The welding rod should be placed into the upper edge of the weld pool. This reduces the possibility of under-

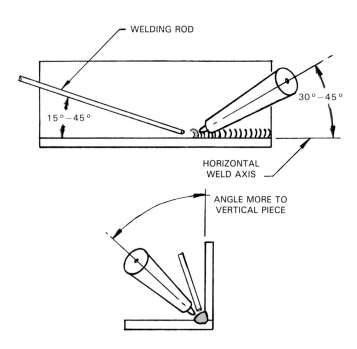

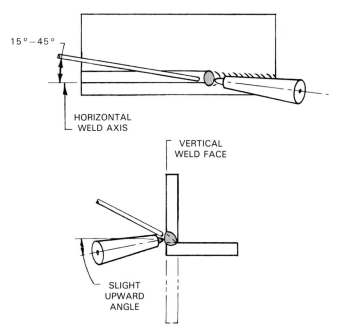

Fig. 7-6. Torch, flame, and welding rod positions for placing a horizontal fillet weld on an inside corner or T-joint. The flame should be directed toward the vertical piece to help oppose the force of gravity on the weld pool.

Fig. 7-7. Torch, flame, and welding rod positions for welding a horizontal butt joint or outside corner joint. The torch and tip are held at a 30° – 45° angle from the surface of the metal. The torch should be angled up slightly to counteract gravity.

cutting on the vertical surface. The rod should be held at a 15° – 45° angle to the weld axis. The welding rod should be used to fill any undercutting. If undercutting does occur, aim the flame a few degrees more toward the vertical piece. The pressure of the welding gases also helps to reduce sagging and undercutting.

Weld pool

A C-shaped weld pool eventually forms as the metal melts. The C-shaped pool shows that the surfaces of both pieces are melting. The welding rod is inserted only after the weld pool forms.

Welding a square-groove, bevel, or V-grooved outside corner joint is similar to welding a butt joint. The edge of one piece of metal and the surface of another piece are heated and fused together. More heat is required to melt the surface, especially when welding thin metal. The flame is usually pointed more toward the surface of the joint than at the edge. See Fig. 7-7.

BUTT JOINT

The pieces of metal for a horizontal butt weld are arranged as shown in Fig. 7-8. The pieces may be tack welded to hold them in correct alignment.

Various techniques are used to create strong welds and well-formed weld beads. Most thicknesses of metal can be welded using the forehand welding method. The downward pull of gravity tries to deform the weld bead in all out-of-position welding. The force of the welding gases is used to overcome the downward pull of gravity. Careful placement of the welding rod into the upper edge of the pool also helps to overcome gravity.

Fig. 7-8. Horizontal butt joint being welded with oxyacetylene welding (OAW). Note the slight upward angle of the torch. The pressure of the welding gases is used to oppose the force of gravity on the weld pool.

EXERCISE 7-3 — SQUARE-GROOVE BUTT JOINT IN THE HORIZONTAL WELDING POSITION

1. Obtain two pieces of mild steel 1/16'' x 1 1/2'' x 6'' (1.6 mm x 38 mm x 152 mm).
2. Clean the edges.
3. Place the pieces in the flat welding position and tack weld in three places.
4. Position the pieces for horizontal welding.

5. Weld the joint, using the suggested torch tip and flame angles.

Inspection:

The completed weld should have a bead with even ripples and bead width. It should also have a convex bead and 100% penetration.

Torch end flame position

The center line of the torch tip and flame are pointed upward toward the weld axis. The upward angle should be about 10° − 30°, as shown in Fig. 7-9. The upward angle of the flame and the force of the welding gases will help to prevent the bead from *sagging* (flowing downward). The angle may be changed to form an acceptable bead. The metal near the center line of the weld on both surfaces must be heated and melted evenly. The torch tip and flame are also positioned between 30° − 45° from the surface of the base metal. See Fig. 7-9. The 30° − 45° angle is used for all welding positions.

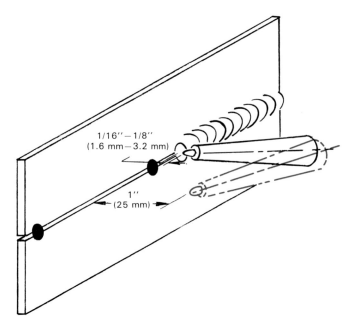

Fig. 7-10. The flip motion of the torch and flame. This motion may be used if the metals tend to overheat. It allows time for the weld pool to cool.

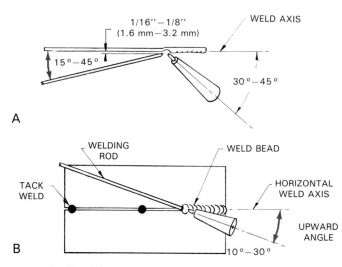

Fig. 7-9. Torch, rod, and flame positions for welding a horizontal butt joint. A—View from above. B—View from front.

Welding rod position

The welding rod is held at a 15° − 45° angle from the base metal. This angle is used for all welding positions. When horizontal welding, the rod is placed into the upper edge of the molten weld pool. This permits the rod to melt and cool the weld pool before the molten metal sags.

Weld pool

The weld pool may become too hot while welding. To cool the pool, withdraw the flame tip about 1″ (25 mm) from the base metal and return it to within 1/16″ − 1/8″ (1.6 mm − 3.2 mm). This rapid up-and-down motion allows the weld pool to cool. See Fig. 7-10. If undercutting occurs on the upper piece, insert the welding rod in the pool more often. This fills the undercut (depression). Increasing the upward angle of the flame also helps to force molten metal higher into the weld pool.

WELDING IN THE VERTICAL WELDING POSITION

Vertical welding is also called welding in the 3G position. Welds in the vertical welding position must meet either of the following sets of conditions.

Condition A

1. Weld axis must be between 80° − 90° from horizontal.
2. Weld face is between 80° − 90° from horizontal.

Condition B

1. Weld axis must be between 15° − 80° from horizontal.
2. Weld face must be between 80° − 280°.
3. Welding is performed from upper side of the joint. For practice welding, place the weld axis as close to vertical as possible.

Vertical welding on any type of joint is generally done using the forehand method. The weld begins at the lowest point on the joint and progresses upward.

The upward pressure of the welding gases keeps the molten metal in the weld pool from sagging. Placing the welding rod into the upper edge of the weld pool also helps to prevent sagging. A pencil-like grip is effective and comfortable for vertical welding. This is an especially good way to hold a light-duty torch.

LAP JOINT

A fillet weld is used on a lap joint. For practice, the weld axis is positioned about 80° −90° from horizontal. This is the 3F position. In a lap joint, the edge of one piece and the surface of another piece are heated.

EXERCISE 7-4 — FILLET WELD ON A LAP JOINT IN THE VERTICAL WELDING POSITION

1. Obtain two pieces of mild steel measuring 1/8″ x 1 1/2″ x 6″ (3.4 mm x 38 mm x 152 mm).
2. Clean the surfaces of both pieces.
3. Place one piece on the other with a 3/4″ (19.1 mm) overlap.
4. Clamp the pieces tightly together.
5. Tack weld in three places on both sides.
6. Remove the clamps and place the weld axis in the vertical position.
7. Wait for the C-shaped weld pool to form before adding the welding rod. Make a fillet weld on both sides of the metal.

Inspection:
The completed welds should have convex beads with an even width and evenly spaced ripples.

Torch and flame position

The center line of the torch tip and flame is held at about a 30° −45° angle from the base metal surface or weld axis. The flame is pointed in the direction of motion. See Fig. 7-11.

The torch and flame are angled so that most of the heat is directed at the surface. This is to avoid excessive melting away of the edge. The tip of the flame should be about 1/16″ −1/8″ from the weld axis.

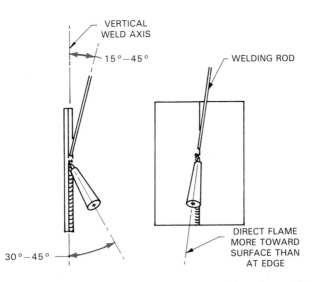

Fig. 7-11. Torch, flame, and welding rod positions for welding a vertical lap joint. Compare the positions with those used for welding a lap joint in the flat welding position.

Welding rod position

The welding rod is usually positioned above the torch and flame. It is held about 30° −45° from the base metal or weld axis. The end of the welding rod should be held close to the weld pool. The end is preheated by keeping it in the reflected heat of the torch flame. See Fig. 7-12.

Fig. 7-12. A vertical lap joint being welded. The flame is aimed more toward the surface piece. The welding rod is added from above.

Weld pool

The welding rod is added when a C-shaped weld pool forms. It is added to the upper edge of the weld pool to prevent sagging. The welding rod should be added to the pool often enough to create a convex bead.

CORNER AND T-JOINTS

A fillet weld is used to weld the inside corner and T-joint on thin metal. A square-groove butt joint may be used for an outside corner joint on metal under 3/16″ (4.8 mm).

The edges of metal over 3/16″ (4.8 mm) thick are cut or ground to provide for better penetration. On thick metal, a fillet weld is commonly added to a bevel-groove weld on the inside corner and T-joint. See Fig. 7-13.

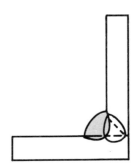

Fig. 7-13. A fillet weld on thick metal. The fillet weld (in red) is on top of the weld made on the bevel-groove inside corner joint.

The weld axis for practice welds is usually between 80°−90° for the vertical position. Practice pieces may be held in a weld positioner, Fig. 7-14, or supported by firebricks or steel blocks.

Fig. 7-14. A T-joint held in the vertical position in an adjustable fixture. The heat is applied equally to both surfaces. The welding rod is added from above.

EXERCISE 7-5 — FILLET WELD ON A T-JOINT IN THE VERTICAL WELDING POSITION

1. Obtain two pieces of mild steel that measure 1/16'' x 1 1/2'' x 6'' (1.6 mm x 38 mm x 152 mm).
2. Clean the surfaces of both pieces.
3. Tack weld the pieces in three places on each side to form a T. This may be done in the flat position.
4. Place the weld axis into a vertical position.
5. Hold the torch tip, flame, and welding rod at the suggested angles. See Fig. 7-15.
6. Melt the two pieces evenly and watch for a C-shaped weld pool to form.
7. When the weld pool forms, add the welding rod to the upper edge of the pool.
8. Continue this procedure to the end of the joint.

Inspection:
The completed weld should not penetrate through either piece. The bead should be convex, even in width, and have evenly spaced ripples.

Torch and flame position

The torch and flame are held at a 45° angle to the metal surfaces when making a weld on an inside corner or T-joint. The torch and flame are also held about

30°−45° up from the weld axis. See Fig. 7-15. The weld is usually made from the bottom and progresses upward using a forehand motion. Heat should be applied evenly, since two surfaces are being joined.

Welding an outside corner joint is similar to welding a vertical butt joint. See Fig. 7-16. However, on the outside corner joint, one piece has an exposed edge and the other piece is a surface. More heat should be aimed toward the surface to avoid melting the edge excessively. See Fig. 7-17. The torch should be held 30°−45° up from the base metal.

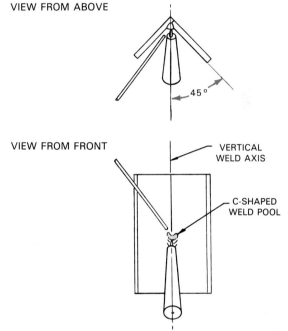

Fig. 7-15. A T-joint being welded in the vertical welding position. The torch tip and flame should be aligned with the weld axis and held at a 30°−45° angle from the weld axis.

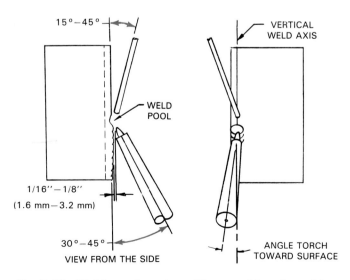

Fig. 7-16. Welding rod, torch, and flame positions for welding an outside corner joint in the vertical welding position.

Fig. 7-17. A weld in progress on an outside corner joint in the vertical welding position. Notice the large tack weld used on these pieces.

Welding rod position

The welding rod should be held at an angle of between 15° – 45° from the base metal. The end of the welding rod should be held near the weld pool to keep it hot or preheated. The rod should be added to the upper edge of the weld pool.

Weld pool

When welding an inside corner or T-joint, a C-shaped weld pool should be formed before adding the welding rod. The C-shaped pool indicates that both pieces have melted. An oval pool is formed when welding a square or bevel-grooved outside corner joint.

BUTT JOINT

When welding a butt joint in the vertical welding position, the weld axis must be 80° – 90° from horizontal. The weld face must be 80° – 90° from a straight down position. See Fig. 7-18.

Both metal surfaces must be heated equally so that the weld pool forms evenly on both sides of the weld axis. On thin metal, very little side-to-side motion is required. On thicker metal, some side-to-side motion is required. A *crescent-shaped motion* works well. See Fig. 7-19.

Torch and flame positions

The center line of the torch and flame should be 30° – 45° up from the metal surface or weld axis. See Fig. 7-20. The tip of the flame should be about 1/16″ – 1/8″ (1.6 mm – 3.2 mm) away from the surface of the weld pool for the greatest amount of heat. If the pool becomes too large, remove the flame momentarily, then return it to its previous height. This flip motion allows for the weld pool to cool and helps prevent sagging of the pool and bead.

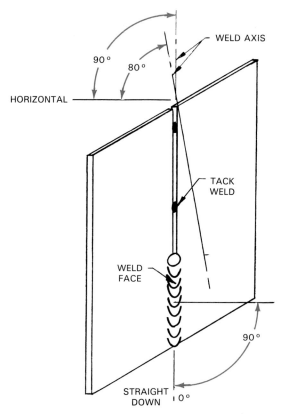

Fig. 7-18. The 80° – 90° vertical or 3G welding position. The weld axis is 90° from horizontal. The weld face is 90° from a straight down position. The weld face may rotate 360°.

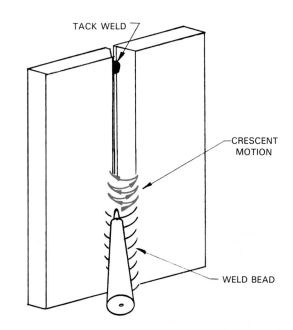

Fig. 7-19. Crescent-shaped weaving motion. This motion is used to make wide beads on thick metal joints.

EXERCISE 7-6 — SQUARE-GROOVE BUTT JOINT IN THE VERTICAL WELDING POSITION

1. Obtain two pieces of mild steel that measure 1/16″ x 1 1/2″ x 6″ (1.6 mm x 38 mm x 152 mm).

2. Clean the edges of both pieces.
3. Tack weld in three places in the flat position.
4. Position the weld axis for vertical welding.
5. Begin welding at the bottom of the joint and progress upwards. Use the suggested torch tip, flame, and welding rod angles. See Fig. 7-20.

Inspection:

The completed weld must have a convex bead with uniform ripples and bead width. 100% penetration is required.

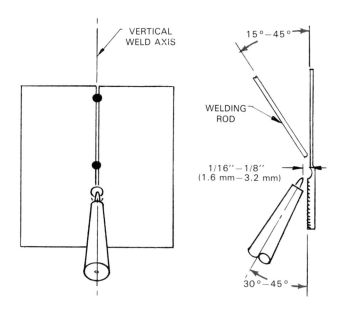

Fig. 7-20. The flame, torch, and welding rod positions for welding a vertical butt joint. Compare the positions to those used in the flat welding position. See Fig. 6-13.

Welding rod position

The welding rod is usually held above the torch and flame, at an angle of 15°−45° from the metal surface. It is preheated by the reflected heat from the welding flame. See Fig. 7-21.

Fig. 7-21. A weld in progress on a vertical butt joint. Note that the welding rod is being added from above.

WELDING IN THE OVERHEAD POSITION

Strong welds and attractive beads can be made in the overhead welding position with care and practice. Welds in the overhead or 4G position must meet the following conditions:

1. The weld axis is between 0°−80°.
2. The weld face is between 0°−80° or 280°−360°.
3. Weld is made from the lower side of the base metal.

Coveralls or jackets should be buttoned at the collar to prevent hot metal from getting inside clothing. Pockets must be buttoned. Flammable items should not be carried in pockets. A cap should be worn. Flame-resistant leather clothing may also be worn for extra protection.

Practice and qualification welds are generally made with the weld axis and weld face in the horizontal or 0° position. See Fig. 7-22.

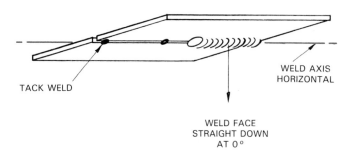

Fig. 7-22. Overhead practice position. The weld axis is horizontal and the weld face is at 360°.

When working in the overhead welding position, a welder usually stands away from the weld axis, similar to flat welding. This allows a good view of the weld in progress. It also decreases the chance of molten metal falling on the welder. Some welders stand in line with the weld axis. This position allows for a good view of the weld and keyhole. However, welding away from the body is not as efficient as welding across the body.

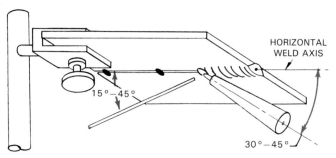

Fig. 7-23. The torch, flame, and welding rod positions for welding a butt joint in the overhead welding position. These positions are mirror images of welding in the flat welding position.

The torch, flame, and welding rod angles are the same for overhead welding and flat welding. However, these angles are rotated 180° into an overhead welding position. See Figs. 7-23 and 7-24.

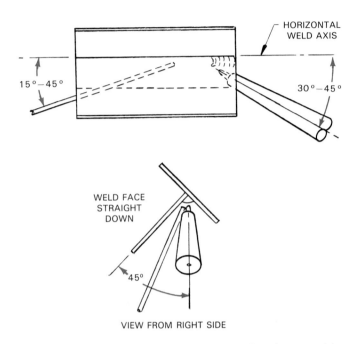

Fig. 7-24. The flame, torch, and welding rod positons used for welding a T-joint in the overhead welding position.

Welds made in the overhead welding position are usually made with a forehand motion. Right-handed welders work from right to left. See Fig. 7-25.

The welding techniques learned in previous positions and joints may be used in the overhead welding position. An overheated weld pool is the greatest problem in overhead welding. If the pool overheats, briefly remove the flame about 1″ (25 mm), then return it. This torch movement permits the weld pool to cool slightly.

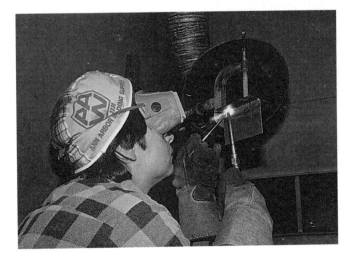

Fig. 7-25. Placing a fillet weld on a T-joint in the overhead welding position.

Making a fillet weld in the overhead position is similar to welding in other positions. The welder must wait for a C-shaped pool to form before adding the welding rod.

EXERCISE 7-7 — FILLET WELD ON A T-JOINT IN THE OVERHEAD WELDING POSITION

1. Obtain two pieces of mild steel that measure 1/16″ x 1 1/2″ x 6″ (1.6 mm x 38 mm x 152 mm).
2. Clean both surfaces on each piece.
3. Place the metal together as shown in Fig. 7-25. Use a weld positioner to place the weld axis into the overhead welding position.
4. Tack weld the pieces in three places.
5. Hold the torch tip, flame, and welding rod at the suggested angles. Refer to Fig. 7-24.
6. Melt the two pieces evenly and watch for a C-shaped weld pool to form.
7. Add the welding rod to the upper edge of the C-shaped pool when it forms.
8. Continue this procedure to the end of the joint.

Inspection:
The completed weld should not penetrate through either piece. The bead should be convex, even in width, and have evenly spaced ripples.

EXERCISE 7-8 — SQUARE-GROOVE BUTT JOINT IN THE OVERHEAD WELDING POSITION

1. Obtain two pieces of mild steel measuring 1/16″ x 1 1/2″ x 6″ (1.6 mm x 38 mm x 152 mm).
2. Clean the edges of the pieces.
3. Tack weld the pieces in three places in the flat welding position.
4. Position the pieces for overhead welding.
5. Weld the joint using the suggested torch tip and flame angles. See Fig. 7-23. Add the welding rod to the upper edge of the molten weld pool.

Inspection:
The completed weld should have a convex bead with even ripples and bead width. It should also have 100% penetration.

REVIEW QUESTIONS

Write your answers on a separate sheet of paper.
1. What is meant by welding out of position?
2. The torch tip and flame should be held between ____° – ____° from the base metal or weld axis for all welding positions. The welding rod should be held between ____° – ____° from the base metal or weld axis for all welding positions.

3. Why should your collar and pockets be buttoned when welding out of position?

4. Name two pieces of protective clothing that are strongly recommended when welding in an overhead welding position.

5. The weld face of a horizontal weld must be within ____° to ____° or ____° to ____°.

6. The weld axis of a horizontal weld must be within ____° of horizontal.

7. List two methods used to overcome sagging of the weld pool, undercutting, and the pull of gravity when making a horizontal weld.

8. List two techniques used to prevent the edge from melting too rapidly when lap welding.

9. What can the welder do to cool the weld pool when welding out of position?

10. What does a C-shaped weld pool indicate to the welder when fillet welding a lap, inside corner, or T-joint?

A welding positioner tilts and rotates the weldment, allowing all welds to be made in the flat position.
(Kioke Aronson, Inc.)

Oxyacetylene cutting is a fast and inexpensive way to cut steel.   (Modern Engineering Co.)

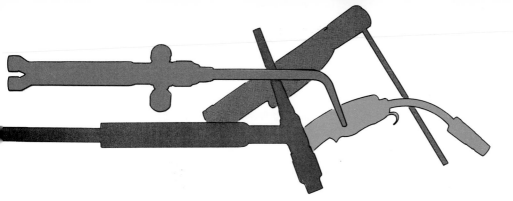

Chapter 8

Oxyfuel Gas Cutting

After studying this chapter, you will be able to:
☐ List fuel gases that are used for oxyfuel gas cutting.
☐ Describe the procedure for turning on and adjusting oxyfuel gas cutting equipment.
☐ Turn on and adjust manual oxyfuel gas cutting equipment.
☐ Perform cuts manually with a cutting torch or cutting torch attachment.
☐ Describe the procedure for shutting down oxyfuel gas cutting equipment.
☐ Identify the basic types of cutting machines.
☐ Describe the procedure for turning on and adjusting oxyfuel gas cutting machines.
☐ Turn on and adjust oxyfuel gas cutting machines.
☐ Perform cuts with an oxyfuel gas cutting machine.
☐ Describe the procedure for shutting down oxyfuel gas cutting machines.

TECHNICAL TERMS

Bell-mouthed kerf, burning, cutting machine, cutting oxygen orifice, cutting torch attachment, drag, electric motor-driven carriage, electronic pattern tracer, ignition temperature, kerf, motor-driven beam-mounted torch, motor-driven magnetic tracer, preheated, preheating orifices, slag, tip nut, track.

OXYFUEL GAS CUTTING PRINCIPLES

Oxyfuel gas cutting (OFC) is a process used to cut metal by rapidly oxidizing it. The heat of a gas flame and pressurized pure oxygen are used in this process. Most materials, including steel, will burn. Oxyfuel gas cutting is also referred to as *burning* by some welders.

If the temperature of paper is raised to its ignition temperature, the paper will burn. The *ignition temperature* is the temperature at which a material will burn if enough oxygen is present. Oxygen must be present for burning to occur. When paper is burned, the oxygen comes from the air.

The ignition temperature for steel is 1500°F (816°C). One or more oxyfuel gas flames are used to heat steel to this temperature when oxyfuel gas cutting. Steel is a bright red or orange-red color at this temperature. When the steel attains a temperature of 1500°F (816°C), a jet stream of high purity (99% +) oxygen is directed at it. This stream of oxygen rapidly oxidizes, or burns, the steel. A continuous cut is made on the steel if the 1500°F (816°C) temperature and oxygen jet are maintained.

Various gases are used as the fuel for oxyfuel gas cutting. These gases are acetylene, hydrogen, natural gas, propane, and MPS (methylacetylene-propadiene). Acetylene and MPS are most commonly used in industry.

OXYFUEL GAS CUTTING EQUIPMENT

Oxyfuel gas cutting and welding equipment is essentially the same. Compare the equipment in Fig. 4-2 and Fig. 8-1. The main differences in the cutting and welding outfits are in the oxygen regulator and the torch. A larger volume and pressure of oxygen may be needed for cutting heavy metal. A pressure regulator with a larger volume capacity and high pressure indications is used.

Cutting torch and tips

The construction of the cutting torch is different from the welding torch. See Fig. 8-2. A cutting torch has an additional oxygen passage built into it. This passage carries the oxygen used to cut, or oxidize, the metal.

Fig. 8-3 shows a cut-away view of an equal pressure type cutting torch attachment. Oxygen and fuel gas enter the torch and are regulated like a welding torch. The gases are mixed in the mixing chamber and travel to *preheating orifices* (holes) in the cutting tip. Pure

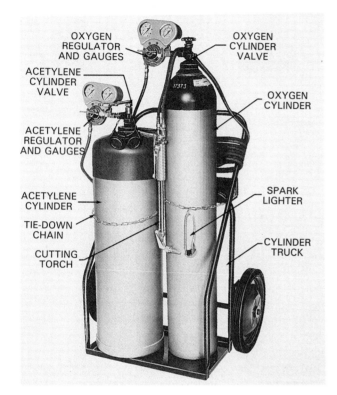

Fig. 8-1. An oxyacetylene cutting outfit. Notice the safety chains that secure the cylinders in an upright position. (Modern Engineering Co.)

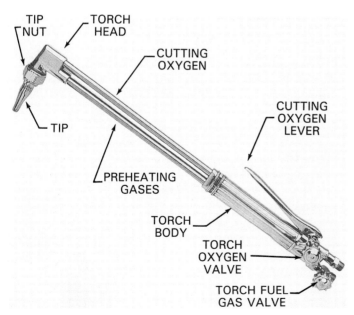

Fig. 8-2. An oxyacetylene cutting torch. The upper tube supplies pure oxygen to the cutting tip.   (Victor Equipment Co.)

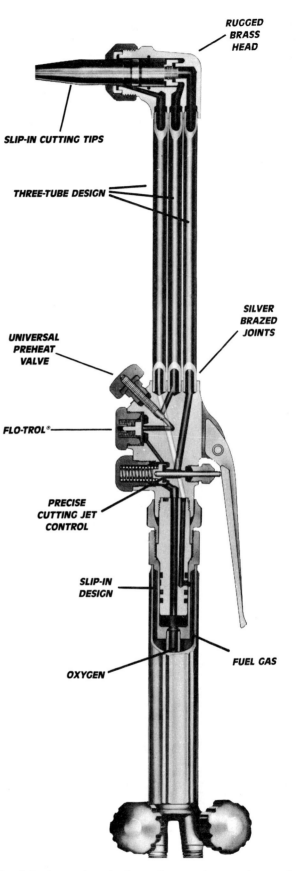

Fig. 8.3. A sectioned view of a cutting torch attachment mounted on a welding torch handle. The preheat gases are mixed in the tip of the torch where it connects into the torch head. Depressing the oxygen cutting lever will allow pure oxygen to travel through the center hole in the cutting tip. (Smith Equipment, Division of Tescom, Inc.)

oxygen is carried through a separate passage to the *cutting oxygen orifice* in the center of the tip. The cutting oxygen is turned on and off by the welder using a quick-acting lever.

Cutting tips are generally made of copper. Sometimes they are chrome-plated to reflect heat. Cutting tips have two or more orifices in the end. The cutting oxygen always flows from the center orifice. Gas for the preheating flames flows from the other orifices. Fig. 8-4 shows several cutting tips with a variety of uses.

Tips used in cutting outfits are always separate from the torch tube. They are commonly unthreaded and held in the cutting torch head by a *tip nut* (large threaded nut). The mating (touching) surfaces of the tip and the torch head are machined smooth. The machined surfaces seal and prevent gas leakage when the tip nut is tightened. See Fig. 8-5.

The flame end of the cutting tip becomes very hot in use. The tips may be easily damaged or pitted. See Fig. 8-6. Never use the cutting tip to hold or pound on anything.

Damaged cutting tips may be cleaned and reformed. Fig. 8-7 shows a tool used to reform the end of a cutting or welding tip. The preheating and oxygen orifices can be cleaned with a set of wire broaches. Fig. 4-30 shows the correct method for cleaning a tip.

NUMBER OF PREHEAT ORIFICES	DEGREE OF PREHEAT	APPLICATION
2	Medium	For straight line or circular cutting of clean plate.
2	Light	For splitting angle iron, trimming plate and sheet metal cutting.
2	Light	For hand cutting rivet heads and machine cutting 30 deg. bevels.
4	Light	For straight line and shape cutting clean plate.
4, 6, 8	Medium	For rusty or painted surfaces.
6	Heavy	For cast iron cutting and preparing welding V's.
6	Very Heavy	For general cutting also for cutting cast iron and stainless steel.
6	Medium	For grooving, flame machining, gouging and removing imperfect welds.
6	Medium	For grooving, gouging or removing imperfect welds.
3	Medium	For machine cutting 45 deg. bevel or hand cutting rivet heads.
6	Heavy	Flared cutting orifices provides large oxygen stream of low velocity for rivet head removal (washing).

Fig. 8-4. A table showing some common cutting torch tips and their uses. (Veriflo Corp.)

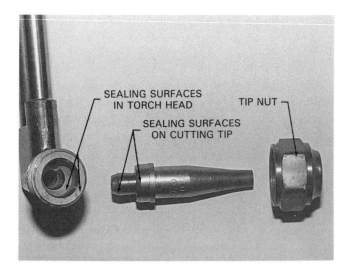

Fig. 8-5. Typical cutting torch, tip, and tip nut arrangement. The smoothly machined surfaces in the torch head and on the torch tip form a gastight seal when the nut is tightened.

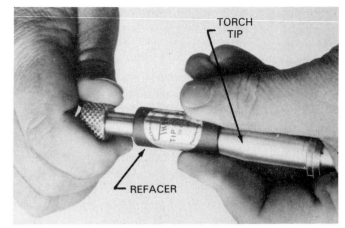

Fig. 8-7. A special tool used to reface the flame end of the cutting tip. This could be used on tip A shown in Fig. 8-6. (Thermacote-Welco Co.)

Cutting torch attachment

A *cutting torch attachment* converts the welding torch body into a cutting torch. See Fig. 8-8. This eliminates the need for buying a separate cutting torch. When the cutting torch attachment is mounted on the torch body, there are four torch valves. Two valves are on the torch body; one is the torch acetylene valve, the other is the torch oxygen valve. A second torch oxygen valve is located on the cutting attachment. The fourth valve is the cutting oxygen lever. It regulates the flow of cutting oxygen.

PREPARING TO CUT

The oxyfuel gas cutting outfit is assembled in the same manner as an oxyfuel gas welding outfit. See

Fig. 8-6. Four cutting tips. The end of tip A can be refaced and the orifices cleaned. Tips B and D are beyond repair. Tip C is in good condition.

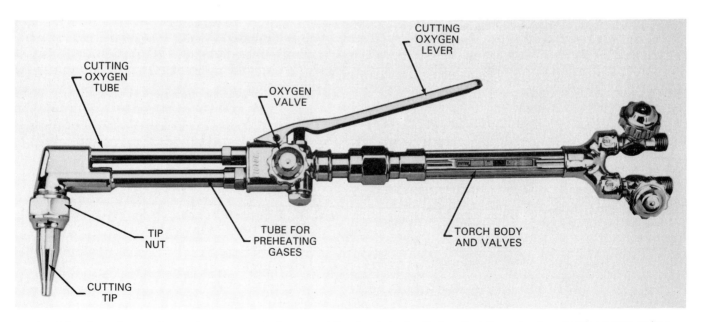

Fig. 8-8. A cutting torch attachment mounted on a welding torch body. Notice that there are two torch oxygen valves. One is opened completely and the other is used to adjust the flame.

Chapter 5 for assembly procedures. Cutting tip sizes are identified by numbers ranging from 00-8. The numbers are stamped on the tip when they are made. Most cutting tip manufacturers use the same tip numbering system. Fig. 8-9 suggests cutting oxygen and acetylene pressures for use with various cutting tips. Suggested tip sizes for use with various metal thicknesses are also shown.

TURNING ON AN OXYACETYLENE CUTTING OUTFIT

Open the torch acetylene and oxygen valves 1/4 – 1/2 turn before setting the working pressures. The cutting torch oxygen valve (lever) should also be open while the oxygen cutting pressure is set. After setting the working pressures for oxygen and acetylene, close the torch valves and release the oxygen cutting lever.

Lighting and adjusting a positive pressure cutting torch

When the correct working pressures have been set, the cutting torch is ready to light. A flint-and-steel sparklighter must be used to prevent severe burns. When lighting the torch, hold it so that the tip is facing downward. Strike a spark while holding the sparklighter approximately 1″ (25 mm) from the tip.

1. Open the acetylene torch valve 1/16 turn.
2. Light the preheating flames with a sparklighter. All flames should light at the same time. If all the flames do not light, shut down the torch and clean any dirty orifices.
3. Continue opening the acetylene torch valve until the smoking stops. The flame must remain in contact with the end of the tip. See Fig. 8-10A.
4. Open the oxygen torch valve until a neutral flame is obtained. See Fig. 8-10B. The cutting oxygen valve must be closed during this adjustment.
5. Open the cutting oxygen valve (lever). The preheating flames should remain neutral. See Fig. 8-10C. If a carburizing flame is obtained, open the torch oxygen valve a little more. Continue to ad-

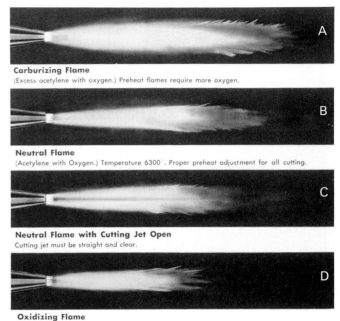

Carburizing Flame
(Excess acetylene with oxygen.) Preheat flames require more oxygen.

Neutral Flame
(Acetylene with Oxygen.) Temperature 6300˚. Proper preheat adjustment for all cutting.

Neutral Flame with Cutting Jet Open
Cutting jet must be straight and clear.

Oxidizing Flame
(Acetylene with Excess Oxygen.) Not recommended for average cutting.

Fig. 8-10. Oxyacetylene cutting flame adjustments. A—Carburizing flame. Excess acetylene with oxygen. B—Neutral flame. C—Neutral flame with cutting oxygen open. D—Oxidizing flame. Acetylene with excess oxygen.

just until neutral preheating flames remain while the cutting oxygen valve is open.

Shutting down a positive pressure cutting torch

The cutting outfit must be shut down when the cutting operation is completed or when the welder must leave the cutting station. The cutting outfit is shut down using a procedure similar to that used for shutting down a welding outfit. Be certain that the correct procedure is followed so that all gauges read zero when the shutdown is completed. Turn the regulator adjusting screws counterclockwise until they feel loose. Use the following procedure to ensure that the outfit is shut down correctly.

Material thickness, inches	1/8	1/4	1/2	3/4	1	1 1/2	2	4	5	6	8	10	12
Recommended tip number	00	0	1	1	2	2	3	3	4	6	6	7	8
Oxygen pressure setting, lb./in.²g	20-25	25-30	30-35	30-35	35-40	40-45	40-45	40-50	45-55	45-55	45-55	45-55	45-55
Acetylene pressure setting, lb./in.²g	3-5	3-5	3-5	3-5	3-7	3-7	5-10	5-10	6-12	7-13	8-14	10-15	10-15
Cutting speed range, in./min.	27-30	26-29	20-24	17-21	14-18	13-17	12-15	8-11	7-9	6-8	5-6	4-5	3-4

Information in table is an approximate guide covering average conditions.
Depth of cut, pressure settings, and cutting speeds will vary slightly for machine torch operation.

Fig. 8-9. Suggested tip sizes for cutting various metal thicknesses with a positive pressure torch. The approximate oxygen and acetylene gas pressures are listed. Also shown is the correct cutting speed. (Goss, Inc.)

1. Turn off the acetylene torch valve.
2. Turn off the oxygen torch valve.
3. Turn off the acetylene cylinder valve.
4. Turn off the oxygen cylinder valve.
5. Reopen the acetylene and oxygen torch valves.
6. Close both torch valves when all gauges read zero.
7. Close both regulators by turning the regulator adjusting screws counterclockwise until they feel loose.

Lighting and adjusting a positive pressure cutting torch attachment

The flame is ready to light after turning the cutting outfit on and setting the correct working pressures. Before lighting the torch, open the torch oxygen valve on the cutting attachment one full turn. This completely opens the torch oxygen valve on the cutting attachment. All oxygen adjustments will now be made at the torch oxygen valve on the torch body.

A flint-and-steel sparklighter must be used to prevent severe burns when lighting the cutting torch attachment. When lighting, point the tip of the torch downward and hold the sparklighter about 1″ (25 mm) from the tip.

1. Open the acetylene torch valve 1/16 turn.
2. Light the preheating flames with a sparklighter. All flames should light at the same time. If all the flames do not light, shut down the torch and clean any dirty orifices.
3. Continue opening the acetylene torch valve until the smoking stops. The flame must remain in contact with the end of the tip.
4. Open the oxygen torch valve until a neutral flame is obtained. The cutting oxygen valve must be closed during this adjustment.
5. Open the cutting oxygen valve (lever). The preheating flames should remain neutral. If a carburizing flame is obtained, open the torch oxygen valve a little more. Continue to adjust until neutral preheating flames remain when the cutting oxygen valve is open.

The flame must always be turned off if the torch is not in the welder's hands. If the torch will not be used for some time, the outfit must be shut down.

Shutting down a positive pressure cutting torch attachment

The cutting outfit must be shut down after the welding operation is completed or when the welder is leaving the welding station. The following shutdown procedure must be used. Make certain that, when the procedure is complete, the gauges read zero and the adjusting screws feel loose.

1. Turn off the acetylene torch valve.
2. Turn off the oxygen torch valve.
3. Turn off the acetylene cylinder valve.
4. Turn off the oxygen cylinder valve.
5. Reopen the acetylene and oxygen torch valves.
6. Close both torch valves when all gauges read zero.
7. Close both regulators by turning the regulator adjusting screws counterclockwise until they are loose.
8. Close the oxgyen valve on the cutting attachment.

MANUAL CUTTING

The edges of metal produced by oxyfuel gas cutting should be smooth and straight. A good cut depends on the welder's skills. Cut quality is also affected by the size of tip and the oxygen pressure used.

Slag is iron oxide. It may form on the underside of the metal being cut. Fig. 8-11 shows a plate that was tipped over to reveal the slag. Slag that is easily removed is acceptable. *Hard slag* (slag which cannot be easily removed) is unacceptable. Enough oxygen pressure must be used to cut through the metal without leaving hard slag.

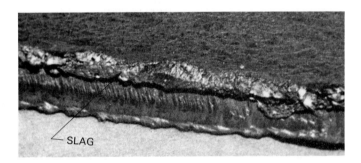

Fig. 8-11. *This plate was turned over to show the slag on the bottom of the piece.*

The slot or opening produced in the metal when cutting is the *kerf.* See Fig. 8-12. The kerf should be as narrow as possible. A wide kerf requires more oxygen to make a good cut. The width of the kerf is determined by the size and shape of the cutting tip. Too much oxygen pressure may result in a *bell-mouthed kerf,* Fig. 8-13.

When oxyfuel gas cutting, the base metal must be *preheated* (heated to its ignition temperature) before cutting oxygen is applied. The ignition temperature for steel is about 1500°F (816°C). The small orifices in the cutting tip supply the preheating flames. These flames preheat the metal ahead of the oxygen stream and also heat the sides of the kerf. The cutting tip should be placed into the torch so that it aligns with the cutting line as shown in Fig. 8-14. A minimum of two preheating orifices should line up with the cutting line. A kerf is formed under the cutting oxygen jet. The quality of the kerf depends on the following:

1. Cutting tip size.
2. Oxygen pressure.
3. Torch forward speed.

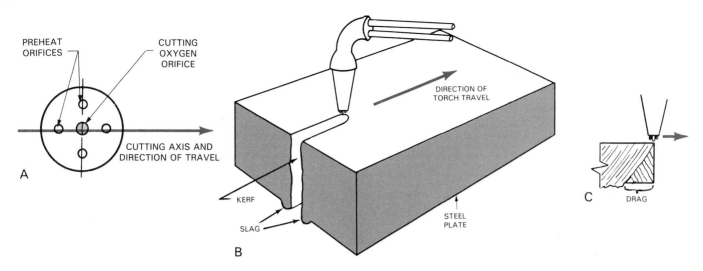

Fig. 8-12. An oxyacetylene cut in progress. A— Four preheat holes; two are aligned with the cutting axis. B—Note the slag at the bottom of the kerf. C—Cutting oxygen slows down as it travels through the metal. Drag results from the oxygen slowing down and the forward motion of the torch.

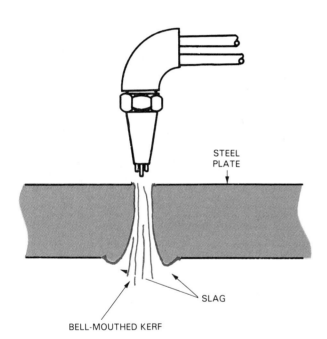

Fig. 8-13. The effect of using too much oxygen when cutting steel. Notice how the kerf widens at the bottom of the plate to create a bell-mouthed kerf.

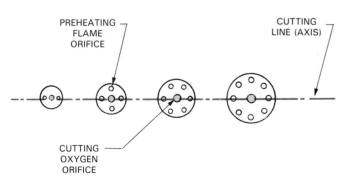

Fig. 8-14. Correct alignment of the preheating orifices on the cutting line. At least two preheating orifices should align with the cutting line.

4. Steady torch movement.
5. Torch tip angle.
6. Distance of preheat flames from the base metal.

Fig. 8-15 shows a good cut and several unacceptable cuts. The characteristics of each cut are also given.

EXERCISE 8-1 — MANUALLY CUTTING STRAIGHT EDGES—FREEHAND

1. Obtain a piece of mild steel that measures at least 1/4" x 6" x 6" (6.4 mm x 152 mm x 152 mm).
2. Lay out three lines on the surface 1 1/2" (38 mm) apart. Use soapstone or chalk. See figure below.
3. Select the correct cutting tip. Use a tip with 4-6 preheating orifices, if possible.
4. Turn on the station and set the correct pressures. Light the torch and adjust for neutral preheating flames.
5. Place your left hand on the table. Rest the torch in the left hand. Slide it through your left hand as the cut progresses. This prevents the torch from shaking.

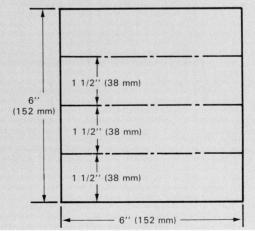

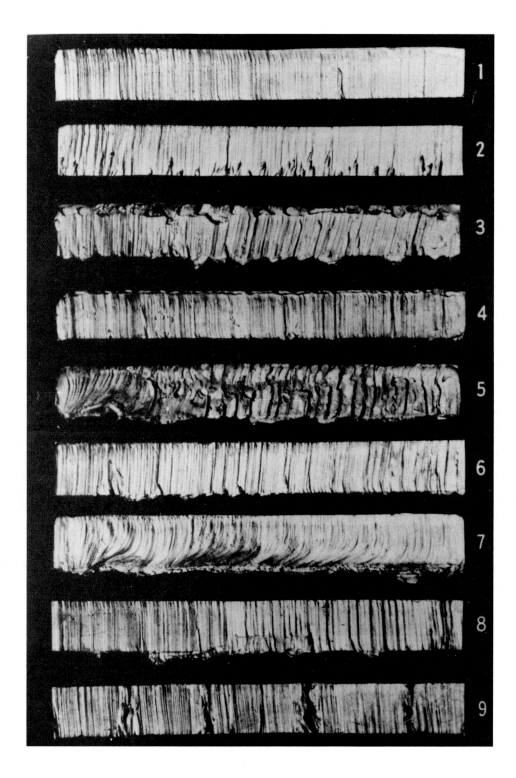

Fig. 8-15. Typical edge conditions resulting from oxyfuel gas cutting operation: (1) good cut in 1'' (25 mm) plate—the edge is square, and the drag lines are essentially vertical and not too pronounced; (2) preheat flames were too small for this cut, and the cutting speed was too slow, causing bad gouging at the bottom; (3) preheating flames were too long, with the result that the top surface melted over, the cut edge is irregular, and there is an excessive amount of adhering slag; (4) oxygen pressure was too low, with the result that the top edge melted over because of the slow cutting speed; (5) oxygen pressure was too high and the nozzle size too small, with the result that control of the cut was lost; (6) cutting speed was too slow, with the result that the irregularities of the drag lines are emphasized; (7) cutting speed was too fast, with the result that there is a pronounced break in the drag line, and the cut edge is irregular; (8) torch travel was unsteady, with the result that the cut edge is wavy and irregular; (9) cut was lost and not carefully restarted, causing bad gouges at the restarting point.   (AWS)

6. For a right-handed welder, start the cut at the right and progress toward your left.
7. Hold the tip 5°—20° from vertical. The center line of the tip should be aligned with the line to be cut. The preheating flames should be about 1/16"—1/8" (1.6 mm—3.2 mm) above the metal.
8. Squeeze the oxygen cutting valve when the metal turns a red-orange color (1500 °F/816 °C). The cut will begin.
9. Progress at a constant rate to complete the cut.
10. Repeat the process for the next two lines.

EXERCISE 8-2 — MANUALLY CUTTING STRAIGHT EDGES USING STEEL ANGLE

1. Obtain a piece of mild steel that measures at least 1/4" x 6" x 6" (6.4 mm x 152 mm x 152 mm).
2. Lay out three lines on the surface 1 1/2" (38 mm) apart. Use soapstone or chalk. See figure below.
3. Select the correct cutting tip. Use a tip with 4-6 preheating orifices, if possible.
4. Turn on the station and set the correct pressures. Light the torch and adjust for neutral preheating flames.
5. Obtain a piece of angle iron that measures at least 1" x 1" x 7" (25 mm x 25 mm x 174 mm).
6. Position the angle just off the line to be cut to allow for the tip thickness. Clamp the angle in place.
7. Hold the tip 5°—15° from vertical. The center line of the tip should be aligned with the line to be cut. The preheating flames should be about 1/16"—1/8" (1.6 mm—3.2 mm) above the metal.
8. Squeeze the oxygen cutting valve when the metal turns a red-orange color (1500 °F/816 °C). The cut will begin.
9. Progress at a constant rate to complete the cut.
10. Repeat the process for the next two lines.

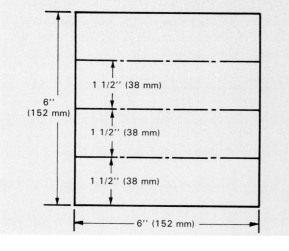

Torch position

The torch angle for cutting will vary with the thickness of the metal. The center line of the cutting tip should be perpendicular to the surface of the metal to form square edges.

The cutting tip should be tilted backward between 5°—20° from vertical when cutting with the correct size tip. See Fig. 8-16. This allows the welder to see into the kerf.

For cutting thick metal, the torch may be vertical as shown in Fig. 8-17. The smallest tip size available

Fig. 8-16. The cutting head and tip are tilted away from the direction of cutting at an angle of 5° to 20° from vertical.

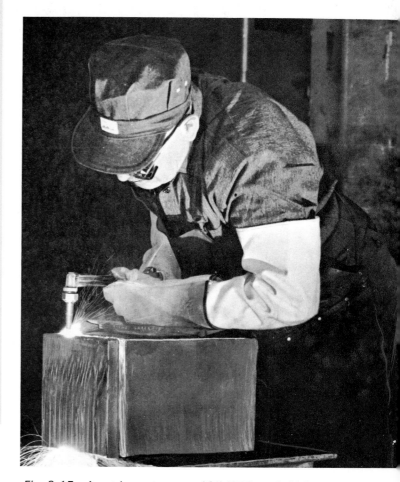

Fig. 8-17. A cut in progress on 10" (250 mm) thick steel. Notice that the torch is held about 90° to the surface of the metal.

may produce too much heat for metal under 1/8″ (3.2 mm) thick. In this case, lower the torch to 15°–20° from horizontal. This angle increases the amount of material being cut and allows a larger tip to produce a clean cut. See Fig. 8-18.

A piece of steel angle (angle iron) may be used as a guide to make straight or beveled cuts. See Fig. 8-20. The angle may be clamped to the part for use as a straight edge. The torch tip is slid along the angle.

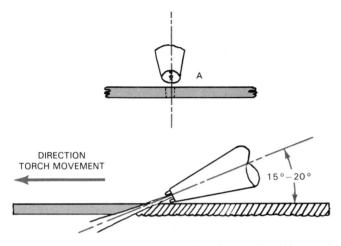

Fig. 8-18. Recommended procedure for cutting thin steel. Notice that the two preheat flames on this tip are in line with the kerf (cut line).

Fig. 8-20. A steel angle (angle iron) used as a guide to make a beveled cut.

Hand position

A cutting torch must be held with two hands. As the cut progresses, the torch is slid through the support hand, as shown in Fig. 8-19A and 8-19B. A 4″–8″ (102 mm–203 mm) cut may be made before moving the support hand.

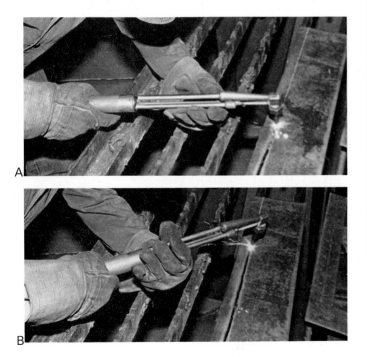

Fig. 8-19. Guiding the cutting torch through the left hand. A–Left hand is held near the torch head. B–Torch is slid through left hand as cut progresses.

EXERCISE 8-3 — MANUALLY CUTTING BEVELED EDGES USING STEEL ANGLE

1. Obtain a piece of mild steel that measures at least 1/4″ x 6″ x 6″ (6.4 mm x 152 mm x 152 mm).
2. Lay out three lines on the surface 1 1/2″ (38 mm) apart. Use soapstone or chalk. See figure below.
3. Select the correct cutting tip. Use a tip with 4-6 preheating orifices, if possible.
4. Turn on the station and set the correct pressures. Light the torch and adjust for neutral preheating flames.
5. Obtain a piece of angle iron that measures at least 1″ x 1″ x 7″ (25 mm x 25 mm x 174 mm).
6. Position the steel angle on its legs, as in Fig. 8-20. Clamp the angle in place.

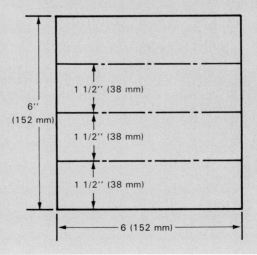

7. Hold the tip 5°−15° from vertical. The center line of the tip should be aligned with the line to be cut. The preheating flames should be about 1/16″−1/8″ (1.6 mm−3.2 mm) above the metal.
8. Squeeze the oxygen cutting valve when the metal turns a red-orange color (1500°F/816°C). The cut will begin.
9. Progress at a constant rate to complete the cut.
10. Repeat the process for the next two lines.

CUTTING MACHINES AND PATTERN TRACERS

Oxyfuel gas *cutting machines* make high-quality cuts at a faster rate than manual flame cutting. Cutting machines are often used to make long cuts. They are also used to cut multiple pieces with close tolerances. The basic types of cutting machines are:
1. Electric motor-driven carriage and track.
2. Motor-driven beam-mounted torch and electronic tracer.
3. Motor-driven magnetic tracer and torch.

Electric motor-driven carriage and track

An *electric motor-driven carriage* uses a variable-speed motor to carry a cutting torch along a straight or curved track. See Fig. 8-21. The speed of the motor can be adjusted to control the speed of the cut. Torch

EXERCISE 8-4 − MACHINE CUTTING

Use any available cutting machine.
1. Obtain a piece of mild steel that measures at least 6″ x 6″ (152 mm x 152 mm). The piece should be at least 1/4″ (6.4 mm) thick. Obtain a pattern, if required.
2. Refer to the procedures for setting up and adjusting your cutting machine.
3. Make cuts as directed by the instructor.
4. When the cuts are completed, shut off the drive motors and turn off the torch(es). Shut down the cutting outfit.

Inspection:

The finished part should be an exact duplicate of the pattern. The edges should be smooth. Slag should not be present.

movement and flame height are consistent because the torch moves along a track. High-quality cuts can be made using a track-mounted carriage. See Fig. 8-22.

The cutting torch used on a motor-driven carriage has a different appearance from a manual cutting torch. The cutting machine has a rack-type gear on it. This allows the torch to be moved up and down to adjust the flame height. The torch has conventional ox-

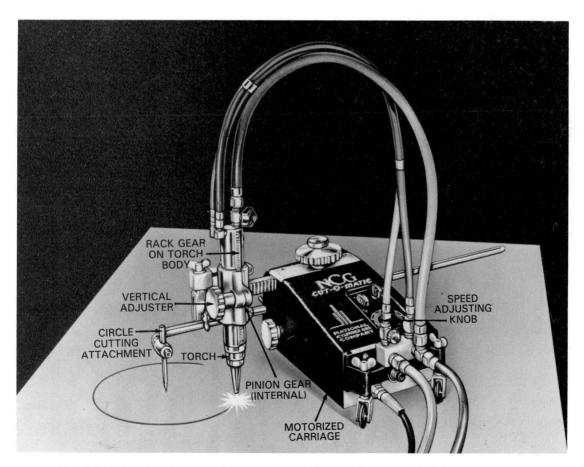

RACK GEAR ON TORCH BODY
VERTICAL ADJUSTER
CIRCLE CUTTING ATTACHMENT
TORCH
PINION GEAR (INTERNAL)
MOTORIZED CARRIAGE
SPEED ADJUSTING KNOB

Fig. 8-21. An electric motor-driven carriage being used to cut a circle in a steel plate.

Fig. 8-22. A motor-driven torch carriage mounted on a track to make long straight cuts.

ygen and acetylene valves. A small lever is used to operate the cutting oxygen valve. The torch can also be adjusted horizontally to align it with the cutting line.

Only one cutting torch is usually mounted on the carriage. Torch angle, flame height, and cutting speed can be adjusted on the carriage. A clutch switch is used to engage and disengage the carriage drive mechanism.

An oxyfuel gas cutting torch mounted on a motor-driven carriage is used as follows:

1. Set up the track or circle cutting attachment.

2. Adjust the torch height so the flames are about 1/16″ − 1/8″ (1.6 mm − 3.2 mm) above the base metal. Adjust the torch to the appropriate angle for the metal being cut. Refer to the Torch position subhead under Manual Cutting. Adjust the forward speed for the carriage. Fig. 8-9 suggests forward speeds to be used with various thicknesses of metal. This speed may be increased, when possible, with a motor-driven carriage.

3. Turn on the cutting outfit using the same procedure as for a manual torch.

4. Light the preheating flames and adjust for a neutral flame.

5. Begin preheating the metal at the edge of the part.

6. Engage the forward drive clutch when the steel becomes orange-red (1500°F/816°C).

7. Disengage the drive clutch when the cut is completed. Disengaging the clutch stops the carriage.

8. Shut off the torch and shut down the cutting outfit using the same procedure as for a manual torch.

MOTOR-DRIVEN BEAM-MOUNTED TORCHES AND ELECTRONIC TRACER

The oxyfuel gas cutting torch is mounted on a strong beam. Several cutting torches may be mounted on the same beam. This allows the machine to cut several parts at the same time. See Fig. 8-23.

Fig. 8-23. Six cutting torches being used to produce duplicate parts. An electronic tracer, near the operator, traces a line drawing and moves the cutting heads. Notice the metal beam connecting all the torches. (ESAB Welding and Cutting Products)

A dense black drawing of the part to be cut is made on a white background. The outline of the drawing is followed by the electronic eye on a motor-driven *electronic pattern tracer*. See Fig. 8-24. As the pattern shape changes, the electric eye follows the edge of the pattern. It will not stray more than .003″ (.076 mm) from the drawing image.

The pattern tracer is electrically connected to two drive motors. As the tracer moves along the outline, it signals the drive motors on the beam. The motors move the beam on which the torches are mounted. All moves that the tracer makes are duplicated by all the torches on the beam.

The beam-mounted torch(es) and electronic tracer are used as follows:

1. Place the drawing on the pattern table under the electronic pattern tracer.
2. Turn on the cutting outfit using the same procedure as for manual cutting.
3. Adjust the light beam on the electronic tracer to the line on the drawing.
4. Light and adjust each torch separately. The flame should be 1/16″ − 1/8″ (1.6 mm − 3.2 mm) above the base metal. The torches should also be perpendicular to the surface of the metal. The drive motor speed is adjusted on the electronic tracer control panel.
5. Turn on the cutting oxygen at each torch and pierce a hole in the metal.
6. Start the beam drive motors.
7. Start the tracer. The torches move as the tracer moves.

8. Shut off the tracer and drive motors when the cut is completed. Turn off all the torches.
9. Shut down the cutting outfit in the same manner as for manual cutting.

MOTOR-DRIVEN MAGNETIC TRACER AND TORCH

The *motor-driven magnetic tracer* and torch is a relatively inexpensive cutting machine. This type of cutting machine works best with shapes that are not too complex. The tracer will not follow sharp angles very well. One torch is generally mounted on this type of cutting machine.

A motor-driven magnetic tracer and cutting machine is shown in Fig. 8-25. A steel pattern must be used when duplicating parts. The pattern is firmly mounted directly above the cutting torch. A cylindrical magnetic tracer, or follower, follows the outline of the steel pattern. The tracer is rotated by a small motor, and held in contact with the pattern by magnetism. The tracer is mounted directly above and on the exact center line of the cutting torch. See Fig. 8-26. Therefore, the torch follows the movement of the tracer as it moves around the pattern. The part that is cut out will be an exact duplicate of the pattern. Parts cut by this method are not as accurate as those cut with an electronic tracer.

The magnetic motor-driven tracer and torch are used as follows:

1. Mount the steel pattern in position.
2. Place the magnetic tracer in contact with the pattern.
3. Adjust the torch angles.

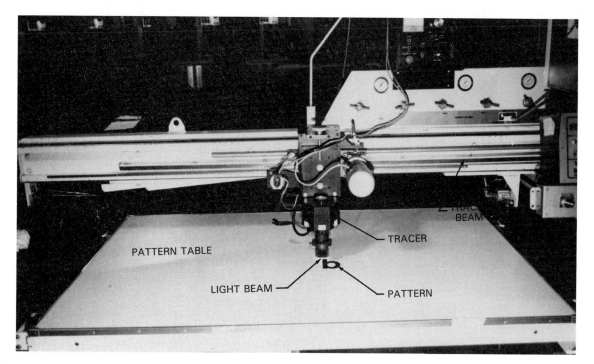

Fig. 8-24. An electronic tracer mounted above a large pattern table. The tracer moves along the tracer beam to trace large patterns.

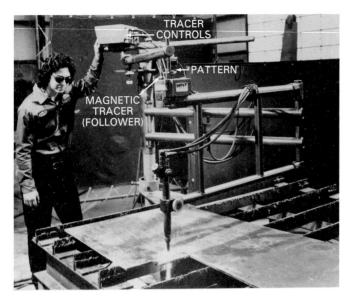

Fig. 8-25. *The magnetic pattern follower is mounted above the cutting tip of a magnetic tracer. As the follower rotates and follows the pattern, the cutting torch travels a duplicate path over the metal.* (ESAB Welding and Cutting Products)

4. Turn on, light, and adjust the torch height, using the same procedure as for manual cutting. The tip of the preheat flames should be 1/16″ − 1/8″ (1.6 mm − 3.2 mm) above the base metal.
5. Start the magnetic tracer motor. The tracer rotates

and follows the pattern. The torch follows the tracer movement.
6. After the cut is completed, shut off the tracer motor. Shut down the cutting torch and cutting outfit.

REVIEW QUESTIONS

Write your answers on a separate sheet of paper.
1. Oxyfuel gas cutting is also known as _____ in the trade.
2. What color is steel at its ignition temperature? What is its ignition temperature?
3. Name four fuel gases that may be used in oxyfuel gas cutting.
4. What is the minimum of preheat orifices shown in Fig. 8-4?
5. Explain the difference between lighting a cutting torch and lighting a cutting torch attachment.
6. The torch tip is normally held at a _____° angle from vertical.
7. Explain why manual oxyfuel gas cutting is a two-handed operation.
8. List three devices used to guide a cutting machine along a line or pattern.
9. The preheating holes in a cutting tip should not be aligned with the line to be cut. True or False?
10. Sketch a bell-mouthed kerf.

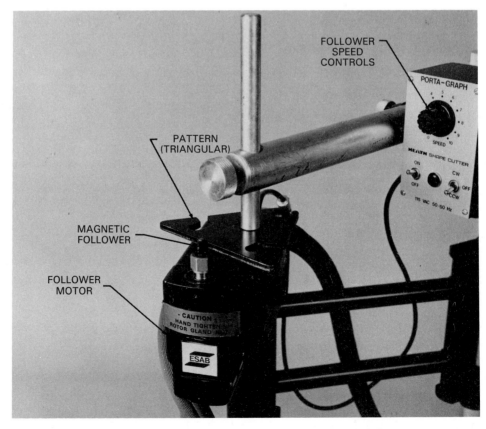

Fig. 8-26. *A close-up view of a magnetic follower and pattern. The magnetic follower rotates around the pattern and the torch (on the frame directly below) traces the same pattern.* (ESAB Welding and Cutting Products)

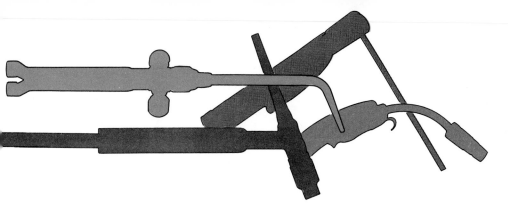

Chapter 9

Brazing and Braze Welding

After studying this chapter, you will be able to:
- ☐ Define brazing and braze welding.
- ☐ Select the correct torch tip, filler metal, and flux for brazing and braze welding.
- ☐ List the safety procedures for brazing or braze welding.
- ☐ Explain the difference between a welded joint and brazed or braze welded joint.
- ☐ Describe the procedure for brazing.
- ☐ Braze metal.
- ☐ Describe the procedure for braze welding.
- ☐ Braze weld metal.

TECHNICAL TERMS

Braze welding, brazement, brazing, brazing rod, cadmium, capillary action, flux, liquidus, solidus.

BRAZING AND BRAZE WELDING PRINCIPLES

The American Welding Society defines *brazing* as "a group of welding procedures that cause materials to join by heating them to the brazing temperature in the presence of a filler metal. The filler metal has a liquidus above 840°F (450°C) and below the solidus temperature of the base metal. The filler metal is distributed between the closely fitting surfaces of the joint by capillary action."

Braze welding is defined by AWS as "a welding process variation in which a filler metal, having a liquidus above 840°F (450°C) and below the solidus of the base metal, is used. Unlike brazing, in braze welding, the filler metal is not distributed in the joint by capillary action."

The major difference between brazing and braze welding is in the fit of the parts. Parts to be brazed must fit very tightly together. *Capillary action* occurs when the filler metal is drawn between tightly fitted mating surfaces. Tightly fitted parts are not required for braze welding.

When fusion welding, the base metal is melted. When brazing or braze welding, the base metal is not melted. The brazing filler metal melts at a temperature above 840°F (450°C) and below the melting point of the base metal. If the filler metal melts below 840°F (450°C), it is a soldering process. Brazing and braze welding on steel is usually done between 840°F−2000°F (450°C−1093°C).

Brazing and braze welding are performed at temperatures lower than required for fusion welding. Therefore, brazing and braze welding processes may be used to join thin metal. Also, because of the lower temperatures, metals may be joined without changing the characteristics produced by heat treatment.

A *brazement* is an assembly of parts to be brazed or braze welded. A brazed or braze welded part is not as strong as a similar part joined by fusion welding. The filler metal used when brazing or braze welding is not as strong as the base metal. In fusion welding, the filler metal is as strong or stronger than the base metal.

Oxyfuel gas welding equipment is commonly used to braze or braze weld. Fuel gases such as acetylene, MPS (methylacetylene-propadiene), propane, natural gas, hydrogen, and city gas may be used. Acetylene and MPS are commonly used because they produce the highest temperatures when combined and burned with oxygen.

Brazing can also be done in a furnace. When furnace brazing, a preshaped piece of brazing filler metal is commonly placed in the joint. The part is then heated in a furnace. The filler metal melts when the part reaches the correct temperature. Hundreds of parts can be brazed simultaneously in this manner. See Fig. 9-1.

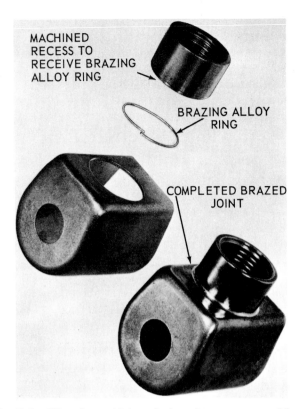

Fig. 9-1. Silver brazed joints designed to use pre-positioned silver alloy rings. The alloy forms almost perfect fillets. Further finishing is not required.
(Handy & Harman/Lucas Milhaupt, Inc.)

Fig. 9-2. Brazing filler metal is available in wire or preshaped forms. The preshaped forms are placed on or in brazements before they are heated.
(Handy & Harman/Lucas Milhaupt, Inc.)

BRAZING FILLER METAL

Brazing filler metal is available in many shapes and sizes. It is offered as wire, metal strips, rods, powder, and preshaped forms to fit specific joints. See Fig. 9-2. Filler metal **brazing rods** are available in various diameters. Rod diameters include: 1/16″ (1.6 mm), 3/32″ (2.4 mm), 1/8″ (3.2 mm), 5/32″ (4.0 mm), 3/16″ (4.8 mm), and 1/4″ (6.4 mm). Rods are usually 36″ (.91 m) long and are sold by the pound (kilogram).

Brazing filler rods are available as bare wire or with a flux coating. The flux coating is applied during manufacture. Flux must be added to bare filler rods while brazing. Flux-coated filler rods are more expensive than bare wire rods. Flux-coated rods are more convenient to use, but must be handled carefully to prevent the coating from coming off. Fig. 9-3 shows a flux-coated filler rod being used to braze a lap joint.

Brazing filler metal can be any metal that melts between 840°F (450°C) and the melting point of the base metal. Aluminum and alloys with low melting points can be used to braze aluminum alloys with higher melting temperatures. Brass is used to braze or braze weld steel. Brass is an alloy of copper and zinc. Silver alloys are commonly used to braze copper, stainless steel, and a variety of other metals. Silver brazing is often referred to, incorrectly, as silver soldering. Silver

Fig. 9-3. Braze welding a lap joint in the flat position. A flux-coated brazing rod is being used.

brazing filler metal melts above 840°F (450°C). Therefore, silver brazing is not a soldering process. Gold alloys are also used as a filler meal when brazing. The table in Fig. 9-4 shows filler metals suggested for use in joining various base metal combinations.

Brazing filler metal is generally an alloy of two or more metals. Fig. 9-5 lists the metals (with their abbreviations) that are combined to create various AWS brazing filler metals. The chemical composition of a filler metal affects its solidus and liquidus temperatures.

	Al & Al alloys	Mg & Mg alloys	Cu & Cu alloys	Carbon & low alloy steels	Cast Iron	Stainless steels	Ni & Ni alloys
Al & Al alloys	BAlSi						
Mg & Mg alloys	X	BMg					
Cu & Cu alloys	X	X	BAg, BAu, BCuP, RBCuZn				
Carbon & low alloy steels	BAlSi	X	BAg, BAu, RBCuZn	BAg, BAu, BCu, RBCuZn, BNi			
Cast Iron	X	X	BAg, BAu, RBCuZn	BAg, RBCuZn	BAg, RBCuZn, BNi		
Stainless steel	BAlSi	X	BAg, BAu	BAg, BAu, BCu, BNi	BAg, BAu, BCu, BNi	BAg, BAu, BCu, BNi	
Ni & Ni alloys	X	X	BAg, BAu, RBCuZn	BAg, BAu, BCu, RBCuZn, BNi	BAg, BCu, RBCuZn	BAg, BAu, BCu, BNi	BAg, BAu, BCu, BNi
Ti & Ti alloys	BAlSi	X	BAg	BAg	BAg	BAg	
Be, Zr, & alloys (reactive metals)	X BAlSi (Be)	X	BAg	BAg, BNi	BAg, BNi	BAg, BNi	BAg, Bni
W, Mo, Ta, Cb & alloys (refractory metals)	X	X	BAg	BAg, BCu, BNi	BAg, BCu, BNi	BAg, BCu, BNi	BAg, BCu, BNi
Tool steels	X	X	BAg, BAu, RBCuZn, BNi	BAg, BAu, BCu, BCuZn, BNi	BAg, BAu, RBCuZn, BNi	BAg, BAu, BCu, BNi	BAg, BAu, BCuZn, BNi

Fig. 9-4. Brazeable metal combinations and suggested filler metals. This table is used when you are brazing two different metals. Locate one metal in the left column and the other metal in the top row of the table. The suggested filler metal is found at the intersection of the vertical column and horizontal row. An ''X'' indicates that brazing this combination is not recommended. ''RB'' means resistance brazing.

FILLER METALS	
BAlSi	Aluminum silicon alloy
BAg	Silver-base alloy
BAu	Gold-base alloy
BCu	Copper-base alloy
BCuP	Copper phosphorous alloy
BCuZn	Copper zinc alloy
BMg	Magnesium-base alloy
BNi	Nickel-base alloy

CHEMICAL ABBREVIATIONS			
Ag	Silver	Mn	Manganese
Al	Aluminum	Mo	Molybdenum
Au	Gold	Ni	Nickel
B	Boron	P	Phosphorous
Be	Beryllium	Pd	Palladium
C	Carbon	Si	Silicon
Cb	Columbium	Sn	Tin
Cd	Cadmium	Ta	Tantalum
Cr	Chromium	Ti	Titanium
Cu	Copper	W	Tungsten
Fe	Iron	Zn	Zinc
Li	Lithium	Zr	Zirconium
Mg	Magnesium		

Fig. 9-5. Listings of filler metals and the chemical abbreviations for base metals and filler metals indicated in Figs. 9-4, 9-6, and 9-7.

The *solidus* is the temperature at which the filler metal begins to melt. The *liquidus* is the temperature at which the filler metal becomes completely liquid.

Generally, when an alloy is formed of two metals, the liquidus of the alloy will be lower than the melting temperature of either of the metals. For example, the liquidus for the alloy BAu-2 in Fig. 9-6 is 1635°F (891°C). The liquidus for gold (Au) is 1945°F (1064°C); for copper (Cu), it is 1981°F (1083°C). When an alloy is formed of three or more metals, the resulting liquidus is not always below the melting temperature of each metal.

Good ventilation should be provided where brazing is performed. Some filler metals contain zinc or cadmium. Both zinc and cadmium are toxic. Also, fumes from some fluxes and metal coatings may be toxic.

PREPARING TO BRAZE OR BRAZE WELD

When brazing or braze welding, wear clothing similar to that worn when oxyfuel gas welding. Coveralls or a cotton shirt and pants are suggested. A cap, gloves, and hard-toed shoes should also be worn. All pockets should be covered and buttoned, if possible. Button

Nominal Chemical Composition (Percentage)

AWS Filler Metal Classification	Ag	Al	Au	Cd	Cr	Cu	Ni	Si	Zn	B	C	Fe	Li	Mg	Mn	Sn	Ti	P	Pd	Zr	Others	Solidus[1] °F	Solidus[1] °C	Liquidus[2] °F	Liquidus[2] °C
BAlSi-2		91.0				.25		7.5	.20			.8		—	.10						.15	1070	577	1135	613
BAlSi-3		84.4			.15	4.0		10.0	.20			.8		.15	.15						.15	970	521	1085	585
BAlSi-4		86.3				.30		12.0	.20			.8		.10	.15						.15	1070	577	1080	582
BAlSi-5		88.4				.30		10.0	.10			.8		.05	.05						.15	1070	577	1095	591
BAlSi-6		88.5				.25		7.5	.20			.8		2.5	.10						.15	1038	559	1125	607
BAlSi-7		87.0				.25		10.0	.20			.8		1.5	.10		.20				.15	1038	559	1105	596
BAlSi-8		85.9				.25		12.0	.20			.8		1.5	.10						.15	1038	559	1075	579
BCuP-1						94.9												5.0			.15	1310	710	1695	924
BCuP-2						92.6												7.3				1310	710	1460	793
BCuP-3	5.0					88.9												6.0				1190	643	1495	813
BCuP-4	6.0					86.6												7.3				1190	643	1325	718
BCuP-5	15.0					79.9												5.0				1190	643	1475	802
BCuP-6	2.0					90.9												7.0				1190	643	1450	788
BCuP-7	5.0					88.1												6.8				1190	643	1420	771
BAu-1			37.5			62.4															.15	1815	991	1860	1016
BAu-2			80.0			19.9																1635	891	1635	891
BAu-3			35.0			64.9																1785	974	1885	1029
BAu-4			82.0			17.9																1740	949	1740	949
BAu-5			30.0			.05	36.0												34.0			2075	1135	2130	1166
BMg-1		9.0							2.0					88.4							.30	830	443	1110	599
BMg-2a		12.0				.05	.005	.05	5.0			.005		82.7	.15						.30	770	410	1050	566
BCu-1						99.9*															.1	1980	1082	1980	1082
BCu-1a						99.0*															.3	1980	1082	1980	1082
BCu-2						86.5*															.5	1980	1082	1980	1082
BAg-1	45.0			24.0		15.0			16.0												.15	1125	607	1145	618
BAg-1a	50.0			18.0		15.5			16.5													1160	627	1175	635
BAg-2	35.0			18.0		26.0			21.0													1125	607	1295	702
BAg-2a	30.0			20.0		27.0			23.0													1125	607	1310	710
BAg-3	50.0			16.0		15.5	3.0		15.5													1170	632	1270	688
BAg-4	40.0					30.0	2.0		28.0													1240	671	1435	779
BAg-5	45.0					30.0			25.0													1250	677	1370	743
BAg-6	50.0					34.0			16.0													1270	688	1425	774
BAg-7	56.0					22.0			17.0							5.0						1145	618	1205	652
BAg-8	72.0					28.0																1435	779	1435	779
BAg-8a	72.0					27.6							.38									1410	766	1410	766
BAg-13	54.0					40.0	1.0		5.0													1325	718	1575	857
BAg-13a	56.0					42.0	2.0															1420	771	1640	893
BAg-18	60.0					29.8										10.0		.025				1115	602	1325	718
BAg-19	92.5					7.2							.23									1435	779	1635	891
BAg-20	30.0					38.0			32.0													1250	677	1410	766
BAg-21	63.0					28.5	2.5									6.0						1275	691	1475	802
BNi-1		.05			14.0		72.4	4.5		3.2	.75	4.5					.05	.02		.05	.50	1790	977	1900	1038
BNi-1a					14.0		73.1	4.5		3.2	.06	4.5						.02				1790	977	1970	1077
BNi-2					7.0		81.3	4.5		3.2	.06	3.0						.02				1780	971	1830	999
BNi-3							91.1	4.5		3.2	.06	.5						.02				1800	982	1900	1038
BNi-4							92.5	3.5		1.8	.06	1.5						.02				1800	982	1950	1066
BNi-5					19.0		70.0	10.2		.03	.10							.02				1975	1079	2075	1135
BNi-6							88.3				.10							11.0				1610	877	1610	877
BNi-7					14.0		74.9	.10		.01	.08	.2						10.1				1630	888	1630	888
BNi-8						4.5	64.7	7.0			.10				23.0			.02				1800	982	1850	1010

1 – Solidus – Melting temperature 2 – Liquidus – Flow temperature * – Minimum

*Fig. 9-6. Brazing filler metals. Solidus and liquidus temperatures are given. **Note: Cadmium (Cd) alloys are listed in red. Cadmium fumes are toxic.***

the top button on coveralls or a shirt to prevent hot metal or sparks from entering. Leathers (jacket, cape, apron, and pants) are recommended when brazing or braze welding in the overhead position.

Cleanliness is one of the most important factors in creating a strong brazed or braze welded joint. The metal in the joint area must be clean. All rust, oil, dirt, and metallic dust should be removed.

Mechanical or chemical cleaning can be used for the joint area. Mechanical cleaning is done with some type of tool or abrasive material. Emery cloth, grit blasting, grinding, or wire brushing are commonly used. The base metal is rinsed and dried after mechanical cleaning. Chemical cleaning includes any process that uses solutions of various chemicals. Chemical cleaning may be done with approved solvents. The metal is cleaned again by the flux during the brazing or braze welding process.

BRAZING AND BRAZE WELDING FLUXES

The most important step in preparing to braze or braze weld is the cleaning of the base metal surfaces. This is done by both mechanical and chemical means, as noted above. Further chemical cleaning is then done using a brazing flux. A *flux* may be defined as a "material used to prevent, dissolve, or facilitate removal of oxides and other undesirable surface substances."

The AWS classifies fluxes into six categories. These flux categories are recommended for use when brazing or braze welding various base metals, including aluminum, magnesium, iron, steel, and nickel. Some

metals require a flux that contains chlorides or fluorides. Other metals require a flux that contains boric acid, borates, fluoborates, or wetting agents. Fig. 9-7 lists the six flux categories and their suggested uses.

Fluxes are available in bottles or cans. Flux may be applied to the surface of the base metal or to the brazing filler rod. A small, clean brush may be used to apply flux to the base metal surfaces. For large areas, flux may be sprayed onto the base metal. A syringe-like applicator can be used to apply flux to a joint.

Flux may also be applied to a base metal filler rod. To apply flux to the rod, heat the last 3″−4″ (75 mm−100 mm) of the brazing rod to a few hundred degrees. A color change should not be seen when heating the rod. Dip the heated end of the rod into the flux container. Flux will stick to the heated section of the rod. When the applied flux is used up, reheat the end of the rod and dip it into the flux container again. This process is continually repeated, as required, while brazing or braze welding.

SELECTING THE TORCH TIP AND FILLER ROD

The recommended tip orifice and rod diameter for braze welding is shown in Fig. 9-8. For brazing, use a tip orifice at least 2-3 numbers larger than recommended for braze welding.

SETTING WORKING PRESSURES

The welding outfit is turned on and lit in the same manner as for oxyfuel gas welding. Chapter 5 provides details for turning on an oxyfuel gas welding outfit. The

AWS brazing flux type no.	Base Metals	Recommended filler metals	Recommended useful temperature range		Flux Ingredients
			°F	°C	
1	All brazeable aluminum alloys	BAlSi	700-1190	371-643	Chlorides, Fluorides
2	All brazeable magnesium alloys	BMg	900-1200	482-649	Chlorides, Fluorides
3A	All except those listed under 1, 2, and 4	BCuP, BAg	1050-1600	566-871	Boric acid, Borates, Fluorides, Fluoborates, Wetting agent
3B	All except those listed under 1, 2, and 4	BCu, BCuP, BAg, BAu, RBCuZn, BNi	1350-2100	732-1149	Boric acid, Borates, Fluorides, Fluoborates, Wetting agent
4	Aluminum bronze, aluminum brass, and iron or nickel base alloys containing minor amounts of Al or Ti, or both	BAg (all) BCuP (copper base alloys only)	1050-1600	566-871	Chlorides, Fluorides, Borates, Wetting agent
5	All except those listed under 1, 2, and 4	Same as 3B (excluding BAg-1 through -7)	1400-2200	760-1204	Borax, Boric acid, Borates, Wetting agent

Fig. 9-7. American Welding Society (AWS) flux classifications for brazing. Consult suppliers of commercial fluxes for specific metals and applications. "RB" means resistance brazing. See Fig. 9-5 for abbreviation meanings.

| METAL THICKNESS | BRAZE WELDING TIP ORIFICE | FILLER ROD DIAMETER | OXYGEN | | ACETYLENE | | SPEED FT/HR |
			PRESSURE (PSIG)	CFH	PRESSURE (PSIG)	CFH	
1/32	74	1/16 in.	1	1.1	1	1	
1/16	69	1/16 in.	1	2.2	1	2	
3/32	64	1/16 in. or 3/32 in.	2	5.5	2	5	20
1/8	57	3/32 in. or 1/8 in.	3	9.9	3	9	16
3/16	55	1/8 in.	4	17.6	4	16	14
1/4	52	1/8 in. or 3/16 in.	5	27.5	5	25	12
5/16	49	1/8 in. or 3/16 in.	6	33.	6	30	10
3/8	45	3/16 in.	7	44.	7	40	9
1/2	42	3/16 in.	7	66.	7	60	8

*Note the tip orifice size as shown in the number drill size. These recommendations are approximate. The torch manufacturers' recommendations should be carefully followed.

Fig. 9-8. Torch tip and rod diameter recommendations. For brazing, use a tip that is 2-3 sizes larger than for braze welding.

oxygen and acetylene working pressures are set to the pressures recommended for the torch tip size, Fig. 9-8.

ADJUSTING THE FLAME

A neutral or slightly carburizing flame is generally used when brazing or braze welding. This is done to prevent the braze joint from oxidizing. The carburizing flame will usually produce a bead with a better appearance. Fig. 5-9 shows a carburizing flame.

BRAZING OR BRAZE WELDING SAFETY PRECAUTIONS

To ensure the welder's safety during brazing and braze welding operations, standard precautions for the safe use of oxyfuel gas welding equipment must be followed. Chapters 4 and 5 include detailed information about the safe use of oxyfuel gas welding equipment. Special precautions that should be followed when brazing or braze welding include:

1. Excellent ventilation is required. Toxic metal and flux fumes are often present when brazing or braze welding.
2. Do not allow brazing fluxes to contact your skin. Many fluxes are harmful to the skin, so care should be taken in handling them. If the fluxes do come into contact with the skin, wash the area thoroughly with soap and water.
3. Only trained personnel should handle acids used for chemical cleaning.
4. Some brazing filler metals contain *cadmium*. When molten, and especially if overheated, they emit cadmium oxide fumes to the atmosphere. Cadmium

oxide fumes are very dangerous if inhaled. The limit value for cadmium oxide fumes is 0.1 milligrams per cubic meter of air for a daily eight-hour exposure. This value represents the maximum tolerance under which workers may be exposed without adverse effects. If concerns regarding cadmium exposure arise, contact the local industrial hygiene department. Cadmium fumes have no odor, and a lethal does not need to be sufficiently irritating to cause immediate discomfort. When the worker has absorbed sufficient quantities, his or her life can be in immediate danger. Symptoms of headache, fever, irritation of the throat, vomiting, nausea, chills, weakness, and diarrhea generally may not appear until some hours after exposure. Injury is primarily to the respiratory passages.

BRAZING

When brazing, the parts are fitted together, the base metal is heated, and filler metal is added to the joint. The filler metal melts at a temperature above 840°F (450°C).

The parts of a brazement must be fitted tightly together. See Fig. 9-9. Only .001″ – .010″ (.025 mm-.254 mm) clearance should be allowed between parts of a brazement.

The area around the joint is heated when brazing, rather than a limited spot as when fusion or braze welding. A larger tip size is used to heat the larger area. Also, the flame is held 1″ – 3″ (25 mm – 75 mm) from the metal. See Fig. 9-10. Do not melt the base metal when brazing. The base metal's melting point must be

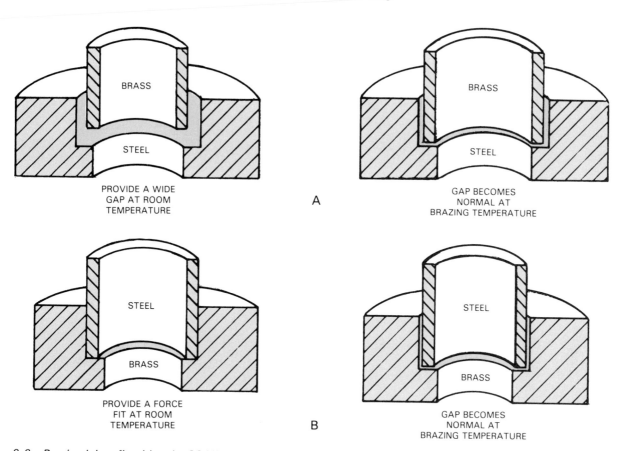

Fig. 9-9. Brazing joints fit with only .001''—.010'' (.025 mm—.254 mm) clearance. The fitting of dissimilar metals must allow for different rates of expansion that occur as the metals are heated. Brass expands at a faster rate than steel. A—Loosely fitted brass in steel fits properly at brazing temperatures. B—Steel force-fitted into brass at room temperature has the correct clearance at brazing temperature.

above the melting point of the filler metal.

The base metal is heated enough to melt the filler metal. The filler rod is touched to the base metal. A filler rod may be held at practically any angle when it is touched to the base metal. The rod should be held

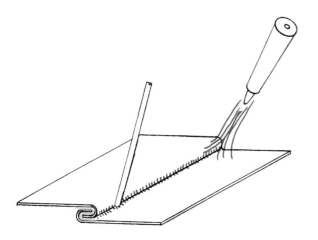

Fig. 9-10. Torch, flame, and rod positions for brazing. The flame is held 1''—3'' (25 mm—75 mm) from the joint. The flame is moved along the joint while touching the rod to the metal. Continue adding the brazing rod until the filler metal is seen throughout the joint. Note: Joint clearances are enlarged in this figure for clarity.

close to the reflected flame to preheat it. As the filler metal melts, it is drawn into the joint. This process of drawing filler metal into a joint is capillary action.

BRAZE WELDING—FLAT POSITION

Braze welding differs from brazing in several ways. Braze welded parts are not fitted together as tightly as brazed parts. Therefore, the filler metal is not drawn into the joint by capillary action. Braze welding is performed on joints similar to those used in fusion welding. See Fig. 9-11.

The torch tip and brazing rod positions for braze welding are generally the same as for oxyfuel gas welding. The tip of the flame is held above the base metal about 1/16''—1/8'' (1.6 mm—3.2 mm). On a butt joint, the torch tip is aligned with the weld axis and held at a 30°—45° angle from the base metal. For a fillet weld on a lap or T-joint, the tip is held at about a 45° angle from the base metal. On a lap joint, the tip should point more toward the surface than toward the edge of the metal. This reduces the tendency for the edge to melt.

When braze welding, the filler rod is generally held at a 15°—45° angle from the base metal surface. The end of the rod should be held close to, but not in, the

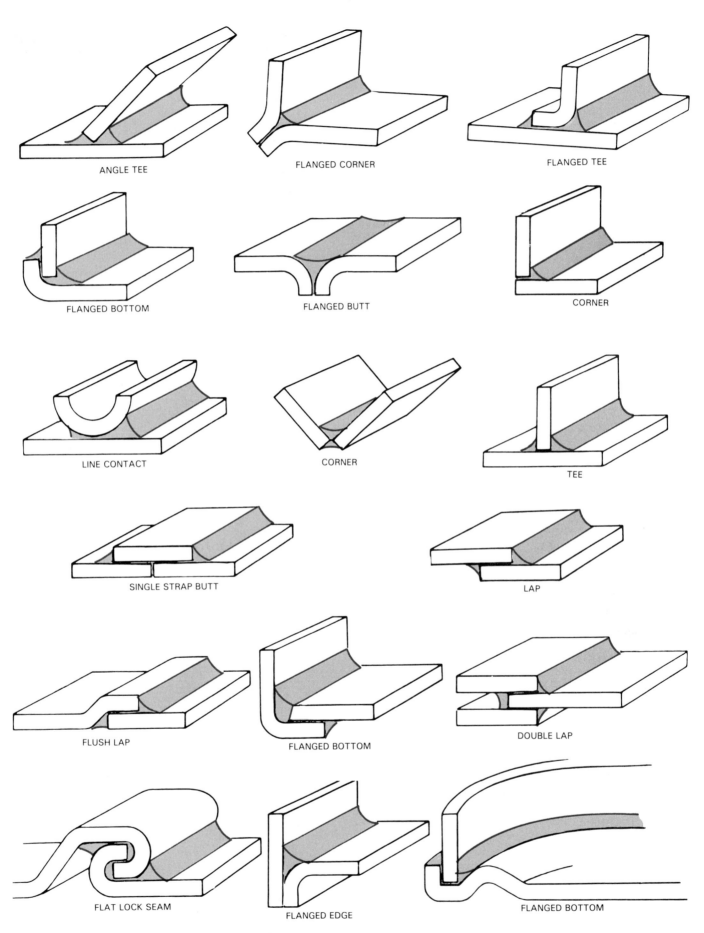

Fig. 9-11. Well-designed joints for braze welding. Note the thickness of the filler metal.   (Aluminum Company of America)

molten pool to preheat it. See Fig. 9-12.

Also, a thick layer of brazing filler metal is added to form a bead on the joint. If the base metal overheats while braze welding, withdraw the flame occasionally to control the bead width. After the base metal cools slightly, return the flame to within 1/16″ − 1/8″ (1.6 mm − 3.2 mm) of the surface.

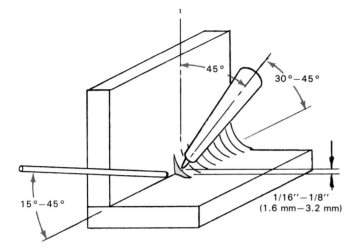

Fig. 9-12. Torch, flame, and brazing rod positions for braze welding. Note the thickness of the bead.

EXERCISE 9-1 — LAYING A BRAZE WELDED BEAD

1. Obtain the following:
   a. A piece of mild steel measuring 1/16″ x 3″ x 6″ (1.6 mm x 76.2 mm x 152.4 mm).
   b. One 36″ (.91 m) long brass uncoated brazing rod of the correct diameter. See Fig. 9-8 for suggested brazing rod diameter.
   c. A container of suitable flux. See Fig. 9-7. Pour some of the flux into a smaller container for use in this exercise.
2. Clean the surface with emery cloth and wipe away dust.
3. Lay out two lines on the surface using a soapstone or chalk. The lines should be 1″ (25.4 mm) apart and 1″ (25.4 mm) from the edges.
4. Select the correct torch tip size. See Fig. 9-8 for suggested tip orifice.
5. Turn on the welding outfit, adjust the working pressures, and light the torch. Adjust for a neutral or slightly carburizing flame.
6. Heat the end of the brazing rod and place it into the small flux container. This coats the end of the rod with flux. Repeat this procedure as often as necessary to keep the rod covered with flux.
7. Heat the metal about 1/8″ (3.2 mm) on each side of the line. Do not melt the base metal. Also, apply heat to only the base metal, not the brazing rod.
8. Touch the brazing rod to the metal occasionally. The base metal is hot enough when the rod melts.

Add the brazing rod to the heated area. Add it often enough to form a convex bead. See figure below.

9. Move the torch and brazing rod ahead at a constant rate of speed, heating the base metal and adding the rod. Continue this procedure to the end of the line.
10. If the bead is too fluid and wide, allow the base metal to cool. Withdraw the flame and allow the metal to cool slightly. Return the flame to about 1/16″ − 1/8″ (1.6 mm − 3.2 mm) above the metal and continue. Withdraw the flame as often as needed to maintain the desired bead width.

Recommendations:
   a. Use a larger tip size if the base metal heats slowly.
   b. Use a smaller tip size if the bead is hard to control or spreads too far from the line.

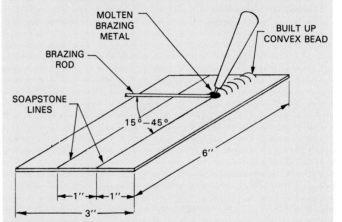

Inspection:
   The bead should be about 1/4″ (6.4 mm) wide. It should have a constant width and evenly spaced ripples.

EXERCISE 9-2 — BRAZE WELDING A FILLET ON AN INSIDE CORNER IN THE FLAT POSITION

1. Obtain the following:
   a. Two pieces of mild steel measuring 1/16″ x 1 1/2″ x 6″ (1.6 mm x 38.1 mm x 152.4 mm).
   b. One 36″ (.91 m) long uncoated brass brazing rod of the correct diameter. See Fig. 9-8 for suggested brazing rod diameter.
   c. A container of the correct flux. See Fig. 9-7. Pour some of the flux into a smaller container for use in this exercise.
2. Clean one surface of each piece with emery cloth and wipe away dust.
3. Arrange the pieces as shown in figure below.
4. Select the correct torch tip for the thickness of metal that you will be braze welding.
5. Turn on the welding outfit and light the torch. Adjust for a neutral or slightly carburizing flame.
6. Heat the end of the brazing rod. Apply flux to the heated end.

103

7. Heat an area about 1/8″ (3.2 mm) on each side of the joint. Do not melt the base metal. Apply heat to the base metal only, not to the brazing rod.

8. Touch the brazing rod to the base metal occasionally. When the rod melts, the base metal is hot enough. Add the brazing rod to the same area until a flat or slightly concave bead is formed. See figure below.

9. Move the torch and rod ahead at a constant rate, heating the metal and adding the filler metal. Continue this procedure to the end of the joint.

10. If the brazed bead is too fluid and wide, allow the base metal to cool. Withdraw the flame and let the metal cool slightly. Return the flame to about 1/16″−1/8″ (1.6 mm−3.2 mm) above the metal and continue. Withdraw the flame as often as necessary to keep the desired bead width.

Recommendations:
   a. Use a larger tip size if the base metal heats slowly.
   b. Use a smaller tip size if the bead spreads too far.

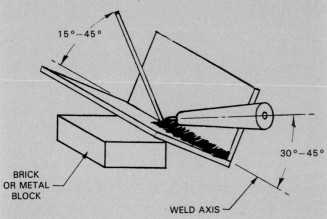

Inspection:
The bead should be about 1/4″ (6.4 mm) wide. It should have a constant width and evenly spaced ripples.

BRAZING AND BRAZE WELDING OUT OF POSITION

Brazing and braze welding are most easily done in the flat position. However, these processes can also be done out of position. Torch and brazing rod angles are the same as for fusion welding out of position. Torch and rod angles are covered in Chapter 7.

Overheating causes the bead to spread and possibly sag downward due to gravity. The brazing filler metal must be kept from overheating. Overheating is controlled as follows:
1. Heat the base metal until it melts the filler metal.
2. Add the filler metal.
3. Remove the flame and allow the bead and base metal to cool slightly.
4. Reheat the base metal and add filler metal again.
5. Continue the heating and cooling process, as required, to the end of the joint.

REVIEW QUESTIONS

Write your answers on a separate sheet of paper.
1. Brazing and braze welding filler metal melts at a temperature above _____°F (_____°C) but below the melting temperature of the base metal.
2. List two reasons why brazing or braze welding is used instead of fusion welding.
3. A _____ is an assembly of parts joined by brazing or braze welding.
4. A brazed or braze welded joint is as strong as a fusion welded joint. True or False?
5. Is "silver soldering" a soldering or brazing process? Why?
6. Refer to Fig. 9-4. What alloy of brazing filler metal is used when brazing carbon steel to aluminum?
7. List four things that can be used to clean the base metal prior to brazing or braze welding.
8. Refer to Fig. 9-7. What type of AWS brazing flux and filler metal is used to braze aluminum alloys?
9. List two types of AWS brazing flux that can be used on steel.
10. List two things that can be done if the bead is too wide and flat while braze welding.

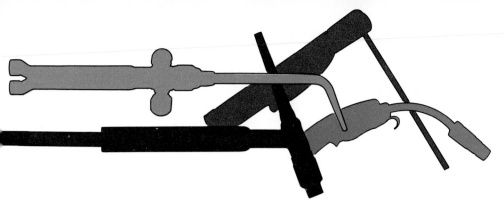

Chapter **10**

Soldering

After studying this chapter, you will be able to:
☐ Describe the principles of soldering.
☐ Select the appropriate filler metal and flux for soldering.
☐ Identify acceptable solders for drinking water systems.
☐ Describe the soldering process.
☐ Solder a lap joint.
☐ Solder a pipe joint.

TECHNICAL TERMS

Air-fuel gas torch, applicator, bridge, inorganic fluxes, liquidus, organic fluxes, rosin fluxes, solder, soldering, soldering alloy, solidus.

SOLDERING PRINCIPLES

Soldering is similar to brazing in many respects. One of the main differences, however, is the temperature at which the operation is performed. *Soldering* is defined as "a group of welding processes that join materials by heating them to the soldering temperature and by using a filler metal having a liquidus *below* 840°F (450°C) and below the solidus of the base metals. The filler metal is distributed between closely fitting surfaces of the joint by capillary action." Brazing is done at temperatures *above* 840°F (450°C).

Parts to be soldered must fit tightly together. The clearance between parts in a soldered joint is commonly .001″−.003″ (.025 mm−.076 mm). The standard clearance between copper tubing and fittings is about .002″−.006″ (.050 mm−.152 mm). When the tube is centered in the fitting, a clearance of .001″−.003″ (.025 mm−.076 mm) exists around the tube. Fig. 10-1 shows several soldered joints.

Before soldering can be done, the base metal near the joint area must be mechanically and/or chemically cleaned. A wire brush or emery cloth is usually used. After mechanical cleaning is completed, a suitable flux

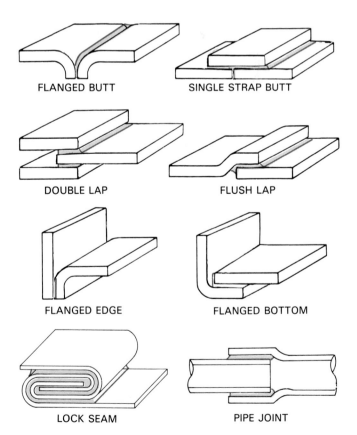

Fig. 10-1. Several solder joint designs. Normal clearance spacing is .001″−.003″ (.025 mm−.076 mm). A thin layer of solder is stronger than a thick layer. A larger surface area produces a stronger soldered joint. (Alcoa)

105

is applied to the area around the joint.

The base metal is heated until it is hot enough to melt the filler metal. Parts to be soldered may be heated using a soldering iron (soldering copper), oxyfuel gas torch, air-fuel gas torch, furnace, or electrical resistance coils. Filler metal in soldering is called *solder.* The solder flows when it is touched to the hot base metal near the joint. The liquid solder is drawn into the tight-fitting joint by capillary action. Liquid solder flows over all surfaces that are heated to the melting point of the solder.

Soldering filler metals, such as lead, tin, and zinc, are not as strong as filler metals used in welding, brazing, or braze welding. However, thin layers of solder in contact with a large enough surface in a joint will make a joint strong enough for its intended purpose. Soldering should not be used when extremely strong joints are required. Thick and wide beads of solder are seldom used, since soldering filler metals are weak. However, thick layers of solder may be used to fill a depression (such as in auto body repair). Thick layers of solder are applied for appearance, not for strength.

Soldering is a low-temperature process. It may be used for the following purposes:

1. Connecting thin metals.
2. Connecting electronic parts, and electrical parts and wires.
3. Connecting metals without affecting their heat treatment characteristics.
4. Jewelry repair and manufacture.
5. Finishing metal surfaces.
6. Connecting plumbing pipes and fittings.

SOLDERING FILLER METALS

Solder, like brazing and braze welding filler metals, is available as wire, paste, and a variety of standard and special preshaped forms. See Fig. 9-2. Most solders are available in 1 pound (.45 kg) spools of wire. Common solder wire diameters are 1/16″ (1.6 mm), 3/32″ (2.4 mm), and 1/8″ (3.2 mm).

Solders contain metals such as aluminum, antimony, cadmium, copper, indium, lead, nickel, silver, tin, or zinc. These metals are often mixed to form *soldering alloys.* The physical properties of these solders differ widely. Fig. 10-2 lists a number of soldering alloys and their composition. The solidus and liquidus temperatures for each solder are also shown. The *solidus* is the temperature at which the solder begins to melt. The *liquidus* is the temperature at which the solder is totally liquid. The solidus and liquidus temperatures are important to know when selecting a solder for a particular job. Solder flows easily into small spaces when the solidus and liquidus are close

SOLDER ALLOY COMPOSITIONS AND MELTING TEMPERATURES												
ALLOY	COMPOSITION (% BY WEIGHT)								SOLIDUS		LIQUIDUS	
	Tin	Lead	Silver	Antimony	Cadmium	Zinc	Aluminum	Indium	°F	°C	°F	°C
Tin-Antimony	95			5					450	232	464	240
Lead-	96		4						430	221	430	221
Tin-	62	36	2						354	180	372	190
Silver	5	94.5	0.5						561	294	574	301
	2.5	97	0.5						577	303	590	310
	1.0	97.5	1.5						588	309	588	309
Tin-	91					9			390	199	390	199
Zinc	80					20			390	199	518	269
	70				95	30			390	199	592	311
	60					40			390	199	645	340
	30				82.5	70			390	199	708	375
Silver-Cadmium		5			40				640	338	740	393
					10							
Cadmium-Zinc						17.5			509	265	509	265
						60			509	265	635	335
						90			509	265	750	399
Zinc-Aluminum						95	5		720	382	720	382
Tin-	50							50	243	117	257	125
Lead-	37.5	37.5						25	230	138	230	138
Indium		50						50	356	180	408	209

Fig. 10-2. The chemical composition of various soldering alloys. Solder begins to melt at its solidus temperature. Solder is completely liquid at its liquidus temperature.

together. Solder fills wide spaces in joints more easily when the solidus and liquidus are farther apart. Compare the temperatures of several alloys in Fig. 10-2.

SOLDERING FLUXES

A soldering flux has two purposes: to clean the metal surface of oxides and other undesirable substances, and to prevent surface oxidation while soldering. Fluxes are available as liquids or pastes. They are sold in leakproof containers intended to keep the flux fresh and clean. See Fig. 10-3. Remove from the container only enough flux to do the job. The container should be tightly resealed after use.

Fig. 10-3. Several types of fluxes and flux containers.

CLASSIFICATIONS

There are three basic classifications of flux. They are organic, inorganic, and rosin fluxes. *Organic fluxes* contain carbon. They have a medium level of cleaning ability. Organic fluxes are corrosive during the soldering operation. They become noncorrosive when the soldering is completed and can be washed away with water.

Inorganic fluxes do not contain carbon. Inorganic fluxes clean better than organic or rosin fluxes. They do not char or burn easily. They can be used for torch, oven, and other soldering methods. Inorganic fluxes are very corrosive. Parts soldered with inorganic flux must be cleaned after soldering. Inorganic fluxes are not used on electrical or electronic parts because they are so corrosive.

Rosin fluxes are the least effective cleaners. They are noncorrosive. Rosin fluxes are recommended for elec-

trical and electronic parts.

Fluxes are used for soldering on a variety of metals.

Some types of fluxes are used for soldering on all metals. Fig. 10-4 shows the recommended fluxes to use for various metals.

Fluxes may be applied in a variety of ways. The most common means of application is with a clean brush. Flux is brushed onto the surface of the joint area. The flux flows around the surfaces of the joint as the part is heated. It is also drawn into tightly fitted joints by capillary action.

RECOMMENDED FLUXES FOR VARIOUS METALS			
Base Metal, Alloy or Applied Finish	**Flux Recommendations**		
	Corrosive (Organic and Inorganic)	Non-corrosive (Rosin)	Special Flux and/ or Solder
Aluminum			X
Aluminum-Bronze			X
Beryllium Copper	X		
Brass	X	X	
Cadmium	X	X	
Cast Iron			X
Copper	X	X	
Copper-Chromium	X		
Copper-Nickel	X		
Copper-Silicon	X		
Gold		X	
Inconel			X
Lead	X	X	
Magnesium			X
Monel	X		
Nickel	X		
Nichrome			X
Palladium		X	
Platinum		X	
Rhodium	X		
Silver	X	X	
Stainless Steel			X
Steel	X		
Tin	X	X	
Tin-Bronze	X	X	
Tin-Lead	X	X	
Tin-Nickel	X	X	
Tin-Zinc	X	X	
Zinc	X		
Zinc Die Castings			X

Fig. 10-4. Recommended fluxes for soldering various metals.

Fluxes may also be applied by means of a large syringe-like *applicator*. In high-volume production, the flux is added in exact amounts. A pressurized flux paste applicator, Fig. 10-5, is used to apply the flux. The flux paste will then flow around the joint when the part is heated.

HAZARDS OF SOLDERS AND FLUXES CONTAINING LEAD

According to the Environmental Protection Agency (EPA) studies, excess lead in drinking water supplies can cause a variety of health problems. On June 19, 1986, President Ronald Reagan signed Public Law

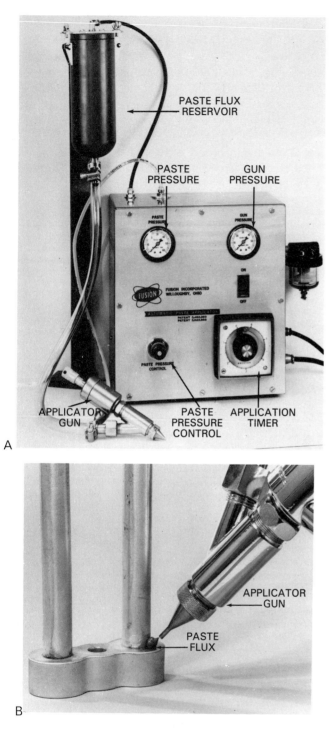

Fig. 10-5. A—Solder flux paste applicator. The amount of flux paste applied is controlled by the paste pressure and by a timer that is accurate to within 1/20 sec. The quantity applied by weight can vary from .001 oz. (0.028g) to 1 oz. (28.4g). For high-volume production soldering, the applicator can be mounted on a robot. B—Close-up of a solder flux paste applicator. (Fusion, Incorporated)

99-339. This law, "The Safe Drinking Water Act Amendment of 1986," banned the use of lead solder, flux, and pipe from use in drinking water installations. Beginning on June 19, 1988, all public water systems were to notify their customers who might be affected by lead contamination of their drinking water. On that same date, all states were to begin enforcing the ban on the use of lead solders, fluxes, and pipes in drinking water systems.

Solders containing lead in excess of 0.2% may still be used in water systems for drainage, heating, fire sprinklers, air conditioning, and machine cooling. Tin and lead alloys are commonly used on parts and systems that do not carry drinking water. These tin-lead alloy solders may contain from 5% tin and 95% lead to 70% tin and 30% lead. The compositions of various tin-lead alloys are shown in Fig. 10-6.

Tin-lead alloy solder is available in rolls of wire. Various combinations of lead and tin are available. The 50% tin and 50% lead (50/50) solder is most common.

ACCEPTABLE SOLDERS FOR DRINKING WATER SYSTEMS

Several companies have developed new lead-free solders. These solders produce joints that are as strong or stronger than those made with tin-lead solders. The new lead-free solders have about $40°F - 60°F$ ($4°C - 15°C$) difference between their solidus and liquidus. This wide range of temperatures makes it possible to apply the new lead-free solders using the same techniques as 50/50 tin-lead solder. The range of temperatures also makes it possible for these solders to *bridge* (fill wide gaps) poorly fitted assemblies. Fig. 10-7 shows the solidus and liquidus of several lead-free solders. Similar equipment and flux is used as when applying 50/50 solder. Lead-free alloys that have been recently developed include:
- 95.5% tin, 4% copper, 0.5% silver.
- 88% tin, less than 2% silver, 4% antimony, less than 2% copper, and 4% zinc.
- 89% tin, less than 2% silver, 5% antimony, 3% copper, and less than 1% nickel.

Several other lead-free soldering alloys are harder to apply because their solidus and liquidus temperatures are closer together. Some of these lead-free solders are:
- 95% tin, 5% antimony
- 96% tin, 4% silver
- 95% tin, 5% silver
- 94% tin, 6% silver
- 97% tin, 3% copper

Lead-free solders are more expensive per pound than tin-lead solders. However, more wire length is available in a one-pound roll, since lead-free solders are lighter. The cost per soldered joint is approximately equal to tin-lead solders.

PREPARING TO SOLDER

For most light soldering jobs, regular work clothes are adequate. Light duty gloves are recommended. A lighter goggles lens, such as a #2 – #4, is adequate.

Cleanliness is the most important factor in creating a good solder joint. Metals to be soldered are cleaned chemically or mechanically. Chemical cleaning, on

ASTM SOLDER CLASSI-FICATION	COMPOSITION (% by weight)		SOLIDUS		LIQUIDUS	
	TIN	LEAD	°F	°C	°F	°C
5	5	95	572	300	596	314
10	10	90	514	268	573	301
15	15	85	437	225	553	290
20	20	80	361	183	535	280
25	25	75	361	183	511	267
30	30	70	361	183	491	255
35	35	65	361	183	477	247
40	40	60	361	183	455	235
45	45	55	361	183	441	228
50	50	50	361	183	421	217
60	60	40	361	183	374	190
*	62	38	361	183	361	183
70	70	30	361	183	378	192

Fig. 10-6. Tin-lead solder compositions. ASTM 60 solder is commonly used for electrical connections. *This is the Eutectic alloy. Its solidus and liquidus temperatures are the same.

most soldering jobs, is done by using an appropriate soldering flux. Mechanical cleaning is done with a wire brush, abrasive cloth or paper, or abrasive blasting. Wire brushes must be clean. Stainless steel brushes are recommended. All particles on the surfaces must be removed by washing and drying the surfaces after mechanical cleaning.

SELECTING THE TORCH TIP AND SOLDER

An oxyfuel gas torch or an *air-fuel gas torch* (acetylene, propane, butane, or MPS) is commonly

LEAD-FREE SOLDERING ALLOY	SOLIDUS		LIQUIDUS	
	°F	°C	°F	°C
95% tin, 5% antimony	450	232	464	240
96% tin, 4% silver	430	221	460	238
95% tin, 5% silver	430	221	473	245
94% tin, 6% silver	430	221	535	279
97% tin, 3% copper	450[1]	232	500[1]	260
Silvabrite 100™ 95.5% tin, 4% copper, 0.5% silver	440	227	500	260
Stay-Safe 50™ 88% tin, <2% silver, 4% antimony, <2% copper, 4% zinc	400	204	440	227
Stay-Safe Bridgit™ 89% tin, <2% silver, 5% antimony, 3% copper, <1% nickel	460	238	630	332
*50% tin, 50% lead	361	183	421	217

Fig. 10-7. Chemical composition, solidus, and liquidus for several lead-free alloys. Notes:[1] These temperatures are approximate. *50/50 tin-lead information is given for comparison only.

used to heat the base metal when soldering. Generally, soldering is done on metal which is less than 1/8″ (3.2 mm) thick. The torch tip used for air-acetylene, air-propane, air-butane, or air-MPS gas generally produces a wide, spreading flame. The same size tip is commonly used for all applications. A tip orifice between #74 and #55 is generally used for oxyfuel gas soldering. However, the correct size is a matter of personal preference.

Solder wire is commonly available in the following diameters: 1/16″ (1.6 mm), 3/32″ (4.2 mm), and 1/8″ (3.2 mm). The size used for an operation depends on the base metal thickness. If the diameter is too small, the solder wire will have to be held on the heated joint longer to completely fill it. Solder with a 1/8″ (3.2 mm) diameter is generally used on copper pipe joints.

ADJUSTING THE FLAME

An air-propane, air-butane, or air-MPS gas torch is started by opening the torch valve slightly and lighting the flame with a sparklighter. The torch valve is adjusted until a neutral flame is obtained.

Air-acetylene and oxyfuel gas cylinders are opened and the pressures set according to the tip size. The flame is adjusted on the air-acetylene torch by using only the acetylene torch valve. Adjust the flame until the correct amount of acetylene is flowing to provide a neutral or slightly carburizing flame. On an oxyfuel gas torch, open the torch acetylene valve until the smoke stops. Then open the oxygen torch valve until the flame is neutral or slightly carburizing. Flame adjustment is detailed in Chapter 5.

SOLDERING SAFETY PRECAUTIONS

Observe all standard safety precautions for welding with oxyfuel gas welding equipment, regardless of the type of soldering equipment used. A number of general safety precautions are presented in Chapter 4. Special precautions that must be followed include:

1. Excellent ventilation is required.
2. Care must be taken when handling fluxes that are harmful to the skin.
3. Only trained personnel should use acids for chemically cleaning metal.
4. Some solders contain cadmium, and emit cadmium oxide fumes when molten or overheated. These fumes are extremely toxic. See Chapter 9 for more information regarding cadmium hazards.
5. The proper protective clothing for oxyfuel gas welding should be worn. A #2 – #4 welding goggle filter lens is adequate for most soldering jobs.

TORCH SOLDERING

Non-production and repair soldering is generally done with a hand-held torch. See Fig. 10-8. An oxyfuel gas flame is generally not needed for soldering. Air-fuel gas torches, which produce lower temperatures, work very well for soldering. Propane, butane, acetylene, or MPS gas can be used to obtain the low-temperature flame.

When soldering, pinpoint heating is not required or recommended. A large torch tip should be used to spread the heat over a broad area. See Fig. 10-9.

The following things must be done to produce an acceptable soldered joint:

1. Metals must be mechanically and/or chemically cleaned. All oxides, grease, and dirt must be removed.
2. The appropriate flux must be used. Flux must be fresh and as chemically pure as possible.
3. The metals must be firmly supported during and after the soldering operation. They should be supported until they cool. As the joint cools, it increases in strength.
4. Metals to be soldered must be heated to the melting temperature of the solder. -
5. Solder should be melted only by the heat of the metals to be soldered and not by the flame. Add only enough solder to complete the job. An excess of solder is unsightly.
6. The soldering operation should be done as quickly as possible to reduce oxidation.
7. Most fluxes should be thoroughly removed from the joint as soon as possible after the soldering operation is completed. This keeps corrosion from continuing.

Fig. 10-9. An autobody repair person who is soft soldering the trunk lid of an automobile. A propane torch with a large tip is used to heat the trunk lid. A wooden stick is used to smooth the molten solder.

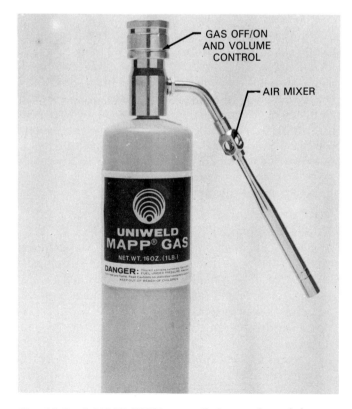

Fig. 10-8. A MAPP (MPS) gas cylinder, torch, and tip combination. This type of torch is used for torch soldering in a variety of applications. (Uniweld Products, Inc.)

Hold the flame about 1″ – 3″ (25 mm – 76 mm) above the solder joint at a 30° – 45° angle from the base metal. The end of the solder wire does not have to be preheated, as in welding or brazing. The solder wire may be held at any convenient angle.

The flame is moved back and forth to heat the entire joint to the soldering temperature at the same time. When soldering pipes or tubes, the flame is moved completely around the joint. As the joint is heated, the solder is touched to it to test the temperature. To prevent overheating the joint as the solder melts, the flame is then withdrawn to about 3″−5″ (76 mm−127 mm) from the joint. The solder wire is placed on the joint and continually added until solder can be seen along the entire length of the joint. If the solder cools before the joint is full, reheat the joint and add more solder. In a well-made joint, solder should not extend more than about 3/8″−1/2″ (9.6 mm−12.7 mm) to either side of the joint center line. If the solder is wider than 1/2″, too large an area of the joint has been heated.

EXERCISE 10-1 − SOLDERING A LAP JOINT

1. Obtain the following:
   a. Six pieces of mild steel that are less than 1/16″ (1.6 mm) thick and measure 1 1/2″ x 5″ (38.1 mm x 127 mm).
   b. A coil of a suitable solder with a 1/8″ (3.2 mm) diameter.
   c. A container of the appropriate flux and a small clean brush to apply the flux. See Fig. 10-4.
   d. An oxyfuel gas welding outfit or air-fuel gas torch.
2. Clean the surfaces of all pieces at least 3/4″ (19.1 mm) from the edges.
3. Use the small brush to apply flux to the two pieces over the area cleaned in step 2.
4. Form a lap joint with the two pieces and clamp the pieces together.
5. Turn on the torch and adjust it to a neutral or slightly carburizing flame.
6. Hold the flame about 1″−3″ (25 mm−76 mm) above the center line of the joint. Move the flame back and forth along the entire length of the joint. Touch the end of the solder wire to the joint occasionally to test the joint temperature. When the solder melts, withdraw the flame to 3″−5″ (76 mm−127 mm). Continue adding solder until it is visible along the entire length of the joint. If the solder cools before the joint is full, reheat the joint and add the solder again. Use the other pieces of metal to complete the exercise by soldering two more lap joints.

Inspection:
   Solder must be seen between the pieces and on each side of the solder joint. The width of the solder on each side of the joint should be uniform and not wider than about 3/8″ (9.5 mm).

EXERCISE 10-2 − SOLDERING A PIPE JOINT

1. Obtain the following:
   a. Two pieces of 3/4″ (19.1 mm) copper pipe approximately 6″ (152 mm) long.
   b. Two 3/4″ x 90° elbows.
   c. A coil of a suitable solder with a 1/8″ (3.2 mm) diameter.
   d. A container of suitable soldering flux and a small clean brush for application of the flux.
   e. An oxyfuel gas welding outfit or other fuel gas torch.
2. Clean the ends of the pipe and the inside of the fittings using emery cloth. The pipe should be cleaned for a distance of at least 1″ (25 mm) from the end.
3. Apply flux to the pipe and fittings using a flux brush.
4. Fit the parts together and support them. A fire brick or metal block may be used.
5. For an oxyacetylene torch, use a tip that is 2−3 number sizes larger than one used for welding this thickness of metal. See Fig. 4-29. If an air-fuel gas (propane, butane, acetylene, or MPS) torch is used, select a large heating tip. See Fig. 10-9.
6. Turn on the fuel gas slightly. Light the gas using a sparklighter. Adjust for a neutral or slightly carburizing flame.
7. Hold the flame 1″−3″ (25 mm−75 mm) from the joint. Heat the area around the joint. Touch the solder wire to the base metal to check the temperature.
8. When the solder melts, continue to heat the base metal and add solder.
9. Control the heat to the joint by withdrawing the torch. This allows the joint to cool. Apply the flame again when the joint cools.
10. Withdraw the flame and solder when the joint appears to be full of solder.
11. Quickly and carefully wipe all excess solder from the joint area using a clean cloth. Do this before the solder becomes firm. Solder the second joint using the same procedure.

Inspection:
   The joints must be filled with solder. They should have a smooth fillet completely around the diameter. Low spots or holes should not be seen in the solder joint. The joint should be wiped clean.

REVIEW QUESTIONS

Write your answers on a separate sheet of paper.
1. Soldering is done below _____°F (_____°C).
2. Strong soldered joints are made with convex beads on loosely fitted joints. True or False?
3. The base metal melts the filler metal when soldering. True or False?
4. Soldered joints are strongest when a thin layer of filler metal is applied to a large surface. True or False?
5. List three applications for which soldering is the best process to use.
6. The temperature at which solder begins to melt is the _____ temperature.
7. For a tightly fitted joint, the solidus and liquidus temperatures should be _____.

8. Refer to Fig. 10-6. What is the solidus and liquidus temperature of a 95/5 tin/silver solder?
9. Refer to Fig. 10-4. What type of flux is recommended for soldering steel?

10. The flame should be held ____″ – ____″ (____ mm – ____ mm) away from the base metal when soldering.

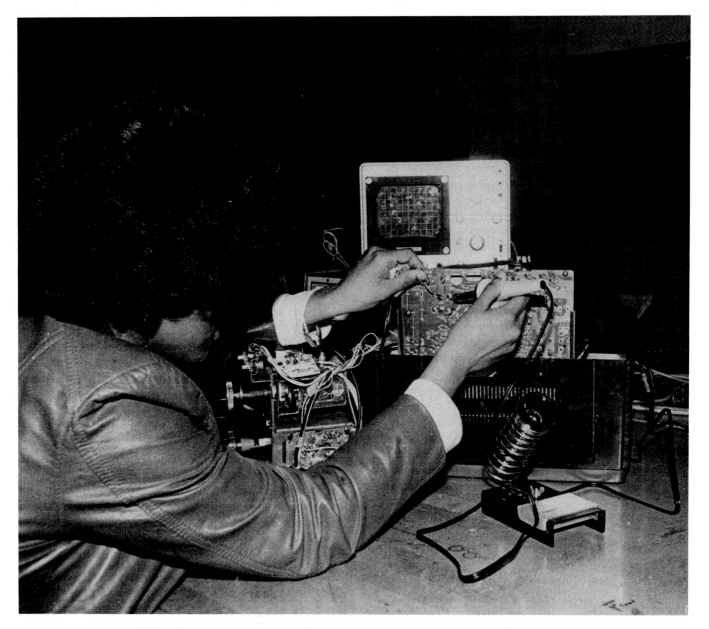

Television repair technician making a repair with a small electric soldering iron.

SHIELDED METAL ARC WELDING

Nederman, Inc.

Shielded metal arc welding (SMAW) being used to construct a storage tank.   (Miller Electric Mfg. Co.)

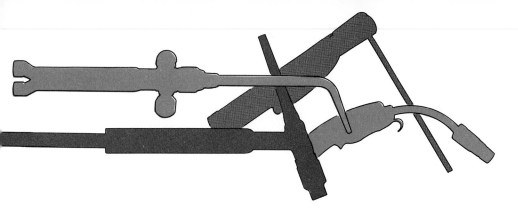

Chapter 11

Shielded Metal Arc Welding Equipment and Supplies

After studying this chapter, you will be able to:
- ☐ Distinguish between direct current (DC) and alternating current (AC).
- ☐ Interpret American Welding Society (AWS) abbreviations regarding welding current polarity.
- ☐ Identify the equipment and accessories used in shielded metal arc welding (SMAW).
- ☐ Identify factors to consider when selecting an arc welding machine.

TECHNICAL TERMS

Alternating current (AC), chipping hammer, constant current (CC), constant voltage (CV), cover plate, cycle, direct current (DC), direct current electrode negative (DCEN), direct current electrode positive (DCEP), direct current reverse polarity (DCRP), direct current straight polarity (DCSP), droopers, duty cycle, electrode holder, electrode lead, filter lens, flash goggles, gauntlet gloves, helmet, infrared rays, input power, lugs, open circuit voltage (OCV), polarity, rated output current, shielded metal arc welding (SMAW), shielding gas, slag, terminals, ultraviolet rays, voltage drop, welding outfit, welding station, wire brush, work booth, workpiece lead.

SHIELDED METAL ARC WELDING PRINCIPLES

Shielded metal arc welding (SMAW) is a welding process in which the base metals are heated to fusion or melting temperature by an electric arc. The arc is created between a covered metal electrode and the base metals. The base metals, arc, electrode, and weld are shielded from the atmosphere while welding. A *shielding gas* is used to protect the weld area from the atmosphere. This gas is not pressurized. As the flux covering on the electrode melts, it creates the shielding gas. The melting electrode wire furnishes filler metal to the weld. See Fig. 11-1.

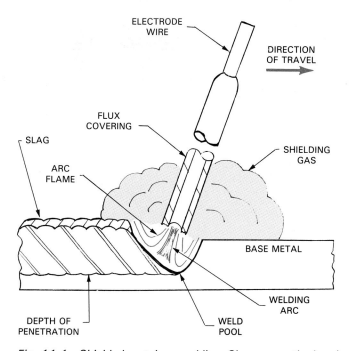

Fig. 11-1. Shielded metal arc welding. Slag covers the bead while it cools. Shielding gas is formed as the flux covering is burned.

The temperature of the arc in SMAW is about 6500°F – 7000°F (3590°C – 3870°C). With the suggested-diameter electrode, the heat created by the arc is enough to melt any weldable metal. Mild carbon steel melts at about 2500°F – 2700°F (1370°C – 1480°C).

Shielded metal arc welding is used to construct buildings, ships, truck chassis, pipe lines, and other weldments. SMAW equipment is fairly inexpensive. It is widely used on farms and in small repair shops.

SMAW CURRENT AND POLARITY

In SMAW, electrical current flows across an air gap between the covered electrode and base metal. An electric arc is formed as the current flows across the air gap. The arc creates the heat required for welding.

The electrical current for welding is supplied by an arc welding machine. Welding machines, also called power sources, produce two types of current: direct current (DC), and alternating current (AC).

An electrical current is actually the flow of electrons within a circuit. Electrons generally flow from a negatively charged (polarized) body to a positively charged body. Direct current may flow from the electrode to the base metal. When the current flows in this direction, the electrode has *negative* polarity and the base metal has *positive* polarity. It is called *direct current electrode negative (DCEN)*. Direct current electrode negative was also known as *direct current straight polarity (DCSP)*. See Fig. 11-2. The current direction may be reversed to flow from the base metal to the electrode. When the current flows in this direction, the base metal has *negative* polarity and the electrode has *positive* polarity. It is called *direct current electrode positive (DCEP)*. Direct current electrode positive is also known as *direct current reverse polarity (DCRP)*. See Fig. 11-3.

The direction selected is determined by the metal thickness, joint position, and type of electrode used. The selection of polarity is covered later in this chapter.

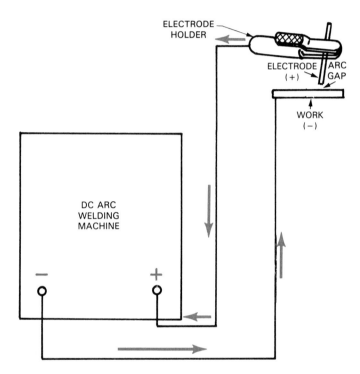

Fig. 11-3. A wiring diagram for a direct current electrode positive (DCEP) arc welding circuit. In DCEP, the current travels from the base metal to the electrode.

DIRECT CURRENT

Direct current (DC) flows in only one direction. Current flows from one terminal of the welding machine to the base metal. It then flows back to the second terminal of the welding machine. The direction of current is reversed at the machine by reversing electrical leads or by flipping a *polarity* switch. See Fig. 11-4. On some welding machines, the electrical leads are manually reversed on the machine terminals. This changes the direction of current flow. On other machines, a polarity switch is flipped to electrically reverse the terminal leads within the machine. Welding must be stopped and the electrode insulated from the circuit whenever the polarity is changed.

The circuit polarity affects the covered electrode when welding. Direct current electrode positive (DCEP) produces deeper penetration than direct current electrode negative (DCEN). DCEN causes the covered elec-

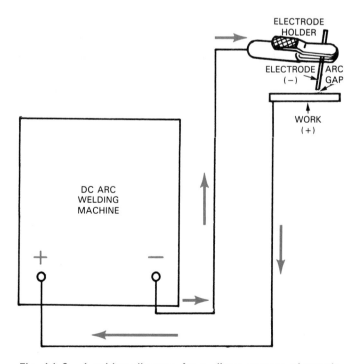

Fig. 11-2. A wiring diagram for a direct current electrode negative (DCEN) arc welding circuit. Notice that the current is traveling from the electrode to the base metal.

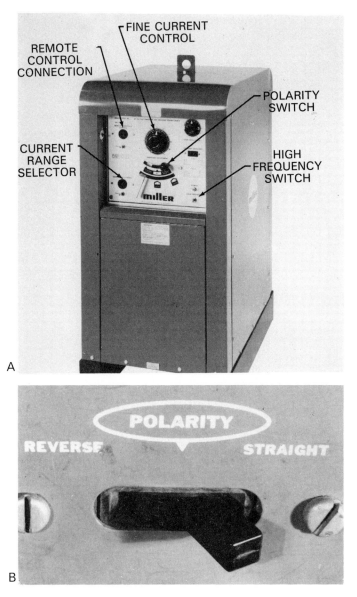

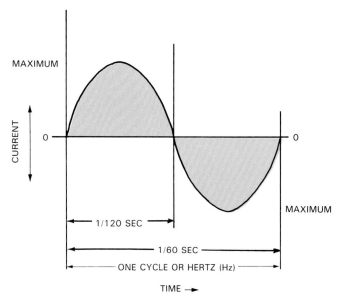

Fig. 11-5. Alternating current plotted against time results in an AC sine wave. The current increases from zero and returns to zero. It then increases in the opposite direction and returns to zero. The complete process is one cycle and takes 1/60 second.

Fig. 11-4. A—An AC/DC arc welding machine. The polarity switch has three positions; DC reverse polarity (DCRP), AC, and DC straight polarity (DCSP). B—Some DC arc welding machines use a toggle switch like this to select either reverse (DCEP) or straight (DCEN) polarity. (Miller Electric Mfg. Co.)

trode to melt faster and deposit filler metal at a faster rate. DCEN heats the base metal more slowly than DCEP. It is commonly used when welding thin metals.

ALTERNATING CURRENT

Alternating current (AC) generally operates at 60 cycles per second in the United States. A *cycle* is a set of repeating events, such as the flow and reversal of flow in an electrical circuit. Alternating current flows in one direction for 1/120 of a second, then reverses its direction for 1/120 of a second to form one complete cycle. See Fig. 11-5. The electrode in an AC circuit changes polarity 120 times per second.

AC-covered electrodes produce a neutral or reducing gas when they are used. These gases protect the weld

area and electrode from oxidation and contamination by preventing the local atmosphere from entering. AC and AC electrodes produce a medium depth of penetration. See Chapter 13 for more information on electrodes.

SMAW OUTFIT

SMAW requires the use of proper protective clothing and various tools, supplies, and equipment. The *welding outfit* includes equipment required to actually create a weld. The *welding station* includes tools, supplies, and other equipment items required to make welding safe and comfortable. A complete SMAW outfit includes the following:
• DC or AC welding machine/power source.
• Electrode and workpiece leads.
• Electrode holder.
A complete SMAW station includes the SMAW outfit plus the following: (See Fig. 11-6.)
• Ventilation.
• Welding table.
• Welding booth with an opaque (solid) or filtered transparent plastic screen.
• Covered electrodes.
• Arc welding helmet and lenses.
• Chipping hammer and wire brush.

ARC WELDING MACHINE

Arc welding machines produce either a *constant current (CC)* or a *constant voltage (CV).* Manual welding

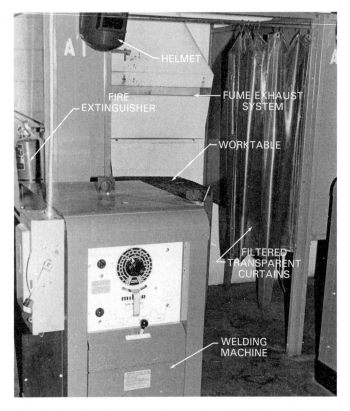

Fig. 11-6. A shielded metal arc welding (SMAW) station. The electrode holder, electrode, chipping hammer, and wire brush are not shown.

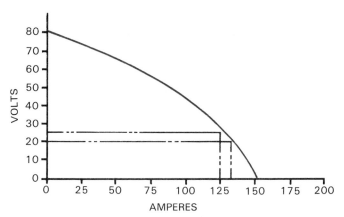

Fig. 11-7. A constant current volt-amp diagram. The constant current (CC) welding machine is called a drooper because of this curve. A 25% change in voltage results in only a 4% change in amperage. The current change is so slight that the current is considered constant.

Fig. 11-8. A tap-type AC arc welding machine. Two amperage ranges are possible. The electrode lead is placed into either the high- or low-range tap.   (Century Mfg. Co.)

processes, such as SMAW, require a constant current (CC) welding machine. If a CC welding machine is not used, large changes in current occur whenever a welder changes the arc length slightly. Since it is impossible to manually maintain a constant arc length without any variation, a CC welding machine is used.

CC welding machines are also called *droopers* or droop curve machines. This is because of the voltage versus amperage curve produced by the machines. Fig. 11-7 shows the curve shape for a CC welding machine.

Examine Fig. 11-7 closely. A 25% increase in voltage is produced by increasing from 20 volts to 25 volts. This change in voltage results in an increase in current from 125 amps to 130 amps. This represents only a 4% current increase. If the welder varies the arc length, it causes a change in voltage. However, the amperage changes very little with a CC welding machine. Although the current varies slightly, these welding machines are considered constant current. Both AC and DC welding machines are available as constant current machines.

Selecting an arc welding machine

AC welding machines meet most welding requirements, are easy to use, and generally cost less than comparable DC welding machines. Therefore, they are widely used in industry, on the farm, and in the home shop. See Fig. 11-8.

DC welding machines are more versatile than AC machines, because the polarity can be changed. The ability to change the polarity allows the welder to make out-of-position welds and to weld thin metal more easily. It also allows the welder to vary the heat applied to the metal. DCEP provides deep penetration for welds on thick metal; DCEN deposits filler metal faster. DCEN is used more easily on thin metal.

Some arc welding machines are combination AC/DC machines. These machines can be used as AC welding machines, then can easily be switched to become DC welding machines. These machines have a greater flexibility in their use. However, they are more expensive than AC or DC machines. See Fig. 11-9.

After selecting an AC, DC, or AC/DC machine, the following variables must also be considered:
1. Input power requirements.
2. Rated output current rating.
3. Duty cycle.
4. Open circuit voltage.

Input power requirements. Welding machines used in school shops, trade schools, and industry are connected to commercially available electric power. Power requirements for a welding machine must be specified to the electrician when the machine is wired into a building. The *input power* of a welding machine must correspond with the type of power that is available. It is fairly expensive to rewire existing 110 V power to 220 V or 440 V.

Engine-driven welding machines are used in the field for pipelines, construction, and other welding operations where electric power is not available. Engines are connected to the welding machine to turn an alternator for AC welding current or a generator for DC welding current.

The input voltage to a welding machine may be 115 V, 230 V, 440 V or higher. These high voltages are reduced by a transformer within the welding machine to the required welding voltage. Welding voltage ranges from 5 V to 30 V.

Rated output current. A nameplate on each welding machine shows the ***rated output current***. See Fig. 11-10. A welding machine must be able to supply the current required for the welds being made. The rated output current depends on the duty cycle.

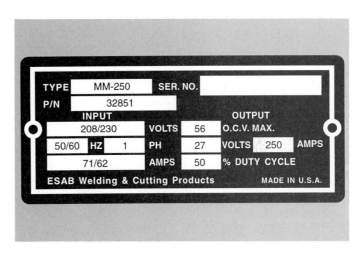

Fig. 11-10. A typical nameplate for an arc welding machine. The output for this machine is 250 amperes at 27 volts, and the duty cycle at that amperage is 50%. (ESAB Welding and Cutting Products.)

Duty cycle. The ***duty cycle*** is a rating that indicates how long a welding machine can be used at its maximum output current without damaging it. Duty cycle is based on a ten-minute time period. A welding machine with a 60% duty cycle can be used at its maximum rated output current for six out of every ten minutes. The welding machine may overheat if the duty cycle is exceeded. At lower current settings, the duty cycle may be increased and the power source used for a longer period of time.

A duty cycle chart should be provided with a new welding machine. In Fig. 11-11, the duty cycle is 20% at the maximum rated output current of 200 amperes (A). The duty cycle is 100% at a rated output current of 100 A.

Open circuit voltage. Welding machines have a maximum voltage or open circuit voltage. ***Open circuit voltage*** (OCV) is the voltage of the welding machine when it is on, but is not being used. The maximum OCV for most manual AC or DC machines is 80 V. This relatively low OCV protects workers from electrical shock. The 80 V OCV is still high compared to the 5 V-30 V closed circuit welding voltage. An OCV of 80 V is required to start the arc easily. It is also necessary to maintain the arc during AC welding.

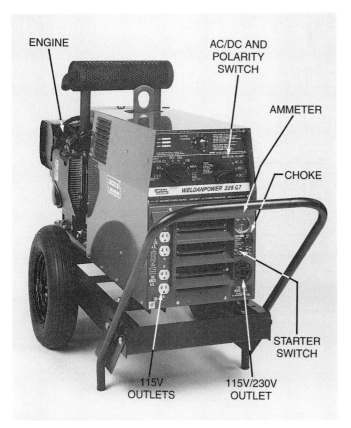

Fig. 11-9. A portable engine-drive AC/DC welding power source. A switch is used to select AC or the desired DC polarity. (The Lincoln Electric Co.)

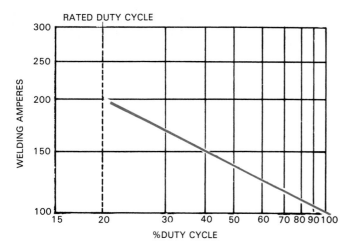

Fig. 11-11. Duty cycle versus welding current graph. This type of graph should be provided with each new arc welding machine. The duty cycle for this machine is 20% at 200 A. At 100 amperes, the machine may be used 100% of the time.

Fig. 11-12. This piece of AWG #1/0 welding lead shows the large number of fine copper wires used. These small wires, woven into larger bundles, give the lead its flexibility. The bundle is wrapped in plastic and several strands of strong cord. Neoprene is then used as an insulator and a wear covering. (Carol Cable Co.)

WELDING LEADS

The electrical cable that connects the electrode holder to a welding machine is the **electrode lead.** The **workpiece lead** (ground) is the electrical cable that connects the base metal to the welding machine.

On large welding machines, leads may be required to carry 600 A or more. Leads must have a large diameter to carry such high current. They must also be flexible so a welder can easily move them. As many as 2500 fine conductors are used in a welding lead to produce the flexibility required. Welding leads must also be well insulated to prevent electrical shock. They are usually insulated with a heavy rubber or neoprene covering. See Fig. 11-12.

Voltage and current are affected when the welding leads are the wrong diameter. Electrical resistance in the lead increases as its diameter decreases or its length increases. **Voltage drop** (voltage loss) occurs when electricity travels a long distance from the welding machine. A larger diameter lead is used to counteract the voltage drop when long leads are required. A greater amount of current flows in a larger diameter lead when voltage drop is held to a minimum.

Welding leads are available in a variety of diameters. The diameter is referred to by number size. These number sizes and their actual diameters are shown in Fig. 11-13. Fig. 11-13 also shows what size lead should be used to carry a given amperage to the weldment and

CURRENT CAPACITY IN AMPERES LENGTH IN FEET AND METERS					
Lead No.	Lead Dia.		Length 0-50 ft. 0-15.2 m	Length 50-100 ft. 15.4-30.5 m	Length 100-250 ft. 30.5-76.2 m
	in.	mm	Amperage	Amperage	Amperage
4/0	.959	24.4	600	600	400
3/0	.827	21.0	500	400	300
2/0	.754	19.2	400	350	300
1/0	.720	18.3	300	300	200
1	.644	16.4	250	200	175
2	.604	15.3	200	195	150
3	.568	14.4	150	150	100
4	.531	13.5	125	100	75
Note: Lengths given are for the total combined length of the electrode and work leads.					

Fig. 11-13. Welding lead size recommendations. Lead sizes range from 4/0 to 4. If the work is 100' (30.5 m) from the machine, 200' (61 m) of leads are needed: 100' (30.5 mm) for the workpiece lead and 100' (30.5 m) for the electrode lead. Size 1/0 leads are required to carry 200 A through 200' (61 m) of leads.

back. The amperages shown are the maximum that can be carried in the various leads over the stated distances.

Lead connections

Welding leads are connected to the welding machine and base metal with lugs, clamps, or special *terminals.* Fig. 11-14 shows several *lugs* that may be used on the machine end of the workpiece and electrode leads. Lug connections can also be used to connect the workpiece lead to the welding table. A special push-and-turn connector is shown in Fig. 11-15A. This quick-connect terminal may be used on both welding leads.

A

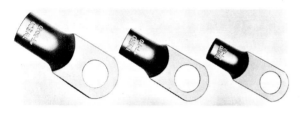

Fig. 11-14. Lugs for welding leads. These three sizes will fit leads from size 4 to 4/0. They are connected to the lead by soldering or mechanical crimping.   (Tweco Products, Inc.)

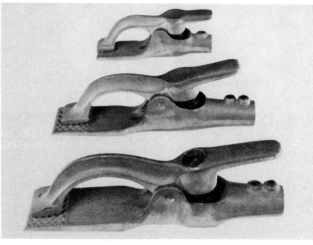

B

Various other types of clamps or special connectors may be used to connect the workpiece lead to the base metal. The lead may be clamped to the base metal using a spring-loaded clamp, as shown in Fig. 11-15B. Occasionally, a part must be welded while it is being rotated. A special rotating workpiece clamp is used for this application, Fig. 11-15C. Lugs, spring clamps, or other special connectors are soldered or mechanically connected to the bared end of the leads.

ELECTRODE HOLDER

The *electrode holder* is held by the welder during the welding operation. The well-insulated handle of the electrode holder protects the welder from electrical shock. An electrode is clamped in the copper alloy jaws of the electrode holder. The jaws provide a good electrical contact for the electrode. See Fig. 11-16. The electrode lead is clamped into the electrode holder. The cable clamp is under the insulated handle.

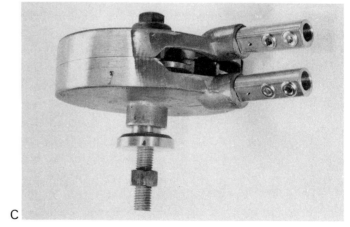

C

Fig. 11-15. Lead connections. A—American and European quick connectors for welding machine leads. The European connector is called a DIN connector. The connectors marked with * are DIN connectors. (Weldcraft Products, Inc.) B— Three sizes of spring-loaded work clamps. The lead may be easily connected or disconnected. (NLC, Inc.) C—A rotating work clamp used to provide a good ground on a rotating part.   (NLC, Inc.)

PROTECTIVE CLOTHING

Pockets on shirts, pants, or coveralls should be covered to prevent sparks from being caught in them. The top collar button should be fastened, especially when welding out of position. Pant legs should not have cuffs.

A cap should be worn to protect your hair. Leather gloves should be worn, with the *gauntlet* type preferred for out-of-position welding. High-top shoes offer added safety from sparks. Steel-toed shoes are recommended when working around heavy metal parts.

Arc welding helmets and lenses

An electric arc creates *ultraviolet* and *infrared rays.* These light rays are harmful to the eyes and cause burns, like sunburn, on bare skin. A welder should cover all bare skin areas while welding.

An arc welding *helmet* serves several purposes. It shields the face and neck from the harmful rays and protects these areas from molten metal that may spatter from the weld area. The filter lens and cover plate

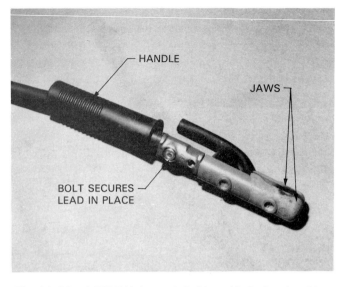

Fig. 11-16. A SMAW electrode holder with the insulated handle slid back to show the lead connection. An electrode is held tightly in the jaws by pressure from a heavy coil spring under the release lever.

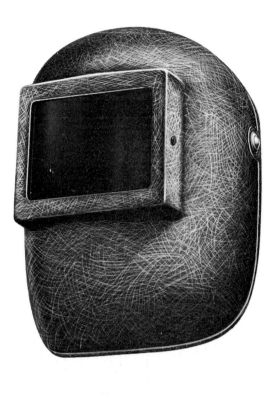

are also held by the helmet. A #10 – #14 *filter lens* should be used for SMAW. A darker lens is required as the brightness of the arc increases. See Fig. 11-17. A good quality filter lens screens out 99.5% of the infrared rays and 99.75% of the ultraviolet rays from the electric arc. Regardless of the filter lens number, the percentage of the rays filtered remains the same for all high-quality filter lenses.

A clear glass or plastic cover plate is also used. The *cover plate* protects the filter lens from arc spatter. Always use a cover plate over a filter lens. Filter lenses are more expensive than cover plates, and should be protected. The cover plate should be replaced regularly.

Filter lenses and cover plates are available in two sizes; 2″ x 4 1/4″ (51 mm x 108 mm) and 4 1/2″ x 5 1/4″ (114 mm x 133 mm). See Fig. 11-18. An arc welding helmet can be adjusted for various head sizes. Adjustable screws secure the head band to the helmet. These screws should be tight enough to prevent the helmet from falling down when it is raised. However,

Application	Lens Shade Number
SMAW (Shielded metal arc welding) Up to 5/32 in. electrodes	10
3/16 - 1/4 in. electrodes	12
5/16 - 3/8 in. electrodes	14
GMAW (Gas metal arc welding) (nonferrous) Up to 5/32 in. electrodes	11
GMAW (Gas metal arc welding) (ferrous) 1/16 - 5/32 in. electrodes	12
GTAW (Gas tungsten arc welding)	10 to 14

Fig. 11-17. Suggested filter lens shade numbers for various arc welding applications.

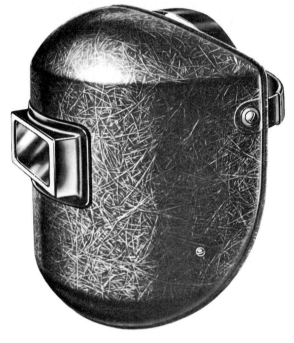

Fig. 11-18. Arc welding helmets are used to protect the eyes and face while welding. The welder can see the weld area clearly while welding. The filter lens, however, reduces the amount of harmful rays that reach the welder's eyes. A cover plate is used to protect the filter lens.   (Fibre-Metal Products Co.)

the helmet should fall into position over the face when the welder's head is nodded.

Arc welders often wear flash goggles under their helmets. *Flash goggles* protect the welder's eyes from flashes from the rear. A #1 – #3 lens is often used in these goggles.

VENTILATION

Adequate ventilation must be provided when performing any type of welding. The size or capacity of the ventilation system for a given area should be calculated by a safety engineer.

Exhaust fumes from an arc welding area can be toxic to a welder. A flexible exhaust pick-up tube is often used to remove the fumes. The pick-up tube is positioned by the welder for maximum performance. Fumes should be picked up and exhausted before they cross the welder's face. See Fig. 11-19.

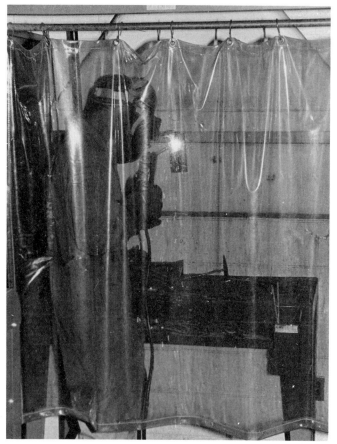

Fig. 11-19. A welding fume exhaust system. The flexible duct allows the welder to position the pick-up hood close to the weld area. (Nederman, Inc.)

WORK BOOTH AND TABLE

The welding table is part of the arc welding station. The workpiece lead may be bolted to the table with a lug or attached to the table with a spring clamp. A weld positioner may be welded or clamped to the table. A shop-fabricated weld positioner is shown in Fig. 7-2.

Welding may be done in a **work booth** or in an open area. In either case, the electric arc must be shielded from the view of the workers who are not wearing eye protection. Solid walls, canvas curtains, or filtered, transparent plastic may be used. Filtered, transparent plastic must filter more than 99% of the ultraviolet and infrared rays from the electric arc. See Fig. 11-20.

Fig. 11-20. Filtered, transparent plastic curtain used on an arc welding booth. The plastic filters out the harmful rays created by the arc welding operation.

Chipping hammer and wire brush

Slag must be removed after each arc weld bead or pass is completed. *Slag* is the hard, brittle material that covers a finished shielded metal arc weld. The next bead will have slag trapped in it if the previous bead is not cleaned thoroughly. Slag is chipped away using a *chipping hammer*. See Fig. 11-21.

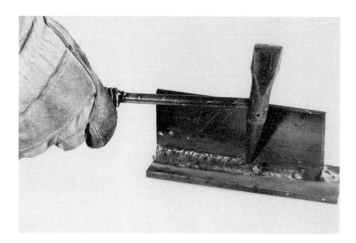

Fig. 11-21. A chipping hammer being used to remove the slag from a partially completed weld.

After chipping the slag, the bead is further cleaned using a *wire brush.* The bristles of the brush are steel. See Fig. 11-23. A wire brush is sometimes combined with a chipping hammer. A rotary wire wheel in a portable electric drill motor may also be used to clean the weld. Always wear goggles when chipping or wire brushing the slag from a weld. This will help to prevent eye damage.

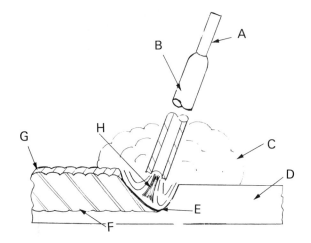

Fig. 11-22. Wire brush is used to clean the weld after the slag has been removed.

REVIEW QUESTIONS

Write your answers on a separate sheet of paper.
1. Identify the parts of a SMAW in progress.
2. Which electrode polarity should be used when welding thin metals?
3. Current that travels from the electrode to the base metal is known as _____ or _____.
4. AC electrodes provide _____ penetration when used for SMAW.
5. List eight items found in an SMAW station.
6. Is a drooper a CC or CV welding machine?
7. Refer to Fig. 11-11. The duty cycle for this welding machine at 125 amperes is _____ %.
8. Refer to Fig. 11-13. A #_____ electrode and workpiece lead should be used to carry 150 amperes of electricity 75 feet to the workpiece and 75 feet back to the welding machine.
9. The temperature of an electric arc in SMAW is between _____ °F – _____ °F (_____ °C – _____ °C).
10. What is the OCV of most AC and DC welding machines?

Chapter 12

Shielded Metal Arc Welding Equipment: Assembly and Adjustment

After studying this chapter, you will be able to:
- ☐ Explain the assembly of a welding machine, leads, and electrode holder.
- ☐ List the steps in connecting an electrode holder to the lead.
- ☐ Describe the procedure for inspecting a shielded metal arc welding (SMAW) outfit.
- ☐ Set the proper amperage and polarity on a welding machine.

TECHNICAL TERMS

Coarse amperage range, jack, quick-connect terminal, range switch, rectifier, tap.

WELDING MACHINE

AC and DC welding machines can be connected to a variety of input power voltages. Input power leads should not be repaired by a welder. A qualified electrician should make needed repairs. A welder may need to change the output power leads, however. The polarity of some DC welding machines is changed by reversing the workpiece and electrode leads.

INPUT POWER

Input power for AC and DC welding machines generally is supplied by commercial sources. Commercial power voltages of 110 V – 115 V, 220 V – 230 V, or higher may be required.

Welding machines using 220 V – 230 V or higher voltages are usually permanently wired (hard-wired) into the building's electrical system. However, some electrical codes permit 220 V – 230 V machines to be plugged into special wall receptacles. See Fig. 12-1. Wiring a 110 V – 115 V or 220 V – 230 V wall receptacle or permanently wiring any welding machine must be done by a qualified electrician.

Engine-driven alternators may be used to produce 115 V or 230 V input power for arc welding machines. The alternators are driven by gasoline or diesel engines.

Engines may also be connected directly to the welding machine. The engines are used to drive an AC

Fig. 12-1. An approved wall-mounted receptacle for use with a special 220 V-250 V plug on some small welding machines.

125

alternator or DC generator. A *rectifier* is used to convert alternating current to direct current in the welding machine. See Fig. 12-2.

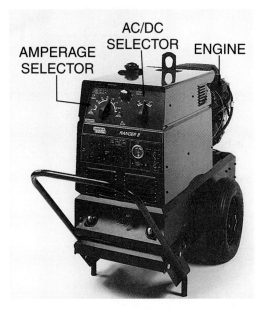

Fig. 12-2. A portable, engine-driven alternator. This power source provides AC, DCRP, or DCSP. (The Lincoln Electric Co.)

OUTPUT CONNECTIONS

AC and DC welding machines have at least two terminals. One terminal is for the electrode lead and the other is for the workpiece lead. See Fig. 12-3. The leads are connected to the threaded copper or brass terminals

Fig. 12-3. The electrode and workpiece lead terminals on a DC arc welding machine. These leads may be reversed to change the polarity.

using hex or wing nuts. Use a proper-size wrench to tighten the connections. The hex or wing nuts must be tight to ensure a good connection.

Amperage range terminals may be used on welding machines. The electrode lead is plugged into one of the amperage range terminals to select a coarse amperage range. Generally, the *coarse amperage range* is one of several ranges of a welding machine that provides electrical current output. Coarse amperage ranges usually extend over 50 A or more, depending on the output current of the machine. See Fig. 12-4. Brass or copper

Fig. 12-4. An AC/DC welding machine. On the AC side, a coarse amperage range may be selected by placing the electrode lead into the 40 A – 230 A or 30 A – 150 A range terminal. The workpiece lead is placed into the workpiece lead terminal. On the DC side, the leads may be reversed to change polarity. Fine amperage adjustments are made using the hand crank on top of the machine.

quick-connect internal/external terminals are commonly used. The external connector is on the end of the electrode lead. The *jack* or *tap* (internal connection) is a socket on the front of the machine. See Fig. 12-5.

ELECTRODE AND WORKPIECE LEADS

The electrode and workpiece leads should be the same diameter. Two variables are used to determine the diameter of the electrode and workpiece leads: maximum amperage and total distance to the work and back. A larger amperage requires larger diameter leads. A longer distance to the work and back requires larger diameter leads. Fig. 11-13 shows a table used to determine lead diameter.

Fig. 12-5. An external quick-connect terminal and internal jack or tap used on some welding machines to connect the workpiece and electrode leads.

Lugs or external quick-connect terminals are attached to one end of each lead. A lug is a heavy-duty electrical terminal that is cylindrical at one end and flat on the other end. A hole is drilled in the flat end. This end is connected to a welding machine terminal with a nut. A *quick-connect terminal* is a heavy-duty electrical terminal that is easy to connect and disconnect from a welding machine. A cam lock device is used to engage the quick-connect terminal. See Fig. 12-6A and B. A short section of insulation is removed from the end of the lead. The lug or connector is then attached to the bare conductors (wires) by mechanical means or soldering.

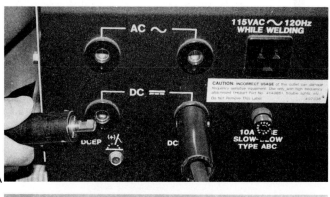

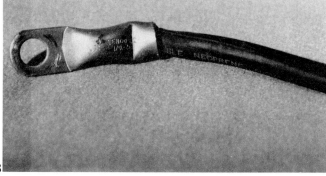

Fig. 12-6. A—An external connector used with a jack or tap on the arc welding machine. B—A copper lug fitting used on welding leads. A mechanically swaged joint was used to connect the lead to the lug.

A lug is attached to the opposite end of the workpiece lead. An area on the work table or workpiece is cleaned to ensure a good electrical connection. The lug may be bolted to the work table or attached to the table or workpiece using a spring-loaded clamp. Magnets or C-clamps may also be used to attach a workpiece lead to the table or workpiece. See Fig. 12-7.

The opposite end of the electrode lead is connected to the electrode holder. The procedure for connecting the electrode holder is discussed later in the chapter.

Leads should not be laid across aisles or other heavy-traffic areas. They may be damaged and cause injuries. For permanent installations, run the leads overhead in conduit or below the floor. For temporary installations on a job site, pass the leads under a large section of C-channel for protection.

Fig. 12-7. A work lead held in contact with the workpiece by a C-clamp. (Jackson Products, Inc.)

CONNECTING THE ELECTRODE HOLDER

One end of the electrode lead is attached to a lug or quick-connect terminal. The opposite end is connected to the electrode holder by mechanical means. See Fig. 12-8. The procedure for making the mechanical connection is as follows:

1. Remove the insulated handle from the electrode holder.
2. Loosen the hex socket cap screws (Allen screws) in the copper or brass socket.
3. Remove the insulation from the electrode lead. The amount of insulation removed should be equal to

the depth of the socket.

4. Wrap copper foil around the bare conductors. This prevents the conductors from unwrapping and also provides a good electrical contact.

5. Insert the wrapped end of the electrode lead with the curved cable pressure plate into the socket.

6. Tighten the hex socket cap screws. This secures the lead tightly between the pressure plate and the socket walls.

7. Reinstall the insulated handle.

The electrode and/or holder should not be in contact with the work table or workpiece when the welding machine is started or damage may occur to the welding machine. Also, do not allow them to be in contact with the workpiece when changing current. To prevent the electrode and/or holder from contacting the workpiece, hang the holder from an insulated hanger in the welding booth when not in use.

INSPECTING THE SMAW OUTFIT

The SMAW outfit should be inspected before welding to ensure that all connections are tight, all insulation is intact, and that all equipment is safe to use. The welding machine must be turned off to prevent electrical hazards when inspections are made.

1. Check to see that the electrode and workpiece leads are tightly attached to the welding machine terminals. Use the proper size wrench to tighten the connections, if required.

2. Visually check the entire length of each lead for damage or wear.

3. Ensure that the electrode lead is connected tightly to the electrode holder.

4. Verify that the workpiece lead is connected tightly to the work table or workpiece.

5. Make sure that the electrode holder or electrode is not in contact with the work table or workpiece.

6. Verify that booth sides and curtains protect people outside the booth from the arc and flying slag.

7. Remove any moisture or standing water from the floor.

8. Make sure all ventilation ports and ducts are not blocked and that the ventilation system is working properly.

ADJUSTING THE MACHINE

The correct polarity and amperage must be determined before adjusting the welding machine. The type of metal, metal thickness, joint design, welding position, and type of electrode should be considered when selecting polarity and amperage.

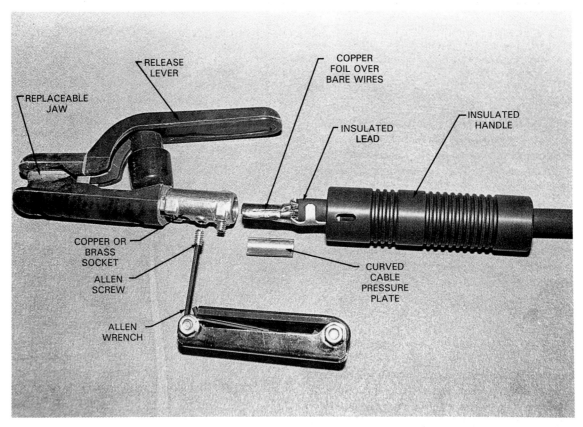

Fig. 12-8. An exploded view of a SMAW electrode holder. Notice the insulated handle and the parts required to connect the welding lead to the electrode holder.

POLARITY

A DC welding machine can be changed from DCEN to DCEP by reversing the workpiece and electrode leads. The polarity can be changed on some machines by flipping a switch. See Fig. 12-9. The welding machine must be turned off, whenever the polarity is changed, to prevent internal damage.

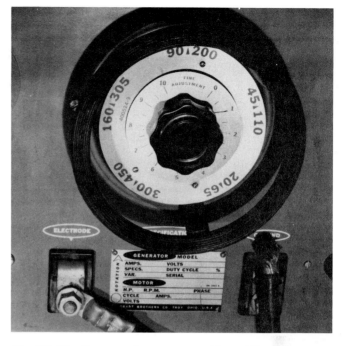

Fig. 12-10. Coarse adjustment on this welding machine is made by turning the large outer handwheel. The fine amperage adjustment uses the smaller knob in the center of the dial.

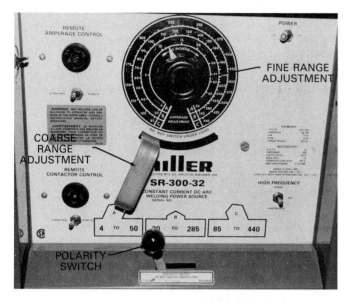

Fig. 12-9. Close-up view of the controls on a DC welding machine. On this machine, a lever is used to select the coarse amperage range. The DC polarity is selected using the lever at the bottom of the photo. Fine amperage adjustments are made by turning the knob at the top. Notice the three coarse and fine range selections.

AC and DC welding machines may use coarse and fine adjustment controls to select the required amperage. The adjustment controls function in the same way, although they may be located in a different place on the welding machine or operate differently. Coarse adjustment provides ranges of amperage settings. Coarse adjustments can be made by changing the position of the leads in the jacks, by moving a lever, or by rotating a handwheel on the face of the machine. See Fig. 12-10. The coarse adjustment lever on a welding machine is called a *range switch.*.

Fine amperage adjustments provide precise control of the current. Fine adjustments are made with a knob, crank, or slide handle. These controls change the amperage within the selected coarse range. The coarse and fine adjustments are generally shown on the face of the welding machine. The welding machine shown in Fig. 12-11 uses high and low amperage taps for the AC coarse adjustment and a sliding handle for fine adjustments.

When all coarse and fine adjustments have been made, the machine may be turned on.

Fig. 12-11. This AC/DC welding machine uses jacks to select a high and low amperage range when using AC. For DC, the jacks are used to select the polarity. Fine adjustments for AC and DC are made using the sliding handle.
(POW-R-MATE, Division of Century Mfg. Co.)

REVIEW QUESTIONS

Write your answers on a separate sheet of paper.
1. A special wall receptacle may be used for _____ V to _____ V welding machines.
2. What is a rectifier used for in a welding machine?
3. A qualified _____ should permanently wire the welding machine to the commercial power source.
4. Lugs are attached to welding leads using _____ or _____.
5. Internal welding lead connections on the face of a welding machine are called _____ or _____.
6. What two variables are used to determine the diameter of the welding leads?
7. List three methods of attaching a workpiece lead to a work table or workpiece.
8. Why shouldn't welding leads be laid across an aisle or other heavy-traffic area?
9. Name two methods used to change DC polarity on a welding machine.
10. List three ways of selecting the fine amperage setting on AC and DC welding machines.

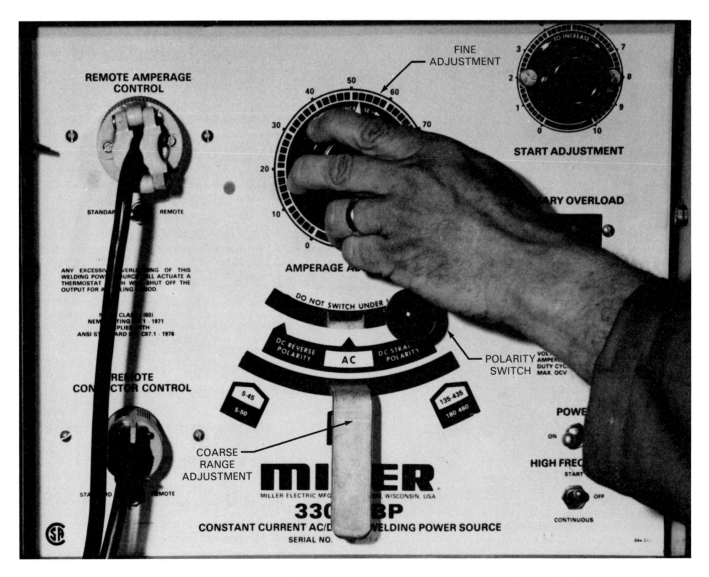

A welder making a fine amperage adjustment. Note the coarse amperage range lever and the polarity selection switch.

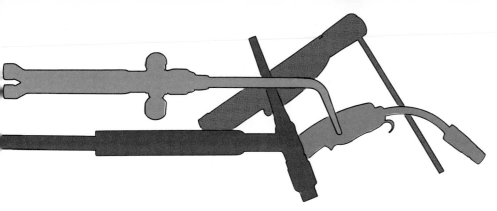

Chapter 13

Shielded Metal Arc Welding Electrodes

After studying this chapter, you will be able to:
- ☐ Identify carbon and low alloy SMAW electrodes.
- ☐ List six purposes of an electrode covering.
- ☐ Interpret the AWS electrode identification system.
- ☐ Determine the trial amperage of a welding machine using the rule-of-thumb method.
- ☐ Select an electrode to meet the requirements of a weld.
- ☐ Identify two means of storing electrodes.

TECHNICAL TERMS

AWS electrode specifications, carbon dioxide, electrode covering, electrode drying oven, electrode identification system, low hydrogen electrodes, reducing gas, welding procedure.

COVERED ELECTRODES

SMAW electrodes are solid, round metal wires covered (coated) with a flux. Electrode wires are made from pure metal or metal alloys. The wire melts when an arc is struck. The melting wire provides filler metal for the weld.

Electrode wires and fluxes for SMAW electrodes are produced under strict manufacturing specifications. These manufacturing specifications were developed by the American Welding Society. *AWS electrode specifications* are published and revised about every five years. Fig. 13-1 shows the AWS electrode specifications for several metals.

ELECTRODE COVERINGS

When manufacturing covered electrodes, the electrode wire is first cut to the desired length. Flux materials are combined into a thick clay-like mixture. The covering is then applied to the electrode wire in a very exact thickness.

An *electrode covering* may serve any or all of the following purposes:
1. Add filler metal to the weld.
2. Create a protective gas shield around the arc and molten metal.
3. Create a flux to clean impurities from the molten metal.
4. Create a hard slag covering to protect the molten bead as it cools.
5. Improve mechanical and chemical properties of the weld by adding alloying elements to the weld metal.
6. Determine the polarity of the electrode.

When an arc is struck, the electrode wire melts due to the heat. The molten electrode wire adds filler metal to the weld crater and forms a weld bead.

The heat also causes the flux covering to burn and create a shielding gas around the weld. The shielding gas may be neutral or a reducing gas such as *carbon dioxide* (CO_2) or hydrogen (H_2). A *reducing gas* is a gas that removes oxygen from the atmosphere by

AWS ELECTRODE SPECIFICATION NUMBER	METAL REFERRED TO IN SPECIFICATION
AWS A5.1	Carbon steels
AWS A5.3	Aluminum and aluminum alloys
AWS A5.4	Corrosion-resistant steels
AWS A5.5	Low alloy steels
AWS A5.6	Copper and copper alloys
AWS A5.11	Nickel and nickel alloys
AWS A5.15	Gray and ductile cast iron

Fig. 13-1. AWS electrode specification numbers for several commonly welded metals.

combining with the oxygen. For example, $CO + O_2$ = CO_2. The shielding gases keep oxygen, dirt, or other airborne contaminants out of the weld area.

The electrode covering forms a flux as it melts. The flux removes impurities from the weld. The impurities and flux float to the surface of the weld bead.

As the flux and impurities cool, they form *slag* (hard crust) over the surface of the bead. The bead cools under the slag, protected from contamination by the thick covering.

The mechanical and chemical properties of the weld metal can also be improved using certain flux coverings. Iron powder may be used in electrode covering compounds. This type of electrode deposits more filler metal into the weld in a given length of time. This reduces welding time and decreases welding costs. Some electrode coverings contain metallic salts. The metallic salts add alloying elements to the molten weld metal as the covering melts. The alloying elements combine with the weld metal to make the weld stronger, more corrosion resistant, or more ductile.

Hydrogen in a completed weld lowers its strength. *Low hydrogen electrodes* deposit a minimum amount of hydrogen into the weld. The electrode identification number for low hydrogen electrodes is discussed later in the chapter.

Chemicals in the electrode covering determine the polarity of the electrode. Certain electrode coverings work well with AC. Other electrode coverings are used with DCEN only or DCEP only. Still, other coverings produce good beads with AC, DCEN, or DCEP.

ELECTRODE SIZES

Covered electrodes are available in a variety of lengths and diameters. They are available in 9″ (229 mm), 12″ (305 mm), 14″ (356 mm), and 18″ (457 mm) lengths.

The diameter of an electrode is the diameter of the uncoated wire. Flux covering thickness is disregarded when determining electrode diameter. AWS covered electrodes are available in the following diameters; 1/16″ (1.6 mm), 5/64″ (2.0 mm), 3/32″ (2.4 mm), 1/8″ (3.2 mm), 5/32″ (4.0 mm), 3/16″ (4.8 mm), 7/32″ (5.6 mm), 1/4″ (6.4 mm), 5/16″ (7.9 mm), and 3/8″ (9.5 mm).

Not all AWS electrodes are produced by every electrode manufacturer in the diameters or lengths shown above. Each electrode manufacturer distributes an electrode guide showing the lengths and diameters they produce. These guides also provide a variety of information regarding electrodes and their suggested uses.

Corrugated cartons or metal cans are used to ship 14″ (356 mm) and 18″ (457 mm) electrodes. They are available in 50-pound containers. The 9″ (229 mm) and 12″ (305 mm) electrodes are shipped in 25-pound (11.4 kg) containers.

ELECTRODE IDENTIFICATION

The proper electrode diameter and classification must be determined in order to produce a good weld. The diameter of the electrode can be measured. The electrode is identified (classified) by a number located near its bare end. See Fig. 13-2.

Fig. 13-2. The American Welding Society (AWS) number is imprinted near the bare metal end of an electrode.

The American Welding Society has developed an *electrode identification system.* Covered electrodes for SMAW low alloy steels and carbon steels are identified with the letter E (for "electrode") followed by 4 or 5 numbers. See Figs. 13-3 and 13-4.

The first two (sometimes three) numerals from the left indicate the minimum tensile strength. Minimum tensile strength is measured in thousands of pounds per square inch or megapascals. In Fig. 13-5A, the minimum tensile strength is 60,000 psi (414 MPa). The minimum tensile strength for the electrode number in Fig. 13-5B is 100,000 psi (689 MPa).

The second numeral from the *right* indicates the recommended welding position and/or type of weld. The numerals represent the following:
1—All positions.
2—Flat welding and horizontal fillet welds.
3—Flat position only.
4—Flat, horizontal, overhead, or vertical down welding.
In Fig. 13-5, the electrodes may be used in all positions.

The last two numerals at the right indicate the polarity or any other special notes. Fig. 13-6 shows a list of the last two numerals and their indicated polarities. In Fig. 13-5A, the electrode requires DCEP. The electrode in Fig. 13-5B requires AC or DCEP. It is also a low hydrogen electrode.

The electrode identification number may also have a suffix to show the alloying elements that are added to the weld metal. Fig. 13-7 shows a list of suffixes and their indicated alloying elements. In Fig. 13-5B, D2 indicates that the alloying elements are composed of .25–.45% molybdenum and 1.25–2.00% manganese.

Low hydrogen electrode have AWS identification numbers that end in 5, 6, or 8. See Fig. 13-8. The numerals 5, 6, or 8 indicate the covering composition and suggested application of the electrode.

ELECTRODE AMPERAGE REQUIREMENTS

The diameter of the electrode wire determines the amount of amperage required to melt it. A larger wire diameter requires a higher amperage. A recommended amperage range is provided by electrode manufacturers for all AWS electrodes and diameters that they produce. Fig. 13-9 suggests amperage ranges for various E60XX and E70XX electrodes and diameters.

A rule-of-thumb method may be used to find a trial amperage when a manufacturer's electrode guide is not available. This method works well for electrode diameters between 3/32" (2.4 mm) and 1/4" (6.4 mm). A 1/8" (3.2 mm) diameter electrode requires 90 A – 120 A. Add 40 A to the 90 A for each 1/32" (.8 mm) of diameter above 1/8" (3.2 mm). Subtract 40 A from

AWS Classification[a]	Type of covering	Capable of producing satisfactory welds in positions shown[b]	Type of current[c]
E70 series — Minimum tensile strength of deposited metal, 70,000 psi (480 MPa)			
E7010-X	High cellulose sodium	F, V, OH, H	DCEP
E7011-X	High cellulose potassium	F, V, OH, H	AC or DCEP
E7015-X	Low hydrogen sodium	F, V, OH, H	DCEP
E7016-X	Low hydrogen potassium	F, V, OH, H	AC or DCEP
E7018-X	Iron powder, low hydrogen	F, V, OH, H	AC or DCEP
E7020-X	High iron oxide	{ H-fillets { F	AC or DCEN AC or DC, either polarity
E7027-X	Iron powder, iron oxide	{ H-fillets { F	AC or DCEN AC or DC, either polarity
E80 series — Minimum tensile strength of deposited metal, 80,000 psi (550 MPa)			
E8010-X	High cellulose sodium	F, V, OH, H	DCEP
E8011-X	High cellulose potassium	F, V, OH, H	AC or DCEP
E8013-X	High titania potassium	F, V, OH, H	AC or DC, either polarity
E8015-X	Low hydrogen sodium	F, V, OH, H	DCEP
E8016-X	Low hydrogen potassium	F, V, OH, H	AC or DCEP
E8018-X	Iron powder, low hydrogen	F, V, OH, H	AC or DCEP
E90 series — Minimum tensile strength of deposited metal, 90,000 psi (620 MPa)			
E9010-X	High cellulose sodium	F, V, OH, H	DCEP
E9011-X	High cellulose potassium	F, V, OH, H	AC or DCEP
E9013-X	High titania potassium	F, V, OH, H	AC or DC, either polarity
E9015-X	Low hydrogen sodium	F, V, OH, H	DCEP
E9016-X	Low hydrogen potassium	F, V, OH, H	AC or DCEP
E9018-X	Iron powder, low hydrogen	F, V, OH, H	AC or DCEP
E100 series — Minimum tensile strength of deposited metal, 100,000 psi (690 MPa)			
E10010-X	High cellulose sodium	F, V, OH, H	DCEP
E10011-X	High cellulose potassium	F, V, OH, H	AC or DCEP
E10013-X	High titania potassium	F, V, OH, H	AC or DC, either polarity
E10015-X	Low hydrogen sodium	F, V, OH, H	DCEP
E10016-X	Low hydrogen potassium	F, V, OH, H	AC or DCEP
E10018-X	Iron powder, low hydrogen	F, V, OH, H	AC or DCEP
E110 series — Minimum tensile strength of deposited metal, 110,000 psi (760 MPa)			
E11015-X	Low hydrogen sodium	F, V. OH, H	DCEP
E11016-X	Low hydrogen potassium	F, V, OH, H	AC or DCEP
E11018-X	Iron powder, low hydrogen	F, V, OH, H	AC or DCEP
E120 series — Minimum tensile strength of deposited metal, 120,000 psi (830 MPa)			
E12015-X	Low hydrogen sodium	F, V, OH, H	DCEP
E12016-X	Low hydrogen potassium	F, V, OH, H	AC or DCEP
E12018-X	Iron powder, low hydrogen	F, V, OH, H	AC or DCEP

a. The letter suffix 'X' as used in this table stands for the suffixes A1, B1, B2, etc. (see Fig. 13-7) and designates the chemical composition of the deposited weld metal.
b. The abbreviations F, V, OH, H, and H-fillets indicate welding positions as follows: F = Flat; H = Horizontal; H-fillets = Horizontal fillets. V = Vertical } For electodes 3/16 in. (4.8 mm) and under, except 5/32 in. (4.0 mm) and under for classifications OH = Overhead } EXX15-X, EXX16-X, and EXX18-X.
c. DCEP means electrode positive (reverse polarity). DCEN means electrode negative (straight polarity).

Fig. 13-3. AWS low alloy steel, covered arc welding electrodes. An explanation of the Xs is found in Fig. 13-7. (American Welding Society)

the 90 A for each 1/32″ (.8 mm) of diameter below 1/8″ (3.2 mm). The amperage that is determined will be near the lower end of the recommended amperage range. For example:

Determine what amperage should be set on the welding machine when using 3/16″ (4.8 mm) electrodes.

Step 1: 1/8″ (4.8 mm) diameter electrode = 90 A

Step 2: Determine the difference between 3/16″ and 1/8″. *3/16″ − 1/8″ = 1/16″*

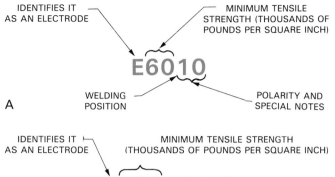

A

B

Fig. 13-5. *Examples of what an AWS electrode number includes. A—E6010 electrode. B—E10016-D2 electrode.*

AWS Classification	Type of covering	Capable of producing satisfactory welds in position shown[a]	Type of current[b]
E60 series electrodes			
E6010	High cellulose sodium	F, V, OH, H	DCEP
E6011	High cellulose potassium	F, V, OH, H	AC or DCEP
E6012	High titania sodium	F, V, OH, H	AC or DCEN
E6013	High titania potassium	F, V, OH, H	AC or DC, either polarity
E6020	High iron oxide	H-fillets	AC or DCEN
E6022[c]	High iron oxide	F	AC or DC, either polarity
E6027	High iron oxide, iron powder	H-fillets, F	AC or DCEN
E70 series electrodes			
E7014	Iron powder, titania	F, V, OH, H	AC or DC, either polarity
E7015	Low hydrogen sodium	F, V, OH, H	DCEP
E7016	Low hydrogen potassium	F, V, OH, H	AC or DCEP
E7018	Low hydrogen potassium, iron powder	F, V, OH, H	AC or DCEP
E7024	Iron powder, titania	H-fillets, F	AC or DC, either polarity
E7027	High iron oxide, iron powder	H-fillets, F	AC or DCEN
E7028	Low hydrogen potassium, iron powder	H-fillets, F	AC or DCEP
E7048	Low hydrogen potassium, iron powder	F, OH, H, V-down	AC or DCEP

a. The abbreviations, F, V, V-down, OH, H, and H-fillets indicate the welding positions as follows:
    F = Flat
    H = Horizontal
    H-fillets = Horizontal fillets
    V-down = Vertical down
    V = Vertical } For electrodes 3/16 in. (4.8 mm) and under,
    H = Overhead } except 5/32 in. (4.0 mm) and under for classifications E7014, E7015, E7016, and E7018.

b. The term DCEP refers to direct current, electrode positive (DC reverse polarity). The term DCEN refers to direct current, electrode negative (DC straight polarity).

c. Electrodes of the E6022 classification are for single-pass welds.

Fig. 13-4. *AWS carbon steel, covered arc welding electrodes. (American Welding Society)*

Last Two Numerals	Polarity and Special Notes
EXX10	DCEP
EXX11	AC or DCEP
EXX12	AC or DCEN
EXX13	AC or DCEN
EXX14	DCEN, DCEP, AC (iron powder)
EXX15	DCEP (low hydrogen)
EXX16	AC, DCEP (low hydrogen)
EXX18	AC, DCEP (iron powder and low hydrogen)
EXX20	DCEN, or AC for horizontal fillet welds; DCEN, DCEP, or AC for flat position welding
EXX24	DCEN, DCEP, or AC (iron powder)
EXX27	DCEN, or AC for horizontal fillet welds; DCEN, DCEP, or AC for flat position (iron powder)
EXX28	AC or DCEP (iron powder and low hydrogen)

Fig. 13-6. *Recommended polarity used with SMAW electrodes. The last two numerals indicate the recommended electrical polarity.*

-A1	1/2% Mo
-B1	1/2% Cr, 1/2% Mo
-B2	1 1/4% Cr, 1/2% Mo
-B3	2 1/4% Cr, 1% Mo
-C1	2 1/2% Ni
-C2	3 1/4% Ni
-C3	1% Ni, .35% Mo, .15% Cr
-D1 and D2	.25-.45% Mo, 1.25-2.00% Mn
-G	.50 min Ni, .30 min Cr, .20 min. Mo, .10 min V, 1.00 min Mn, .80 min Si (Only one of the listed elements is required for the G classification).

Fig. 13-7. *AWS electrode suffixes. When a suffix is used, it indicates that alloys have been added to the electrode. Example: E8016-B2.*

Step 3: Convert 1/16″ to 32nds of an inch. *1/16″ = 2/32″*

Step 4: Multiply the number of 32nds by 40 A. *2 × 40 = 80*

Step 5: Add the 90 A from Step 1 to the answer in Step 4. *90 A + 80 A = 170 A*

Answer: *The welding machine should be set for 170 A.*

The amperage obtained by using the manufacturer's electrode guide, the rule-of-thumb method, or from

the tables in Fig. 13-9 may need to be modified. The position of the weld and the skill of the welder determine the final amperage setting. Inspect the weld to determine whether the amperage needs to be raised or lowered. The need for amperage changes will be covered in more detail in Chapter 14.

CHOOSING AN ELECTRODE

A *welding procedure specification* is written to specify welding requirements for commercial jobs such

RIGHT HAND DIGIT	COVERING COMPOSITIONS	APPLICATION (USE)
5 E-7015	Low hydrogen sodium type.	This is a low hydrogen electrode for welding low carbon, alloy steels. Power shovels and other earth moving machinery require this rod. The weld files or machines easily. Use DCEP (DCRP) only.
6 E-7016	Same as ''5'' but with potassium salts used for arc stabilizing.	It has the same general application as (5) above except it can be used on either DCEP (DCRP) or AC.
8 E-7028	Iron powder (low hydrogen), flat position only.	For low carbon alloy steels, use DC or AC.
E-8018	Iron powder plus low-hydrogen sodium covering.	Similar to (5) and (6) DCEP (DCRP) or AC. Heavy covering allows the use of high speed drag welding. AC or DCEP (DCRP may be used.

Fig. 13-8. Low hydrogen (Lo-Hi's) electrode covering compositions and applications. Lo-Hi's withstand higher temperatures and can be used with higher amperages.

Suggested Metal Thickness		Electrode size		E6010 and E6011	E6012	E6013	E6020	E6022	E6027
in.	mm	in.	mm						
1/16 & less	1.6 & less	1/16	1.6		20-40	20-40			
1/16-5/64	1.6-2.0	5/64	2.0		25-60	25-60			
5/64-1/8	2.0-3.2	3/32	2.4	40-80	35-85	45-90			
1/8-1/4	3.2-6.4	1/8	3.2	75-125	80-140	80-130	100-150	110-160	125-185
1/4-3/8	6.4-9.5	5/32	4.0	110-170	110-190	105-180	130-190	140-190	169-240
3/8-1/2	9.5-12.7	3/16	4.8	140-215	140-240	150-230	175-250	170-400	210-300
1/2-3/4	12.7-19.1	7/32	5.6	170-250	200-320	210-300	225-310	370-520	250-350
3/4-1	19.1-25.4	1/4	6.4	210-320	250-400	250-350	275-375		300-420
1 - up	25.4 - up	5/16	8.0	275-425	300-500	320-430	340-450		375-475

Suggested Metal Thickness		Electrode size		E7014	E7015 and E7016	E7018	E7024 and E7028	E7027	E7048
in.	mm	in.	mm						
5/64-1/8	2.0-3.2	3/32*	2.4*	80-125	65-110	70-100	100-145		
1/8-1/4	3.2-6.4	1/8	3.2	110-160	100-150	115-165	140-190	125-185	80-140
1/4-3/8	6.4-9.5	5/32	4.0	150-210	140-200	150-220	180-250	160-240	150-220
3/8-1/2	9.5-12.7	3/16	4.3	200-275	180-255	200-275	230-305	210-300	210-270
1/2-3/4	12.7-19.1	7/32	5.6	260-340	240-320	260-340	275-365	250-350	
3/4-1	19.1-25.4	1/4	6.4	330-415	300-390	315-400	335-430	300-420	
1 - up	25.4 - up	5/16*	8.0*	390-500	375-475	375-470	400-525	375-475	

Note: When welding vertically up, currents near the lower limit of the range are generally used.
   * These diameters are not manufactured in the E7028 classification.

Fig. 13-9. A—Suggested amperage ranges for use with various E60XX electrodes and diameters. B—Suggested amperage ranges for various E70XX electrodes and diameters. Electrode guides furnished by manufacturers also give amperage ranges.

as bridges, pipelines, or steel structures. A welding procedure specification outlines exactly how each weld is to be made. It also lists the diameter and type of AWS electrode to be used. Welding procedure specifications are generally written by engineers.

Most fabrication, repair, and craft welding does not require a welding procedure. The welder or shop owner selects the electrode to be used for these situations.

Several factors must be considered when selecting an electrode.
1. Type of metal.
2. Metal thickness.
3. Groove design.
4. Joint alignment.
5. Available welding current.
6. Welder skill.
7. Welding position.
8. Deposition rate (rate of weld deposit).
9. Depth of penetration.
10. Weld bead finish.
11. Additional metal properties.

After considering all the factors, an electrode selection can be made. An electrode manufacturer's guide may also be helpful when selecting electrodes.

Type of metal

The tensile strength of the deposited weld metal should be as strong as that of the base metal. The tensile strength of the steel being welded must be determined. The steel supplier can provide you with this information. If the steel has a tensile strength of 60,000 psi (414 MPa), an E60XX or E70XX electrode can be used. An E120XX electrode is required for steel with a tensile strength of 120,000 psi (827 MPa).

Metal thickness

One pass may be enough for metal up to 1/4″ (6.4 mm) thick. More than one pass is needed for metal over 1/4″ thick. A pass should not be more than 1/4″ (6.4 mm) thick. A 1/4″ − 3/8″ wide bead is required on 1/4″ (6.4 mm) thick steel. Wider beads are used on thicker metal.

A weld bead should be 2 − 3 times the diameter of the electrode. A 1/4″ (6.4 mm) wide bead requires an electrode 3/32″ − 1/8″ (2.4 mm − 3.2 mm) in diameter. A 3/8″ (9.6 mm) wide bead requires an electrode with a 1/8″ − 3/16″ (3.2 mm − 4.8 mm) diameter.

Certain electrodes work well on sheet metal because they do not penetrate deeply. E6012 and E6013 electrodes work well on thin pieces of metal.

Groove design

In a wide V-groove joint, a small diameter electrode may be used for the root pass. Larger diameter electrodes are then used for the filler and cover passes. Smaller diameter electrodes may be used for a narrow bevel-, V-, J-, or U-groove joint.

Joint alignment

Poorly designed or assembled joints may have large openings that are hard to fill. Certain electrodes are designed to span large root openings. The E6012, E6013, and E7014 electrodes are designed for use on poorly assembled joints with large root openings.

Available welding current

A wide variety of electrodes can be used if direct and alternating welding current is available. The selection is more limited if only an AC or a DC welding machine is available.

Welder skill

A large-diameter electrode and high amperage may be used successfully by a highly skilled welder. The same electrode may not provide proper penetration for a less-skilled welder. An E7024 electrode can be used by an inexperienced welder to produce good-quality welds in the flat position.

Welding position

The welding position must also be considered when selecting an electrode. All electrodes can be used for welds done in the flat welding position. Welds made in the overhead welding position are done with EXX1X or EXX4X electrodes. (The Xs can represent any number.)

Deposition rate

Deposition rate is the amount of filler metal deposited in the joint in one minute. If a high deposition rate is required, a larger electrode diameter, higher amperage, and DCEP may be required.

Depth of penetration

E6010 and E6011 electrodes penetrate more deeply than E6012 and E6013 electrodes. E6010 and E6011 electrodes provide good penetration on thick metal.

Weld bead finish

An E6013 or E7024 electrode produces a strong bead with a good appearance in butt and corner joints. An E6010 electrode produces strong welds in the same joints. The appearance of the bead will not be as good, however.

Additional metal properties

Electrodes such as the E10016-D2 contain alloying elements in their coverings. The alloying elements are combined with the weld metal during welding. Alloying elements improve the strength, corrosion resistance, or other characteristics of the finished weld.

Low hydrogen electrodes contain chemicals that eliminate hydrogen in the weld. Low hydrogen electrodes generally produce welds that contain fewer gas inclusions. This produces a completed weld that is stronger and tougher.

CARE OF ELECTRODES

The electrode covering should not be damaged. Electrodes should not be bent. Keep electrodes in their shipping containers until they are used. The shipping container provides safe storage for the electrodes after it is opened.

Electrode coatings that are damp will not work properly. Water (H_2O) contains hydrogen (H_2). Hydrogen deposited in a weld weakens it. Welders should store their electrodes in an *electrode drying oven.* An adequate supply of electrodes can be kept under ideal conditions in such an oven. See Fig. 13-10.

Low hydrogen electrodes must be kept especially dry to prevent the addition of hydrogen to the material in the covering.

All electrodes must be used within a specified time after they are removed from the drying oven. The amount of time varies with the type of electrode. The amount of time out of the oven is extremely critical for high-strength, low hydrogen electrodes such as E10018 and E12018. When the specified time has expired, all electrodes should be returned to the oven and re-baked.

An *electrode dispenser,* Fig. 13-11, may be used for temporary storage of electrodes. The dispenser protects the electrodes and keeps them relatively dry.

REVIEW QUESTIONS

Write your answers on a separate sheet of paper.
1. Which AWS electrode specification number provides information about electrodes to weld nickel and nickel alloys?
2. List four purposes of the electrode covering.

Fig. 13-10. An electrode drying oven. A—Note the temperature gauge and latch on the door. B—Different electrode types and diameters are stored in separate bins inside the oven.

Fig. 13-11. Electrode dispenser. This dispenser protects the electrodes from damage and keeps them relatively dry while they are out of the oven. The electrodes are loaded into the dispenser and the sealing top installed. Moving the handle up pushes one electrode through the top seal for easy removal. (Red-D-Arc, Ltd.)

3. The slag on a SMAW bead contains _____ and _____.

4. Metallic salts in the electrode covering may cause the weld metal to become stronger, more corrosion-resistant, or more ductile. True or False?

5. Describe what is indicated by each of the characters in the E6013 electrode designation.

6. What does B2 represent in an AWS E8016-B2 electrode designation?

7. Use the rule-of-thumb method to determine the trial amperage for a 7/32″ (5.6 mm) electrode. Show your work.

8. The electrode type and diameter is given in a welding procedure. True or False?

9. An E7013 electrode would be a good electrode to use for steel with a tensile strength of 88,000 psi. True or False?

10. A 1/2″ (12.7 mm) wide stringer bead requires an electrode between _____″ and _____″ diameter.

Shielded metal arc welding being done in the flat welding position on the thick steel plates of a ship during construction.
(Miller Electric Mfg. Co.)

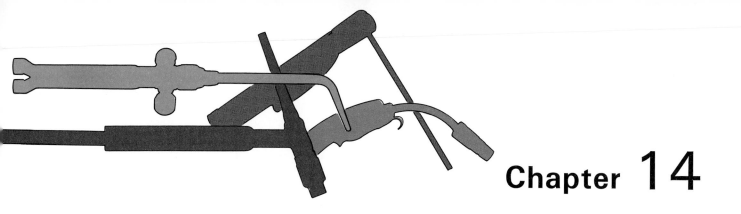

Chapter 14

SMAW: Flat Welding Position

After studying this chapter, you will be able to:
- ☐ Identify the safety rules required for arc welding.
- ☐ Demonstrate how an arc is started.
- ☐ Read a weld bead.
- ☐ Make a fillet weld on a lap joint in the flat welding position using shielded metal arc welding.
- ☐ Make a fillet weld on an inside corner joint in the flat welding position using shielded metal arc welding.
- ☐ Make a fillet weld on a T-joint in the flat welding position using shielded metal arc welding.
- ☐ Make a square-, J-, and V-grooved weld on a butt joint in the flat welding position using shielded metal arc welding.

TECHNICAL TERMS

Arc blow, arc length, backward arc blow, bullet-shaped ripples, chipping, chipping goggles, closed arc, drag welding, fire watch, forward arc blow, keyhole, layer, magnetic field, open arc, pass, phosgene gas, reading the bead, restarting the arc, root pass, running a bead, stringer bead, toxic, under load, weaving bead.

PREPARING TO WELD

Comfortable, yet fire-resistant, clothing should be worn for SMAW. Protect all areas of your skin from burns caused by arc rays or molten metal. Shirts and jackets should be buttoned at the collar. Wear a cap to protect your head and hair. Gloves should be worn to protect your hands and forearms. Hard-toed, high-top shoes or boots should also be worn. Pants without cuffs should be worn to prevent hot metal or sparks from being caught in them. Do not wear clothing with ragged edges or loose threads. The edges and threads may catch fire easily.

Make a visual safety inspection of your welding outfit or station. Refer to Chapter 12 for information detailing the inspection of a welding outfit.

Select the appropriate type and size of electrode for your job. Chapter 13 covers information regarding electrodes, such as suggested amperage and polarity. Guides published by electrode manufacturers also provide polarity and amperage information.

SMAW SAFETY PRECAUTIONS

Arc welding presents dangers of electrical shock, fumes and gases, hot metal, arc rays, and fire. The importance of safety precautions cannot be stressed enough. The application of proper safety precautions prevents injury to personnel and damage to equipment.

Safety precautions that may be applied to all forms of arc welding include:

1. Always have installation, maintenance, and repair work on equipment and electrical circuits done by qualified people.
2. The electrode and work circuits are HOT (current is flowing) when the welding machine is on. Do not touch the electrode, workpiece, or work table with bare skin, wet gloves, or wet clothing.
3. Do not weld on damp or metal floors. If welding must be done in these locations, be certain you are insulated from them.
4. Keep the welding machine, welding cables, electrode holder, lugs, and clamps in safe working order.
5. Connect the work table or workpiece to a good electrical ground.
6. Never make polarity or current changes on the welding machine while the machine is *under load* (while welding or with the electrode holder touching the table or workpiece).

7. Avoid breathing hazardous fumes or gases while welding. Fumes should be removed from the weld area before they pass the welder's face.

8. Use extra ventilation, or wear supplied air breathing apparatus, when welding on lead, cadmium, or galvanized (zinc-coated) metals. These metals produce *toxic* or poisonous fumes.

9. Do not weld near degreasing or cleaning chemicals that contain chlorinated hydrocarbons. Arc welding rays and the heat of the arc can react with these solvents to produce phosgene gas. *Phosgene gas* is a highly toxic and poisonous gas. Parts cleaned with chlorinated hydrocarbons must be thoroughly rinsed to remove these cleaning agents.

10. Always wear clothing that is flame resistant for protection from fire, molten metal, and harmful arc rays. Wear clean, oil-free clothing such as leathers. Wear pants without cuffs, a cap, gloves, and high-top shoes.

11. Never look at an arc. Arc rays can cause skin burns and damage to the eyes.

12. Always use an arc welding helmet with the proper filter lens in place.

13. Nonflammable screens should be set up around weld areas. The screens protect others in the area from arc rays and metal spatter.

14. Remove all flammable or explosive materials from welding areas. Fire extinguishers should be available.

15. Place a person on *fire watch.* Sparks and hot metal can go through cracks into other areas or fall onto other floors of a building.

16. Do not heat, cut, or weld on tanks or other containers until appropriate safety steps have been taken. Containers may have been used to store explosive or chemically toxic materials. Specific steps for cleaning must be followed to ensure that explosions do not occur. The local fire or industrial safety department can advise you on how to properly clean a container prior to welding.

17. Wear ear plugs to keep sparks out of your ears when welding out of position.

18. Wear safety glasses when chipping. Flash goggles should be worn under the welding helmet for protection from reflected rays.

19. Engine-driven welders must be used in an open, well-ventilated area. Vent exhaust fumes outside.

STRIKING AN ARC

An electrode holder is generally held with one hand. It can be gripped like a hammer or screwdriver. The electrode lead may be draped over the lower arm to make the electrode holder feel lighter. See Fig. 14-1.

In order to strike an arc, the electrode must first touch the base metal. The electrode should only remain in contact with the base metal momentarily. This causes electricity to start flowing. The electrode is then pulled

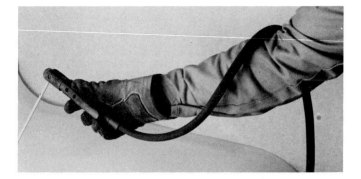

Fig. 14-1. The electrode holder can be held like a hammer. Draping the electrode lead over the forearm makes the holder feel lighter.

away from the base metal a short distance. Current continues to flow across this gap, resulting in an arc.

Most electrodes have a relatively thin covering of flux. When an arc is struck using a thinly covered electrode, the arc can be seen. An *open arc* is an arc that can be seen.

Two methods are used to strike an arc. A welder may scratch the electrode on the metal and then withdraw it. A straight up-and-down or pecking motion may also be used. See Fig. 14-2. After the arc is struck and stabilized, it is brought down to the correct arc length.

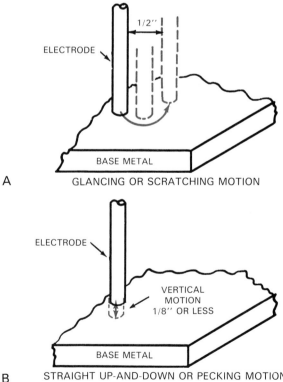

Fig. 14-2. A—The scratch method of striking an arc. Notice that this is a very short movement. To maintain the arc, the end of the electrode must remain within 1/8" (3.2 mm) of the base metal. B—The straight up-and-down or pecking method of striking an arc. The electrode must remain within 1/8" (3.2 mm) of the base metal to maintain the arc.

The *arc length* is the distance between the electrode and the base metal. This distance should be approximately equal to the diameter of the electrode.

The electrode may become welded to the base metal if the arc is not struck properly. To release the electrode, proceed as follows:

1. Keep your welding helmet down and release the electrode from the electrode holder.
2. Lift your helmet.
3. Place the electrode holder on an insulated hook.
4. Grasp the electrode near the base metal. Bend the electrode back and forth until it breaks free.

ARC BLOW

A *magnetic field* is created whenever electricity travels in a wire or electrode. When current travels in an electrode, a magnetic field is created around the electrode. The magnetic field changes direction as the current changes direction. The arc is also deflected from its normal path by magnetic forces. This deflection is *arc blow*. See Fig. 14-3.

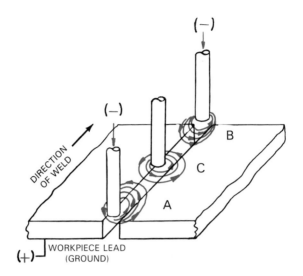

Fig. 14-4. The magnetic field around the electrode is deflected at the ends of a weld joint (A and B). The field attempts to flow in the metal and not through the air. The concentration of magnetic flux at the ends of the metal forces the arc toward the center of the base metal. The arc "blows away" from the area directly under the electrode at the ends of the weld. Notice that the magnetic field is not distorted in the center areas of the weld joint at C.

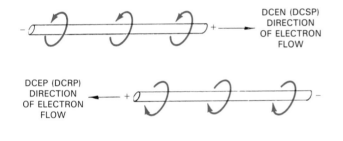

Fig. 14-3. The magnetic field around a wire. The magnetic field around a wire with DCEN (DCSP) rotates in the opposite direction to the field around a wire carrying DCEP (DCRP).

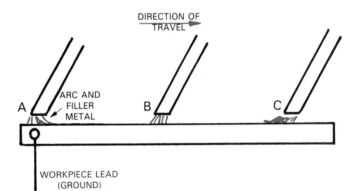

Fig. 14-5. The effects of direct current arc blow on the arc and electrode wire. As the arc is started (A), the arc is blown toward the center of the weld joint. In the center (B), the arc travels straight down. As the arc approaches the end of the joint (C), the arc and filler metal are again blown toward the center of the weld joint.

When alternating current is used, the magnetic field is continually cancelled as the current changes direction. However, when direct current is used, the current flows in one direction and a strong magnetic field may develop. A magnetic field prefers to travel in a conductive material such as metal. Air causes the magnetic field to weaken. See Fig. 14-4.

Arc blow is greatest at the ends of the weld joint. The magnetic field tries to stay in the metal. This causes the filler metal to blow toward the center of the joint. See Fig. 14-5. Arc blow that occurs at the beginning of a joint is *forward arc blow*. *Backward arc blow* occurs at the end of the joint. The least amount of arc blow occurs in the center area of the weld or joint.

Arc blow may affect the quality of a weld. If arc blow does occur, one or more of the following steps may be taken:

1. Place the ground connection as far from the weld joint as possible.
2. If forward arc blow is a problem, connect the workpiece lead (ground) near the end of the weld joint.
3. If backward ard blow is a problem, place the workpiece lead near the beginning of the weld.
4. Reduce the welding current to reduce the strength of the magnetic field.
5. Position the electrode so that the arc force counteracts the arc blow force.
6. Use the shortest arc possible to produce a good bead. A short arc permits the filler metal to enter the arc pool before it is blown away. A short arc

also permits the arc force to overcome the arc blow force.

7. Use run-on or run-off tabs on the joint. See Fig. 14-6.

8. Wrap the electrode lead around the base metal in the direction that will counteract the arc blow force.

9. Change to an AC machine and electrodes.

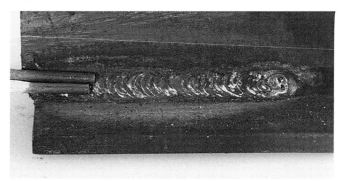

Fig. 14-7. The width of a SMAW stringer bead should be two to three times the electrode diameter. After welding for a short distance, check the bead width with two of the electrodes being used.

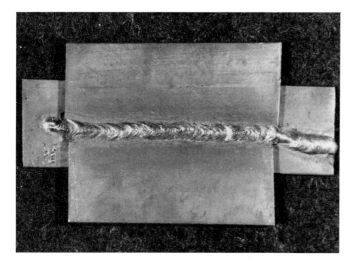

Fig. 14-6. Run-on and run-off tabs used on a butt joint. The tabs allow a welder to strike an arc on the tab. The arc is completely stabilized by the time it reaches the actual base metal. At the end of the joint, the bead is continued past the base metal and onto the tab. A strong weld from beginning to end is possible by using these tabs.

RUNNING A BEAD

Making a weld bead is known as *running a bead.* A weld pool forms after the arc is started. Filler metal melts from the electrode and is deposited in the weld pool. The bead now begins to form.

The arc length determines the voltage and amperage across the arc. Therefore, the arc length must remain constant throughout the entire weld. The normal arc length is about equal to the electrode diameter. A beginning welder should practice holding a constant arc length with one hand.

Two types of beads are used when welding: the stringer bead and the weaving bead. A *stringer bead* is one made along a line without any side-to-side motion. The width of a stringer bead should be two to three times the electrode diameter. See Fig. 14-7. Its height should be about one-eighth of the bead's width. For example, a stringer bead made with a 1/8″ (3.2 mm) diameter electrode should be about 1/4″−3/8″ (6.4 mm−9.5 mm) wide. The height of the bead should be approximately 1/32″−3/64″ (0.8 mm−1.2 mm). As the bead reaches the correct width, the electrode is moved forward at a constant speed. A *weaving bead* is one made along a line while the electrode is moved

from side to side. See Fig. 14-8. Weaving beads should not be wider than six times the electrode diameter. Weaving beads wider than this may cool before trapped impurities and air can reach the surface of the weld. Weaving beads are often used as the final *pass* of a multiple-pass weld.

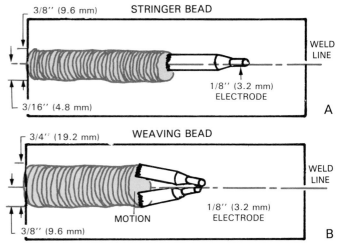

Fig. 14-8. Suggested dimensions for stringer and weaving beads. The stringer bead is two to three times as wide as the electrode diameter. A weaving bead should not be wider than six times the electrode diameter.

A weld is generally made from left to right by a right-handed welder. A left-handed welder commonly progresses from right to left. The electrode may be held vertically or tipped about 20° in the direction of travel. See Figs. 14-9 and 14-10.

Watch the shape of the ripples that form at the rear of the pool. The ripples should be bullet-shaped, and be closely, yet evenly, spaced. Pointed ripples indicate that the forward speed is too fast. Flattened ripples indicate that the travel speed is too slow.

The following must be kept constant while welding:
1. The arc length.

142

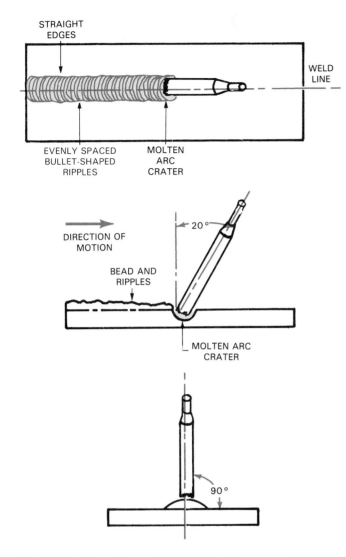

Fig. 14-9. Three views of a SMAW bead in progress. Note that the electrode is tipped 20° in the direction of travel. The completed bead should have straight edges, evenly spaced ripples, and a uniform height.

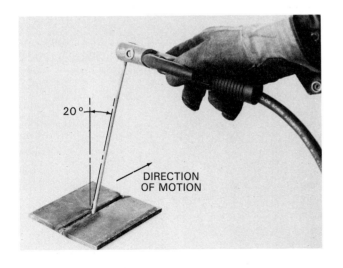

Fig. 14-10. A butt joint being welded with the shielded metal arc. Note that the electrode is tipped 20° in the direction of travel. This helps to pile up the filler metal behind the weld pool.

2. The electrode angle.
3. The bead width.
4. The forward welding speed.

The arc will stop if the arc length is allowed to become too large or if the electrode touches the base metal.

RESTARTING THE ARC

An arc may stop due to faulty welding techniques. It may also be stopped intentionally to change electrodes or make other adjustments. A weld bead that is stopped for any reason will end with a large arc crater. The arc must be restarted so that the crater is completely filled. The ripples in the weld bead must also appear evenly spaced. When *restarting the arc* proceed as follows:

1. Restrike the arc 1/2″ − 1″ (12.7 mm − 25.4 mm) ahead of the old crater.
2. Move backwards rapidly until the rear of the new weld pool touches the rear of the old crater.
3. Move forward as soon as they touch to complete the weld. See Fig. 14-11.

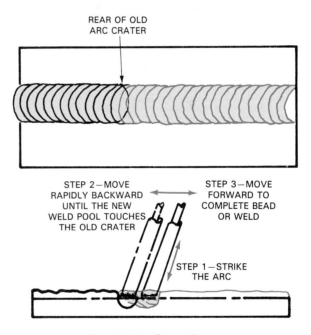

Fig. 14-11. Restarting an arc.

DRAG WELDING

Iron powder electrodes, such as the E7024, have extremely thick coverings. Electrodes with thick coverings can be used for drag welding. *Drag welding* is a welding process in which the electrode covering is in contact with the base metal. Drag welding ensures a constant arc length. The electrode covering does not conduct electricity. As a result, the arc will not stop when the electrode covering touches the base metal.

When striking an arc with a thickly covered electrode, you must hit the base metal hard enough to chip away some of the covering. This exposes the electrode wire and allows the arc to begin. The covering is kept in contact with the base metal after the arc is struck. The arc is enclosed by the thick electrode covering. This type of arc is known as a *closed arc*. Running a bead with the drag method is the same as running a bead with an open arc. It is easier, however, because the arc length is kept constant by dragging the electrode on the flux covering. See Fig. 14-12.

Fig. 14-13. Cleaning the slag from a fillet weld using the pick end of the chipping hammer. The wide end of the hammer can be used on flat beads and *in* larger areas.

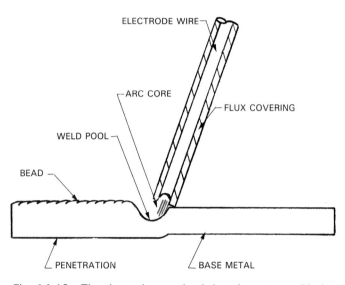

Fig. 14-12. The electrode covering is kept in contact with the base metal as the electrode is pulled along the metal surface.

CLEANING THE WELD

Each weld bead or pass must be cleaned by *chipping* and brushing. This cleaning prevents slag and dirt from mixing with the following passes. Cleaning is also necessary in order to examine a completed bead.

Chipping should be done after the weld has cooled for a few minutes. *Chipping goggles* must be worn while chipping and wire brushing. Chipping should be done where other workers are not endangered by flying particles. The cooled slag coating is usually hard and brittle. It is removed with a chipping hammer. Small areas of the weld are cleaned with the pick end of the chipping hammer. See Fig. 14-13. A wire brush is used to clean the bead thoroughly after chipping.

READING THE BEAD

A welder must be able to visually determine whether a weld is made properly. This can be done by looking at the weld while it is being made and after it is completed. Visual inspection of a weld bead is referred to as *reading the bead*. The process of examining or reading the weld bead can provide information on the following welding variables:
1. Amperage setting.
2. Arc length.
3. Travel speed.

Fig. 14-14 shows weld beads that have been made under different welding conditions. The sketches and photographs show the welds as viewed from above and from the end. If the correct amperage is used, a completed stringer bead will have an even width. It will also have evenly spaced, *bullet-shaped ripples*. The bead width and height will also be correct, Fig. 14-14A. If the welding amperage is too low, the bead will be narrow, built up, and have poor penetration. See Fig. 14-14B. Excessive current produces a weld that is wide, low, and has a great deal of spatter on the metal's surface. See Fig. 14-14C.

A built-up bead with poor penetration is generally caused by a short arc. Overlap is usually present. See Fig. 14-14D. When the arc length is too long, the bead will be too low. It will also have poor penetration and have undercutting at the weld toes, Fig. 14-14E.

Fig. 14-14F shows a weld that was made too slowly. The bead is too wide and built up too much. Also notice the flatter shape of the ripples. A weld that is made too fast will be low and narrow as shown in Fig. 14-14G. The ripples are usually pointed.

These weld characteristics can be seen while the weld is in progress. Change the arc length and welding speed immediately. The effects of these changes should be seen as a more-perfect bead is formed. If the current needs to be changed, the weld must be stopped. Hang the electrode holder on an insulated hanger before changing the amperage. The weld is then restarted after the amperage is changed.

EXERCISE 14-1 — RUNNING A BEAD USING AN OPEN ARC

1. Obtain one low carbon steel plate measuring 1/4'' x 3'' x 6'' (6.4 mm x 76 mm x 152 mm).
2. Also obtain five—1/8'' (3.2 mm) diameter E6013 electrodes.
3. Mark three 6'' (152 mm) long lines on the plate,

SMAW: Flat Welding Position

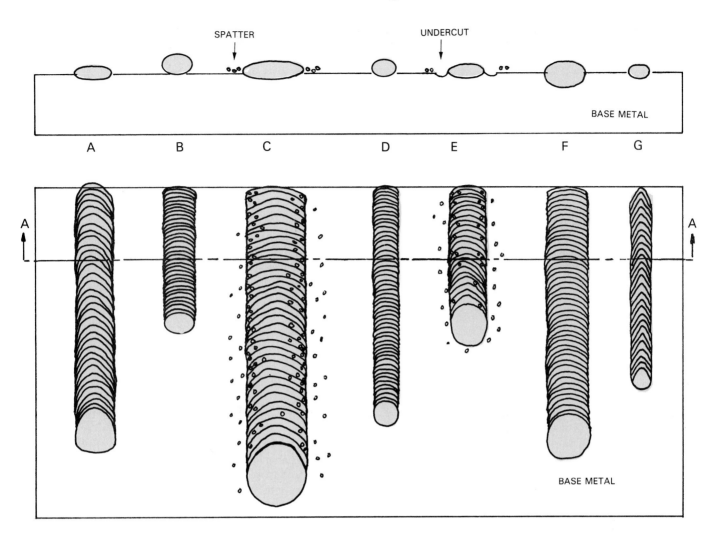

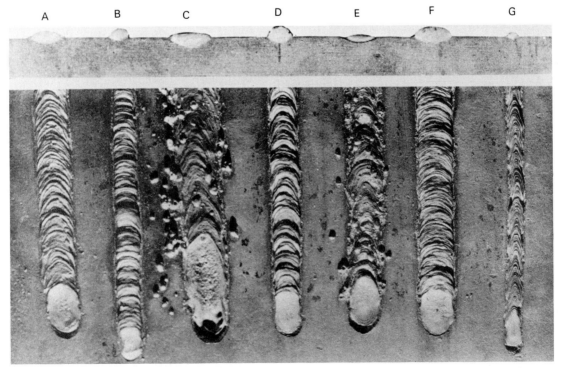

Fig. 14-14. The effects of current, arc length, and travel speed on covered SMAW electrode beads. A—Correct arc length, travel speed, and current. B—Amperage too low. C—Amperage too high. D—Arc length too short. E— Arc length too long. F—Travel speed too slow. G—Travel speed too fast. (American Welding Society)

each 3/4'' (19.1 mm) apart.

4. Determine the amperage range and polarity for this electrode. Refer to Chapter 13 or an electrode manufacturer's guide.
5. Make a safety inspection of the arc welding outfit or station.
6. Set the amperage near the low end of the suggested range.
7. Make certain that the work table or practice plate is attached to the workpiece lead.
8. Strike the arc using the scratching or pecking method.
9. Run a bead of correct width and height along the marked line.
10. Chip and wire-brush the bead. Chipping goggles must be worn.
11. Read the finished bead and compare it to Fig. 14-14.
12. Change your amperage, arc length, or travel speed as required. Run two additional beads. Read each bead while it is being made. Clean the completed welds and make changes in your welding method as required.
13. Make three additional beads on the opposite side of the metal about 3/4'' (19.1 mm) apart.

Inspection:

Each bead should be the proper width and height. The ripples should be bullet-shaped and evenly spaced. Each bead should improve as errors are read and corrected.

FILLET WELDING

A fillet weld may be made on a lap, inside corner, or T-joint. The legs of the fillet are usually equal. Normally, the size of each leg is at least equal to the thickness of the metal.

LAP JOINT

In the lap joint, the weld is made along an edge of one piece and the surface of another piece. The electrode is tilted about 20° from vertical in the direction of travel. The electrode should point more toward the surface when welding thinner metal. This is not as necessary when welding thick metal. The electrode should be held at a 45° angle to the base metal surface when welding metal over 1/4'' (6.4 mm) thick. See Fig. 14-15.

When the arc is struck, lay a stringer bead deep into the root of the joint. A C-shaped weld pool should be created. A "C" shape at the leading edge of the weld pool indicates that both the edge and surface are melting. Reduce the travel speed if the "C" shape does not form. A welder can determine the correct forward speed by watching the rear of the weld pool. It should be bullet-shaped.

Thick metal may require more than one weld pass. Each pass must be cleaned before the next pass is made.

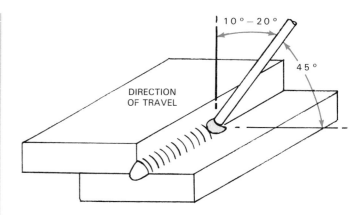

Fig. 14-15. The first bead on a multiple-pass fillet weld on thick metal. The electrode is held at 45° to the base metal surface. It is tipped in the direction of travel at 20°.

The correct sequence of the passes in a multiple-pass weld is suggested in Fig. 14-16. A multiple-pass weld generally contains more than one *layer.* Each layer may contain more than one bead or pass. The final pass may be a weaving bead if it is not too wide.

INSIDE CORNER OR T-JOINT

A fillet weld for an inside corner or T-joint is made along two adjacent surfaces. The electrode is tilted 20° in the direction of travel. It is held at a 45° angle to the base metal surfaces. The first pass must be made deeply into the root of the weld.

The leading edge of the weld pool must be "C" shaped. This ensures that fusion is occurring on both surfaces. Each pass or bead must penetrate the base metal and/or the previous beads that it touches. See Fig. 14-17. A weaving bead may be used for the final pass.

EXERCISE 14-2 — FILLET WELD ON A T-JOINT IN THE FLAT WELDING POSITION

1. Obtain three—1/8'' (3.2 mm) diameter E6010 electrodes, six—1/8'' (3.2 mm) E6013 electrodes, and three—5/32'' (2.4 mm) iron powder electrodes, such as E6027.
2. Also obtain four pieces of mild steel measuring 1/4'' x 2'' x 6'' (6.4 mm x 51 mm x 152 mm).
3. Determine and set the correct amperage range and polarity for each electrode.
4. Form two T-joints. Tack weld each T-joint three times on each side using E6010 electrodes.
5. The leg size of each fillet is to be about 1/4''—5/16'' (6.4 mm —7.9 mm). Make a fillet weld on each side of both T-joints as follows:
   a. Use E6010 electrodes on one weld.
   b. Use E6013 electordes on two welds.
   c. Use iron powder electrodes (E6027) on one weld.

Note: Clean each pass before making the next one.

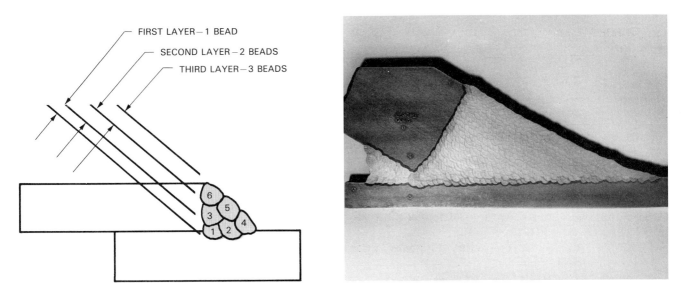

FIRST LAYER—1 BEAD
SECOND LAYER—2 BEADS
THIRD LAYER—3 BEADS

Fig. 14-16. A—Multiple-pass fillet weld on thick metal. Six passes are used in this example. Notice that there are three layers of welds. Each layer contains one or more weld beads. Note the sequence of the beads in each layer. B—Macrophotograph of a multiple-pass weld made on very thick metal using flux cored arc welding (FCAW). Note the number of passes and layers required to complete this weld. (Highland Fabricators, Scotland)

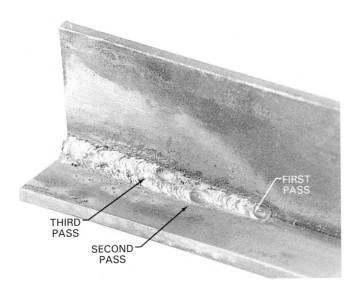

THIRD PASS
SECOND PASS
FIRST PASS

Fig. 14-17. Three passes were used on this fillet weld. Each bead was stopped at a different point to show their placement. The first pass was laid in the corner. The second pass penetrates one-half on the first bead and one-half on the horizontal part. The third pass was laid from the second pass up to the vertical part. The second pass acts as a ledge to keep the third pass from sagging.

Inspection:

The fillet welds must be the correct size. A convex bead must be formed with smooth and evenly spaced ripples. No undercut is permitted.

BUTT JOINT

Metal less than 1/4″ (6.4 mm) thick is welded using a square-groove joint. A bevel-, V-, J-, or U-groove joint is used on thicker metal. The *root pass* is the most important pass. Complete penetration to the opposite side of the metal can only occur on this pass. A *keyhole* must be seen throughout the root pass to ensure penetration. Refer to Fig. 14-18. Additional passes must fuse with the base metal and the previous beads.

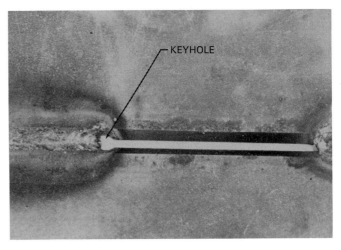

KEYHOLE

Fig. 14-18. A keyhole as seen looking down into a bevel-groove butt joint.

The weld toes must fuse with the surface of the base metal. No undercut should be seen. A butt weld may be welded from one or both sides. See Fig. 14-19. Double-bevel-groove welds are used to ensure 100% fusion and penetration when welding thick sections. Less electrode material and welder time is required to weld a double-bevel-groove than to make a single-bevel-groove weld.

147

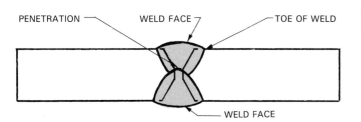

Fig. 14-19. A double-bevel-groove butt joint. Notice the groove preparation. The weld on top was made first and then the weld on the bottom was made. Notice how the second bead penetrates into the root of the first bead.

EXERCISE 14-3 — V-GROOVE BUTT JOINT IN THE FLAT WELDING POSITION

1. Obtain two—1/8" (3.2 mm) diameter E6010 electrodes and two—1/8" (3.2 mm) diameter E6011 electrodes.
2. Also obtain two pieces of mild steel measuring 3/8" x 3" x 6" (9.5 mm x 76 mm x 152 mm).
3. Flame cut or grind a 45° bevel on the 6" (152 mm) edge of both pieces.
4. Form the V—groove butt joint with a 1/16"—3/32" (1.6 mm—2.4 mm) root opening. Tack weld the pieces in three places.
5. Run the root pass using the keyhole method. This ensures 100% penetration.
6. Clean the root pass and run additional passes to complete the weld.

Inspection:
The completed weld must have 100% penetration visible on the reverse side. The face of the bead must be properly shaped. Bead ripples must be evenly spaced. No undercut should be visible.

REVIEW QUESTIONS

Write your answers on a separate sheet of paper.
1. Welders must not weld in damp places or with damp gloves. They must not touch the work and electrode at the same time. Why?
2. The heat of the arc will cause chlorinated hydrocarbons to form _____, a deadly gas.
3. Why is it advisable to wear ear plugs for out-of-position welding?
4. Where is DC arc blow the greatest while welding?
5. Name the two methods used to strike a SMAW arc.
6. What size arc gap is suggested with a 5/32" (4.0 mm) diameter electrode?
7. How wide should a stringer bead be for a 1/8" (3.2 mm) diameter electrode? What is the maximum width for a weaving bead when using a 1/8" (3.2 mm) electrode?
8. The arc maintained within the covering during drag welding is called a _____ arc.
9. How far ahead of the previous crater do you strike an arc when restarting it?
10. What should be changed if the bead is narrow, undercut, and spattered?

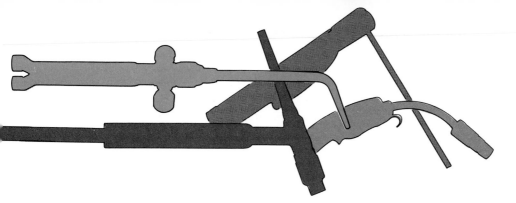

Chapter 15

SMAW: Horizontal, Vertical, and Overhead Welding Positions

After studying this chapter, you will be able to:
- ☐ Identify the proper protective clothing to be worn when welding out of position.
- ☐ Weld in the horizontal welding position.
- ☐ Identify the characteristics of a good weld.
- ☐ Weld in the vertical welding position.
- ☐ Describe the procedure for welding uphill and downhill.
- ☐ Identity special protective equipment that should be worn for welding in the overhead welding position.
- ☐ Weld in the overhead welding position.
- ☐ Identify weld defects.

TECHNICAL TERMS

Downhill, inclusions, oscillating, overlap, undercut, uphill, weld toe, whip motion.

PREPARING TO WELD

A welder should dress properly for out-of-position welding. Flame-resistant clothing should be worn. All pockets should be covered and/or buttoned. A cap should be worn to protect hair from hot metal and sparks. Ear plugs or ear muffs should be worn to prevent hot metal from spattering into your ears. Gauntlet gloves and leathers are recommended, since you will be handling hot metal. High-top, hard-toe shoes should be worn. Pull your pant legs over the top of your shoes. See Fig. 15-1. Observe all the arc welding safety precautions discussed in Chapter 14.

An electrode used for out-of-position welding should have a smaller diameter than an electrode used for flat

Fig. 15-1. An arc welder wearing a leather jacket, leather trousers, hard-toed shoes, gauntlet gloves, a cap, and an arc helmet. Notice that the jacket is buttoned at the collar.

position welding. A smaller-diameter electrode provides the following advantages:

1. Lower amperage can be used.
2. Smaller weld pool is created.
3. The molten metal in the pool is easier to control.

The slag deposit from the electrode should be fast setting (quick-freezing). A fast-setting slag deposit prevents the weld metal from sagging. Some AWS electrodes that have fast-setting slag deposits are E6010, E6011, E6012, and E6013.

The electrode should be placed in the electrode holder at an angle convenient to the welder. Electrodes should not be bent—their flux covering may chip, resulting in a loss of shielding gas. The electrode may be placed perpendicular to the electrode holder, Fig. 15-2. It may also be placed at an angle or parallel to the length of the electrode holder, as shown in Fig. 15-3.

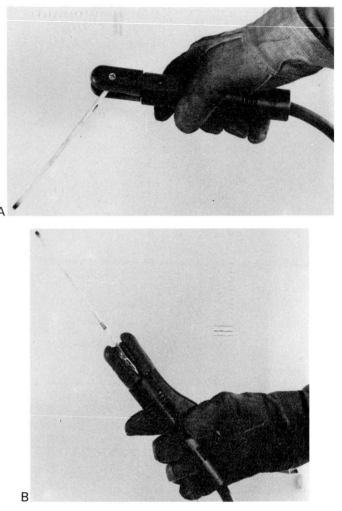

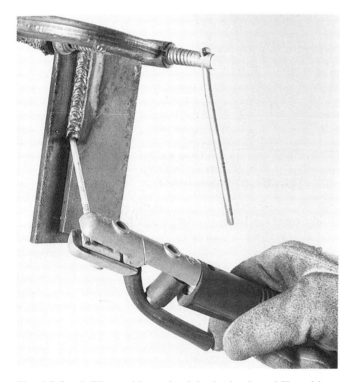

Fig. 15-2. A fillet weld on a lap joint in the downhill position. Note that the electrode holder is held like a hammer or screwdriver with the electrode pointing up.

The electrode holder should be held like a hammer or screwdriver. A firm, but not tight, grip should be used. Refer to Fig. 14-1. The electrode lead may be draped over your arm to prevent fatigue.

HORIZONTAL POSITION

When making practice fillet welds in the horizontal position, the weld axis is generally horizontal and the weld face is 45° from horizontal. The molten metal tends to sag downward out of the weld when welding in this position.

Fig. 15-3. Two ways of clamping the electrode into the electrode holder. A—Recommended for flat, horizontal, and vertical welding positions. B—Recommended for the overhead welding position.

FILLET WELDS

Fillet welds are made in the horizontal welding position on the lap, inside corner, and T-joints. A stringer or weaving bead can be used to make a fillet weld.

Fillet welds must penetrate deeply into the root of the joint, as shown in Fig. 15-4. The *weld toe* should be fused properly into both pieces of the weldment.

Multiple-pass welds may be required on thick metal sections. The lowest pass in a layer should be made first to provide a base for the other beads. See Fig. 15-5.

Fillet welds may be made by drag welding. Certain electrodes, such as the E6027, are designed for making fillet welds in the horizontal welding position.

Lap joints

A fillet weld on a lap joint must fuse the edge of one piece and the surface of another piece. Watch the weld toes as they blend into the base metal to ensure proper fusion. Increase the welding amperage or slow the weld speed if the weld is not fused properly. Bullet-

150

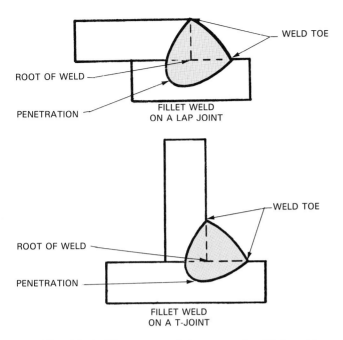

WELD TOE

ROOT OF WELD

PENETRATION

FILLET WELD
ON A LAP JOINT

WELD TOE

ROOT OF WELD

PENETRATION

FILLET WELD
ON A T-JOINT

Fig. 15-4. The names of the parts of a fillet weld.

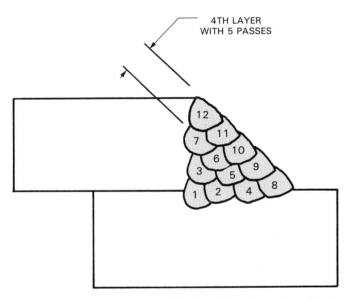

4TH LAYER
WITH 5 PASSES

Fig. 15-5. The suggested order of welding the various beads in a multiple-pass fillet weld. The order of the passes is very important. Notice that the second pass acts as a step for the third pass to rest on. The fourth pass likewise is a step for the fifth pass and so on.

shaped ripples at the rear of the weld pool indicate the correct welding speed. There should not be an overlap or undercut on a fillet weld on a lap joint.

EXERCISE 15-1 — FILLET WELD ON A LAP JOINT IN THE HORIZONTAL WELDING POSITION

1. Obtain two pieces of mild steel measuring 1/4'' x 1 1/2'' x 6'' (6.4 mm x 38.1 mm x 152.4 mm).

2. Obtain five-5/32'' (4.0 mm) diameter E6012 electrodes.
3. Determine the amperage and polarity. Set the welding machine to the correct amperage and polarity.
4. Overlap the two pieces about 3/4'' (19.1 mm). Tack weld the pieces in three places on each side.
5. Lay a fillet weld on both sides. The legs of the fillets should be 1/4'' x 1/4'' (6.4 mm x 6.4 mm). Use a crescent-shaped motion. Bullet-shaped ripples indicate the correct welding speed.

Inspection:
A convex bead is required. The beads should be bullet-shaped and have a consistent width. No undercutting should be visible.

Inside corners and T-joints

A fillet weld in the horizontal welding position is usually made with the electrode positioned 45° from horizontal. The end of the electrode in the electrode holder should be tilted about 20° in the direction of travel. See

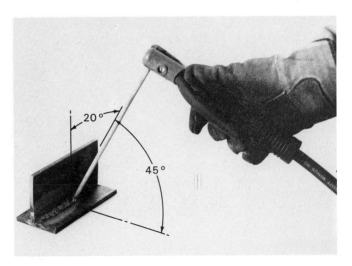

20°

45°

Fig. 15-6. The correct electrode angles for welding a fillet weld in the horizontal welding position. The electrode is tilted 20° in the direction of travel and 45° from the metal surface. On a lap joint, the electrode may point more toward the surface and away from the edge.

Fig. 15-6. A 20° angle ensures that the force of the arc deposits filler metal behind the weld pool. This, in turn, helps to form a uniformly shaped bead. The bead will build up excessively if more than a 20° angle is used.

A short *oscillating* (forward and back) motion also helps in forming a good bead. The oscillating motion, combined with a slight crescent-shaped motion as shown in Fig. 15-7, forces the molten metal upward and backward. This motion prevents the filler metal from sagging. The electrode motion should be stopped briefly each time the electrode reaches the toe of the weld. This helps to prevent undercutting at the weld toes.

151

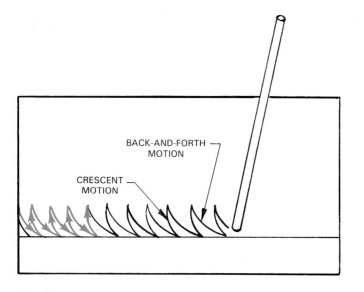

Fig. 15-7. A short back-and-forth arc motion combined with a crescent shape movement of the electrode may be used. This motion will help to push the filler metal up and back in the bead.

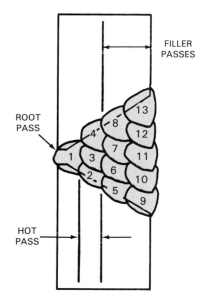

Fig. 15-8. The suggested order of welding the passes (beads) in a multiple-pass V-groove butt weld in the horizontal welding position. The lowest bead in each layer should be welded first to provide a base for the next bead to build on.

EXERCISE 15-2 — FILLET WELD ON A T-JOINT IN THE HORIZONTAL WELDING POSITION

1. Obtain two pieces of mild steel measuring 1/4″ x 2″ x 6″ (6.4 mm x 50.8 mm x 152.4 mm).
2. Obtain three 5/32″ (4.0 mm) diameter E6013 electrodes and three 5/32″ (4.0 mm) diameter E6027 electrodes.
3. Set the amperage and polarity on the welding machine.
4. Form a T-joint with the two pieces and tack weld in three places on each side.
5. Make a normal open arc fillet weld on one side of the T-joint using the E6013 electrodes.
6. Make a fillet weld on the other side using E6027 electrodes and the drag weld technique.

Inspection:

Both beads should be even in width, smooth, and have evenly spaced ripples. The beads should be convex with bullet-shaped ripples. Undercutting is not permitted.

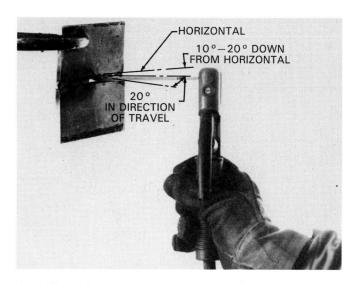

Fig. 15-9. The suggested electrode angles for welding a butt joint in the horizontal welding position.

BUTT JOINTS

Multiple-pass welds are required for butt welds on metal over 1/4″ (6.4 mm) thick. Small weld pools are used to prevent sagging. The first bead laid down, which fills the root opening, is called the root pass. See Fig. 15-8. A stringer bead is used for the root pass. The keyhole welding method should be used to ensure 100% penetration.

A butt weld in the horizontal welding position is usually made with the electrode at an upward angle of 10°−20° from horizontal. The upward angle prevents the molten metal from sagging. The end of the electrode in the electrode holder should be tilted about 20° in the direction of travel. See Fig. 15-9.

EXERCISE 15-3 — V-GROOVE WELD ON A BUTT JOINT IN THE HORIZONTAL WELDING POSITION

1. Obtain two pieces of mild steel measuring 1/4″ x 2″ x 6″ (6.4 mm x 50.8 mm x 152.4 mm).
2. Obtain five 3/32″ or 1/8″ (2.4 mm or 3.2mm) diameter EXX1X electrodes.
3. Prepare the edges for a 90° V-groove butt joint.
4. Tack weld the pieces in three places in the flat position.
5. Position the joint for horizontal welding.
6. Weld the joint with one or two passes using the keyhole welding method. The root pass must penetrate 100%.

Inspection:
The weld must penetrate 100%. A convex bead is required. The bead should be consistent in width and have evenly spaced, bullet-shaped ripples.

VERTICAL WELDING POSITION

In the vertical position, the weld axis is generally within 10° of vertical. The face of the weld is also close to being vertical.

Welding in the vertical position is done either downhill or uphill. A *downhill* weld begins at the top of the joint and ends at the bottom of the joint. See Fig. 15-10. The electrode is tipped in the direction of travel about 10°−20°. Slag should be prevented from entering the weld pool. The correct welding speed keeps the weld pool ahead of the slag. Slag in the weld causes *inclusions* (foreign particles) that will weaken the completed weld. The first bead laid downhill should be a stringer bead. Weaving beads may be used for the following passes.

Fig. 15-11. A butt joint being welded with the uphill method. The electrode is tilted up in the direction of travel about 10°−20°.

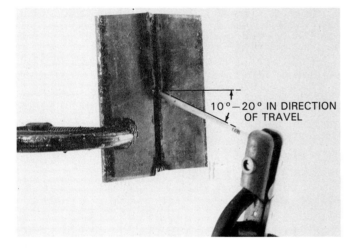

Fig. 15-10. A butt joint being welded with the downhill method. The electrode is tipped about 10°−20° in the direction of travel.

An *uphill* weld begins at the bottom of the joint and ends at the top of the joint. The electrode is tipped in the direction of travel about 10°−20°. See Fig. 15-11. The greatest problem in vertical welding is keeping the molten metal in the weld pool. A *whip motion* is used to control the temperature of the pool. A whip motion prevents the bead from sagging by allowing the weld pool to cool for a short time. A whip motion can be used for welding any type of joint. Generally, it is only used when welding uphill or overhead.

When using a whip motion, the electrode tip is moved ahead of the weld pool about 1″ (25 mm). The electrode is then pulled away from the base metal while still maintaining a long arc. The electrode is then moved back to the weld pool and a normal arc length. The back-and-forth movement should occur within 1−2 seconds. Fig. 15-12 illustrates the whip motion. The molten weld pool cools during the whip motion. The whip motion should occur with a regular rhythm as the bead is laid.

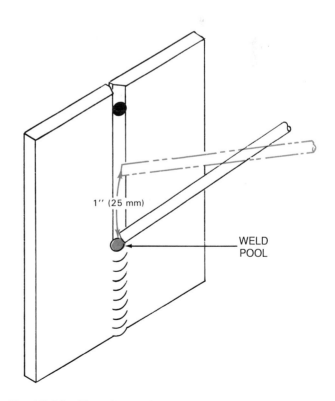

Fig. 15-12. The whip motion used to control the heat of the weld pool. The electrode is moved ahead of the pool about 1″ (25 mm) and returned to the original weld pool. This entire action should take no more than 1−2 seconds. While the electrode is moving, the weld pool can cool.

153

FILLET WELDS

Fillet welds may be made using a stringer or a weaving bead. A whip motion will help to control weld pool temperature while welding in the vertical welding position.

Lap joints

On a lap joint, the electrode should be aimed slightly more toward the surface than toward the edge. This will prevent excessive melting of the edge. Greater weld strength is obtained when welding uphill. The electrode should be inclined about 10° − 20° in the direction of travel.

EXERCISE 15-4 — FILLET WELD ON A LAP JOINT IN THE VERTICAL WELDING POSITION

1. Obtain two pieces of mild steel measuring 1/8" x 2" x 6" (3.2 mm x 50.8 mm x 152.4 mm).
2. Obtain five 1/16" or 3/32" (1.6 mm or 2.4 mm) diameter E6011 electrodes.
3. Form a lap joint with a 3/4" (19.1 mm) overlap on each side.
4. Tack weld the pieces in three places on each side.
5. Place the weldment in a vertical position.
6. Place a fillet weld on both sides. The fillet legs should be 1/8" (3.2 mm) long. One weld should be made uphill. A whip motion may be used. The other weld should be made downhill.

Inspection:
The completed welds should have convex beads. No overlap or undercut is permitted. The bead should have smooth, evenly spaced ripples.

Inside corner and T-joints

When making a fillet weld in the vertical welding position, the electrode should be tipped in the direction of travel about 10° − 20°. It should be held at 45° to the surfaces of the base metal. A weaving bead is commonly used for a fillet weld. A crescent-shaped movement of the electrode is often used. Hesitate at the weld toes when using a weaving bead to prevent undercutting. See Fig. 15-13.

OVERHEAD WELDING POSITION

Special safety precautions must be followed when welding overhead. Hot metal or slag may fall from the weld area, creating unique problems for the welder. Refer to the safety precautions listed in Chapter 14 for specific information.

When welding in the overhead welding position, work is done more comfortably and efficiently across the body. Welding can be done from right to left or left to right. Position yourself so that you can see into the root of the weld.

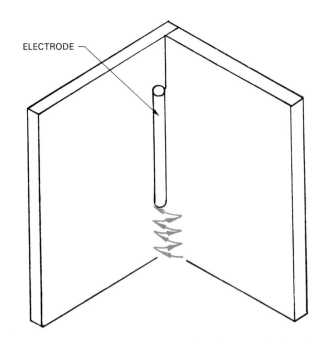

ELECTRODE

Fig. 15-13. A crescent motion used with a weaving bead on a fillet weld. The angles for the electrode should be the same as flat welding.

The electrode should be placed in the holder at a comfortable angle. Some welders clamp the electrode so it is parallel to the electrode holder.

A small weld pool should be maintained to keep the metal from sagging. Welding with a smaller diameter electrode and a lower amperage will help in keeping a small weld pool. A stringer bead is used for the root pass. The keyhole method ensures 100% penetration. A short oscillating motion is recommended to control the pool and bead shape. See Fig. 15-14. A whip motion can be used to control the molten metal.

FILLET WELDS

The electrode angle between the base metal and weld axis on all types of joints is the same as when welding in the flat position. The electrode is held at about 45° from the base metal. It is tilted in the direction of travel

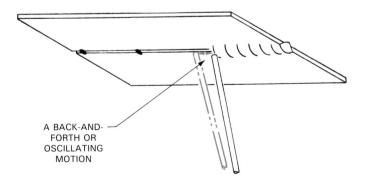

A BACK-AND-FORTH OR OSCILLATING MOTION

Fig. 15-14. A short back-and-forth swing or oscillating motion may be used to control the temperature of the metal in the weld pool.

at a 10° – 20° angle. The electrode is aimed more toward the surface piece and away from the edge. This prevents excessive melting of the edge as may occur in flat welding.

have evenly spaced ripples. The fillet legs must be 3/8'' (9.5 mm) long. The completed weld should have a straight or convex weld face.

EXERCISE 15-5 — BEVEL-GROOVE WELD ON A BUTT JOINT—UPHILL

1. Obtain two pieces of mild steel measuring 1/4'' x 2'' x 6'' (6.4 mm x 50.8 mm x 152.4 mm).
2. Obtain five 1/16'' or 3/32'' (1.6 mm or 2.4 mm) diameter E60XX electrodes.
3. Prepare the edges of the metal for a 45° bevel-groove butt joint. Leave a 1/16'' (1.6 mm) root thickness.
4. Form a bevel-groove butt joint with a 1/16'' (1.6 mm) root opening.
5. Tack weld the pieces in three places.
6. Place the weldment in a vertical position.
7. Use the keyhole method on the root pass to ensure 100% penetration.
8. Weld the following passes as stringer or weaving beads. Use a whip motion, if necessary, to control the molten weld pool.

Inspection:
The completed weld should be convex with 100% penetration. The bead ripples should be smooth and evenly spaced.

EXERCISE 15-7 — V—GROOVE WELD FOR A BUTT JOINT IN THE OVERHEAD WELDING POSITION

1. Obtain two pieces of mild steel measuring 1/4'' x 2'' x 6'' (6.4 mm x 50.8 mm x 152.4 mm).
2. Obtain five 1/16'' or 3/32'' (1.6 mm or 2.4 mm) diameter E6010 electrodes.
3. Prepare the edges of the pieces for a 90° V-groove butt joint. Leave a 1/16'' (1.6 mm) root thickness.
4. Form a butt joint with a 1/16'' (1.6 mm) root opening. Tack weld in three places.
5. Place the joint into an overhead position.
6. Make a root pass using the keyhole method to get 100% penetration. An oscillating or whip motion is suggested to control the arc crater.
7. Complete the weld with stringer or small weaving beads.

Inspection:
The weld should have a convex face and have 100% penetration. The beads should be uniform in width and have evenly spaced ripples. No undercut or overlap should be visible.

BUTT JOINTS

When welding a groove-type joint in the overhead position, the electrode is aligned with the weld axis. A 10° – 20° forward tilt of the electrode is recommended. A multiple-pass weld is cleaned between passes. The weld area will cool sufficiently between passes to help prevent overheating. If overheating occurs, the weld pool can be controlled by using an oscillating or whip motion.

EXERCISE 15-6 — FILLET WELD A T-JOINT IN THE OVERHEAD WELDING POSITION

1. Obtain two pieces of mild steel measuring 1/4'' x 2'' x 6'' (6.4 mm x 50.8 mm x 152.4 mm).
2. Obtain five 1/16'' or 3/32'' (1.6 mm or 2.4 mm) diameter E6012 electrodes.
3. Form a T-joint with the two pieces. Tack weld in three places on each side of the joint in the flat welding position.
4. Place the weldment in an overhead position.
5. Wear protective clothing recommended for overhead welding. Make a fillet weld with 3/8'' (9.5 mm) legs, using several passes.
6. Make a fillet weld with 3/8'' (9.5 mm) legs on the other side of the T-joint.

Inspection:
The weld should be even in width. Each bead should

WELD DEFECTS

Defects that can be seen include overlap, undercut, voids in the weld face, incorrect weld size and shape, and lack of penetration on the groove-joint weld.

Overlap occurs when the weld toe is not fused into the base metal. Fig. 15-15A and B shows examples of overlap.

Undercut occurs when a weld crater is not filled completely by the electrode filler metal. See Fig. 15-15C and D for two examples of undercut.

Voids in the weld face may be low spots where the electrode was not added. This is generally a result of improper electrode motion or improper restarting of the weld. Gases in the cooling weld may leave a void in the weld face.

Welds that are too wide or too low can be found by visual inspection. An incorrect bead contour is easily seen. Fig. 15-15E and F shows two defects in bead shape.

Complete (100%) penetration is generally required only on a weld on a groove-type joint. A bead with 100% penetration will show a small, uniform bead-like bump on the root side of the weld.

Fillet welds seldom require 100% penetration. They do, however, require enough to penetrate past the intersection of the two pieces being welded. The welding symbol often specifies a leg size and weld size large enough to ensure adequate penetration. Poor penetration on a fillet weld cannot be inspected visually.

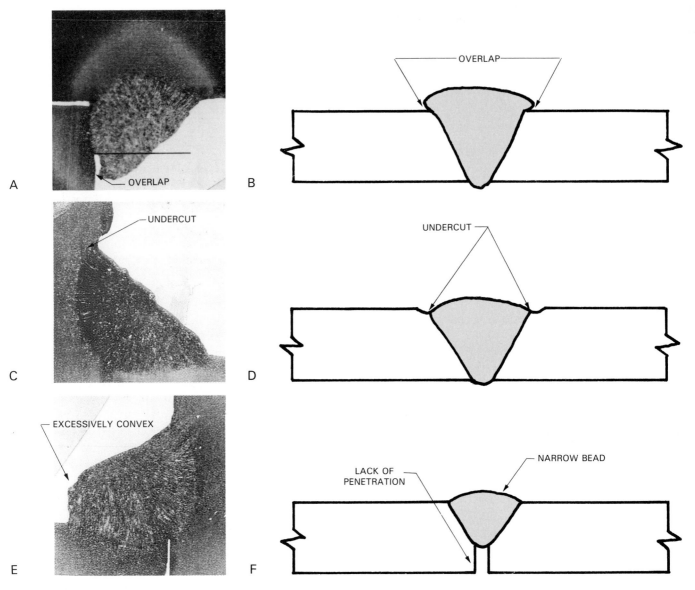

Fig. 15-15. Cross section of weld defects. A—Overlap on a fillet weld. B—Overlap on a groove weld on a butt joint. C—Undercut on a fillet weld. D—Undercut on a groove weld on a butt joint. E—A poorly contoured fillet weld. F—A poorly contoured groove weld on a butt joint.

REVIEW QUESTIONS

Write your answers on a separate sheet of paper.
1. Name five pieces of protective clothing that are to be worn for out-of-position welding.
2. List three advantages of using smaller diameter electrodes for out-of-position welding.
3. Why is the lowest bead in each layer made first on a multiple-pass V-groove butt weld when welding horizontally?
4. Identify the defects shown in the sketch below.

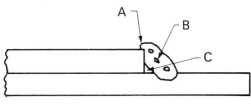

5. Name the part of the weld shown in the sketch.

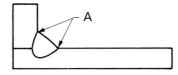

6. An electrode especially designed for making horizontal fillet welds is the _____.
7. Why must the weld pool be kept ahead of the slag when welding downhill?
8. The electrode is moved ahead about _____" (_____ mm) and up to a long arc during the whip motion.
9. The entire whip motion should take only _____ seconds.
10. What type of motion is recommended to control the molten metal during an overhead weld?

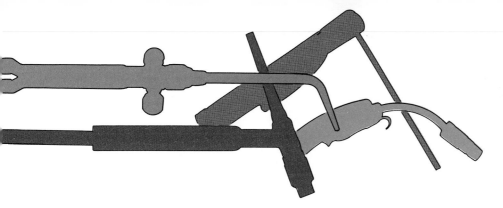

Chapter 16

Surfacing

After studying this chapter, you will be able to:
☐ List reasons for surfacing a part.
☐ Identify the various surfacing processes.
☐ List reasons for wear that occurs in parts.
☐ Define characteristics of surfacing electrodes.
☐ List two means of testing material hardness.
☐ Describe abbreviations used when specifying surfacing electrodes.
☐ Select the proper surfacing electrode.
☐ Surface a part.

TECHNICAL TERMS

Abrasion, abrasion resistance, Brinell hardness test, buildup process, buttering, chemical corrosion, cladding, corrosion resistance, fatigue, grit blasting, hardfacing, hardness, hot hardness, impact strength, machinability, metal transfer wear, metal-to-metal wear resistance, oxidation resistance, Rockwell C hardness test, surfacing, thermal spraying, vapor degreasers.

SURFACING PRINCIPLES

Surfacing is defined by the American Welding Society as "the application by welding, brazing, or thermal spraying of a layer(s) of material to a surface to obtain desired properties or dimensions, as opposed to making a joint." A company generally spends less money surfacing a part than buying a new one. Surfacing also usually reduces the downtime on the equipment. There are several surfacing processes. They include hardfacing, buttering, cladding, and buildup.

Hardfacing is a surfacing process in which hard materials are applied to the surface of a part. This method is used to reduce wear or loss of materials by impact or abrasion. See Fig. 16-1. Hardfacing a part also results in fewer repairs due to wear.

Hardfacing beads may be laid in a basket weave or dot pattern. See Fig. 16-2. These patterns are used when a sticky material, such as dirt or mud, comes in contact with the part. The sticky material accumulates in the depressed areas of the basket weave pattern. The build-up of dirt or mud helps to protect the metal from further abrasion.

In the *buttering* process, one or more layers of easily welded material are applied to the surface of a part that has poor welding characteristics. This process is used to form a transition layer when welding dissimilar metals.

The *cladding* process is used to apply surfacing materials that will improve the corrosion or heat resistance of a part. When a part is worn, the surface

Fig. 16-1. A hammermill head that is used in a stone-crushing machine. This head has been rebuilt with hardfacing electrodes. (Stoody Co.)

Fig. 16-2. Close-up views of hardfacing material applied to the edge of power shovel buckets. A — The basket weave pattern is often used. Abrasive materials, such as sand and mud, will build up between the beads and provide an additional layer of protection for the steel. (Cronatron Welding Systems, Inc.) B — The dot pattern is often used to reduce the possibility of stress cracking in the base metal.
(Cronatron Welding Systems, Inc.)

materials. These materials may be applied using the oxyfuel gas, arc, or plasma process. Solid surfacing rods may be used when surfacing with oxyfuel gas. Special covered electrodes are used for surfacing with the shielded metal arc process (SMAW). Surfacing materials are enclosed in a hollow wire and applied using the gas metal arc (GMAW) process.

Surfacing materials may also be applied by *thermal spraying.* In this process, powdered or finely ground materials are fed into a flame or arc and carried to the surface of the part. Surfacing materials fed from a hopper into the oxyfuel gas flame may be used. See Fig. 16-3. Thermal spraying may also be done using the plasma arc. See Fig. 16-4.

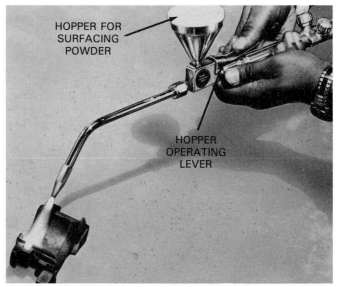

Fig. 16-3. An oxyfuel gas torch equipped with a hopper for surfacing material. A silicon-boron-nickel powder is being used to repair a container mold. (Wall Colmonoy Corp.)

may be returned to its original dimensions by using the *buildup process.*

Surfaces of parts wear for several reasons. *Chemical corrosion* may eat away the surface. *Metal transfer wear* occurs when metal leaves one surface and fuses or sticks to another surface. This commonly occurs on bearing surfaces. Surfaces may be rubbed together constantly and wear down due to friction or *abrasion.* *Fatigue* may occur when surfaces are constantly struck, stretched, compressed, or bent. Metals will eventually weaken and wear down or fracture.

A metal surface is commonly repaired when it wears down. The cause of the failure must be determined to properly repair the part. A surfacing material is then selected to resist the type of wear that occurred on the original part.

Surfacing materials are applied to a part using solid, powder, or finely ground metallic or nonmetallic

Fig. 16-4. A thermal spraying torch with a built-in wire feed mechanism. This torch is applying surfacing material to a worn crankshaft bearing journal. The torch is mounted in a lathe cross-feed for better control of the spray thickness. The journal will be ground to size after it is rebuilt.

SURFACING ELECTRODES

Surfacing electrodes are used to apply surfacing materials to metal parts. The surfacing materials reduce wear to a part or build up its dimensions. The following characteristics or qualities should be considered when selecting a surfacing electrode:
• Hardness.
• Hot hardness.
• Impact strength.
• Oxidation resistance.
• Corrosion resistance.
• Abrasion resistance.
• Metal-to-metal wear resistance.
• Machinability.

Hardness is the ability to resist penetration. Hardness is measured by several methods including the popular Rockwell C and Brinell hardness tests. Fig. 16-5 shows a comparison between the scales used for Rockwell C and Brinell hardness tests. The *Rockwell C hardness test* is generally used for hard materials.

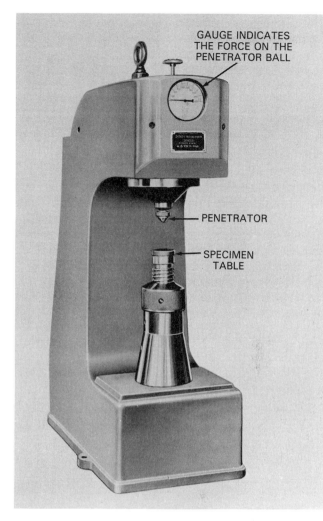

Fig. 16-6. A hydraulic Brinell hardness testing machine. (Detroit Testing Machine Co.)

| HARDNESS TEST COMPARISONS ||
Rockwell C	Brinell
69	755
60	631
50	497
40	380
30	288
24	245
20	224
10	179
0	143

Fig. 16-5. Comparison of hardness numbers in the Rockwell C and Brinell scales.

A special piece of equipment forces a pointed diamond into the metal surface. The diamond penetrates deeper into the surface of a soft metal. A gauge indicates the hardness number. The Rockwell C hardness scale ranges from 0 to approximately 70. A higher number indicates a harder material. A *Brinell hardness test* is used to evaluate the hardness of softer metals like copper, brass, bronze, and aluminum. In this test, a 10 mm diameter steel ball is forced into the surface of the test piece. The ball penetrates deep into the surface of a soft material. A gauge indicates the hardness number, Fig. 16-6. Brinell numbers range from approximately 100 to 800. A higher number indicates a harder material.

Hot hardness is the ability of a metal to retain its strength at high temperatures. The hardness of a metal generally decreases as the temperature is increased. A metal that retains its hardness at high temperatures has good hot hardness.

Impact strength is the ability of a material to withstand impact or hammering forces without cracking or breaking. Stone-crushing hammers and forging dies have high impact strength.

Oxidation resistance is the ability of a material to withstand the corrosive effects of atmospheric oxidation. Steel with high oxidation resistance will not rust easily.

Corrosion resistance is the ability of a material to resist corrosion from chemicals such as salt spray and acids. Corrosion may cause fatigue and failure of parts.

Abrasion resistance is the ability of a material to resist wear from scratching by sharp objects. Rocks, sand grains, or sharp metal are common causes of abrasion.

Metal-to-metal wear resistance is the ability of a material to resist wear from metal-to-metal contact. Bearing surfaces require a high amount of metal-to-metal wear resistance.

Machinability is the ability of a part to be machined or ground to size. The hardness of a material must be considered if the part is to be machined after surfacing. Some surfacing materials are so hard that they may be difficult to machine.

ELECTRODE CLASSIFICATIONS

Surfacing materials can be applied using a variety of welding processes. The oxyfuel gas welding, flux cored arc welding (FCAW), and gas tungsten arc welding (GTAW), processes may be used. Solid, bare surfacing wires and rods are used with OFW and GTAW. Hollow, alloy-filled electrodes are used with FCAW.

Solid, covered surfacing electrodes are used with the shielded metal arc process. Hollow surfacing electrodes may also be used with this process. The hollow electrodes are filled with alloying materials such as tungsten carbide crystals. The tungsten carbide crystals create an extremely hard surface when applied to a metal surface. Surfacing electrodes are classified by the primary material contained in the electrode. See Fig. 16-7.

Surfacing electrodes for the shielded metal arc process are available in the following diameters; 1/8″ (3.2 mm), 5/32″ (4.0 mm), 3/16″ (4.8 mm), 1/4″ (6.4

NEW CLASSIFICATION	FILLER METAL	OLD DESIGNATION
Fe5	High-speed steel	IA5
FeMn	Austenitic Manganese	IB2
FeCr	Chromium steel	IC1
CoCr-A	Cobalt and Chromium	IIA
CoCr-C	Cobalt and Chromium	IIB
CuZn	Copper and Zinc (brass)	IVA
CuSi	Copper and Silicon	IVB
CuAl	Copper and Aluminum	IVC
NiCr	Nickel and Chromium	VB

Fig. 16-7. New and old AWS surfacing filler metal classifications. See Fig. 16-8 for the chemical abbreviations.

mm), 5/16″ (8.0 mm), 3/8″ (9.5 mm). The electrodes are available in 8″ − 28″ (203 mm − 711 mm) lengths. The 14″ (356 mm) and 18″ (457 mm) lengths are the most widely used.

Most surfacing electrodes are designed for use in the flat position. Several are produced for use in all positions.

Specifications for surfacing electrodes are commonly abbreviated. The abbreviations can generally be divided into three parts. See Fig. 16-8. The first part indicates whether it is to be used as an electrode or a welding rod. The second part indicates the type of filler metal to be used. The third part indicates a subclassification of the filler metal.

Characteristics of various surfacing electrodes and rods are shown in Fig. 16-9. Suggested applications for the surfacing materials are also shown. For specific corrosion-resistant applications, consult AWS A5.13 or a corrosion resistance expert.

SURFACING PROCESS

The surfacing process requires a few essential steps. Each part to be surfaced must be thoroughly cleaned.

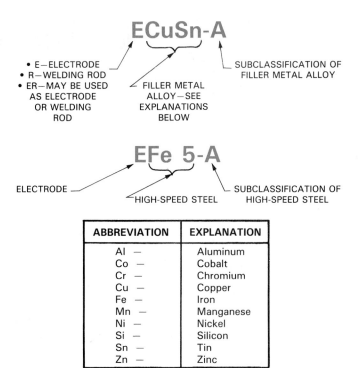

Fig. 16-8. Explanation of electrode and rod specifications. The chemical abbreviations used with AWS surfacing filler metal classifications are also shown.

This ensures that the surfacing material can adhere to the part. Preheating of some metals is necessary. Preheating can be done using an oven, torch, electrical resistance heater, or any other suitable heat source. The final step is to choose the desired process and equipment and apply the surfacing materials to produce an acceptable new surface.

CLEANING THE SURFACE

Parts must be cleaned to ensure the proper bonding of the base metal and surfacing materials. The type of base metal determines the cleaning method to be used. Mechanical cleaning may be done with abrasive or **grit blasting.** Aluminum and magnesium are usually grit blasted with aluminum oxide or quartz crystals. Steam, liquid or **vapor degreasers,** and industrial solvents may be used to chemically clean the parts. Steel may be cleaned with steam or degreasers. Porous metals like cast iron are heated to 400°F − 600°F (204°C − 315°C). Heating to these temperatures vaporizes grease and oil in the porous areas.

PREHEATING

Some metals must be preheated prior to surfacing. Preheating to a specific temperature prepares the metal for surfacing. It ensures a complete bond between the base metal and surfacing material. Preheating recommendations for several metals are listed below. If the type of base metal cannot be determined, preheat to

Classification	Hardness	Hot Hardness	Impact Strength	Oxidation Resistance	Corrosion Resistance	Abrasion Resistance	Metal-to Metal Wear Resistance	Machinability	Applications
Fe5-A,B,C	C55-60[1]	H<1000°F	H[2]	L	L	M	H[3]	H	Cutting tools, dies, guides
FeMn-A & B	450-550BHN[6]	H<500-600°F	H	L	L	M	H	L	Soft-rock-crushing equipment
FeCr-A	C51-62	H<800-900°F	L	H	L	L-H[7]	L-H[7]	L	Farm equipment
CoCr-A	C42	H>1200°F	M	H	H[4]	L	H	L	Engine exhaust valves
-B	C46	H>1200°F	M	H	H[4]	L	H	L	
-C	C55	H>1200°F	L	H	H[4]	M	H	L	
CuAl-A2	115-140BHN[5]	H<400°F	M	M	M[4]	L	H	H	Bearings and gear surfaces
CuAl-A3	140-180BHN[5]	H<400°F	L	M	M[4]	L	H	H	
-B	140-180BHN[5]	H<400°F	L	M	M[4]	L	H	H	
-C	180-220BHN[5]	H<400°F	L	M	M[4]	L	H	H	
-D	230-270BHN[5]	H<400°F	L	M	M[4]	L	H	H	
-E	280-320BHN[5]	H<400°F	L	M	M[4]	L	H	H	
CuSi	80-100BHN	H<800°F	M	M	M[4]	L	H	H	Bearing surfaces
CuSn-A	70-85BHN	H<800°F	L	L	M[4]	L	H	H	Bearing surfaces
-C	85-100BHN	H<800°F	L	L	M[4]	L	H	H	
NiCr-A	C28	H<800°F	M	H<1750°F	4	L-H[7]	H	M	Seal rings, cams, screw conveyors
-B	C38	H<800°F	M	H<1750°F	4	L-H[7]	H	M	
-C	C45	H<800°F	L	H<1750°F	4	L-H[7]	H	M	

KEY:
C - Rockwell C hardness number
BHN - Brinell hardness number
L - Low
M - medium
H - high

1. Anneal to Rockwell C30
2. After tempering
3. After annealing
4. Consult authority for each application
5. Hardness depends on welding process used
6. After work hardening
7. Depends on application

Fig. 16-9. A summary of the characteristics of various surfacing materials. Refer to AWS A5.13 or a corrosion authority for specific applications.

500°F − 700°F (260°C − 371°C). Be certain, however, that the part is not made of manganese steel.

- Cast iron: 500°F − 700°F (260°C − 371°C).
- Low and mild carbon steel: preheating is not required except for heavy sections.
- Manganese steel (steel containing 12 − 14% manganese): 200°F (93°C). Do not heat above 500°F (260°C).

SURFACING WITH A SHIELDED METAL ARC

Direct current electrode positive (DCEP) polarity is generally used when surfacing with a shielded metal arc. The equipment used for surfacing is the same as that used for shielded metal arc welding. Different electrodes are used, however.

Surfacing materials may be applied to an entire surface. Surfacing materials may also be applied as long beads or may completely cover small areas of a part. Surfacing materials are applied as stringer or weaving beads or may be sprayed onto the surface. Surfacing materials are applied to the surface of the base metal. Deep penetration of the surface is not required or desired. Fig. 16-2 shows examples of patterns used for surfacing materials.

The surface must be thoroughly cleaned to ensure a good bond of the surfacing materials. After cleaning, the metals to be surfaced may require preheating. Preheating ensures a better bond between the base metal and surfacing materials.

The process of depositing surfacing materials with an electrode is the same as SMAW. An arc is struck in the same manner as used for SMAW. A long arc is used for surfacing. Such an arc is needed to spread the heat over a large area. Penetration is prevented by spreading the heat out. To maintain the arc, a higher

amperage than recommended should be used for a given electrode diameter.

A surfacing electrode is held at approximately 90° to the base metal. It is tilted in the direction of travel at approximately a 45° angle. An oscillating (forward and back) motion is recommended for stringer beads. See Fig. 16-10. A weaving or circular motion is used to create wider beads. See Fig. 16-11. A weaving or circular bead should not be wider than six times the electrode diameter.

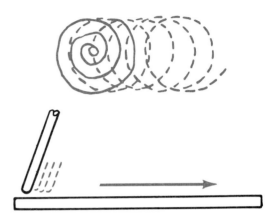

Fig. 16-11. A circular motion is suggested for beads that are from 3/4″ − 1 1/4″ (19 mm − 32 mm) wide.

Each bead of surfacing material should be between 1/16″ − 1/4″ (1.6 mm − 6.4 mm) thick.

Beads that are laid side by side should overlap by about 1/4 − 1/3 the bead width. See Fig. 16-12. If more than one layer is required, the beads should also overlap. Fig. 16-12 shows the suggested overlap of beads and layers.

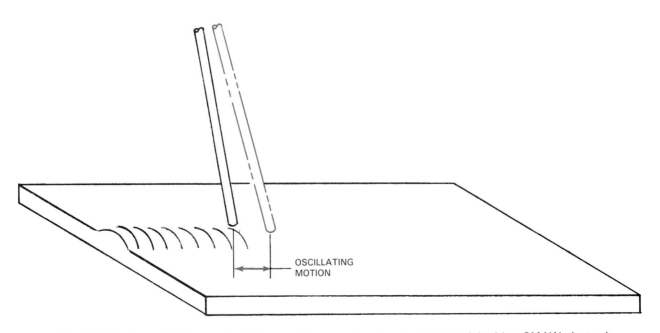

OSCILLATING
MOTION

Fig. 16-10. An oscillating motion being used to apply hard surfacing material with a SMAW electrode.

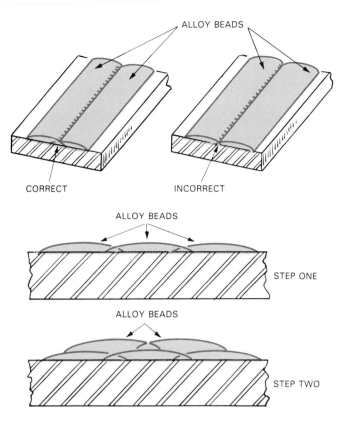

CORRECT   INCORRECT

ALLOY BEADS

STEP ONE

ALLOY BEADS

STEP TWO

Fig. 16-12. The suggested method for overlapping surfacing beads across a wide surface.

EXERCISE 16-1 — SURFACING WITH A SHIELDED METAL ARC

1. Obtain a piece of mild steel measuring 1/4" x 4" x 6" (6.4 mm x 101.6 mm x 152.4 mm).
2. Also obtain four 1/8" (3.2 mm) Fe5 hardfacing electrodes.
3. If a Rockwell hardness tester is available, test the hardness of the mild steel plate.

4. Lay three or more overlapping stringer beads or weaving beads on the surface of the metal. Apply only one layer of material. The beads should be 1" (25.4 mm) wide and about 5" (127 mm) long.

Inspection:

The beads should overlap 1/4—1/3 the width of the bead. The beads should have a uniform width, and smooth, evenly spaced ripples. If possible, test the Rockwell C hardness of the surfaced area. The hardness should be C55-60.

REVIEW QUESTIONS

Write your answers on a separate sheet of paper.

1. List at least two reasons for surfacing or hardfacing a part.
2. List three causes of wear.
3. What type of surfacing process uses powdered surfacing material?
4. A pointed diamond is forced into the metal surface when using a _____ hardness tester.
5. The ability to resist wear from sharp objects is called _____ resistance.
6. Describe the meaning of the abbreviations used in specifying surfacing materials.

$$\underset{A}{\text{ER}}\underset{B}{\text{Cu}}\underset{C}{\text{Si}}-\underset{D}{\text{A}}$$

7. Refer to Fig. 16-9. What type of surfacing electrode could be used if important considerations were metal-to-metal wear resistance and excellent machinability?
8. Why should cast iron be preheated prior to applying a surface material?
9. What polarity is normally used with surfacing electrodes?
10. Why is a long arc used when surfacing?

163

Thermal spraying process being used to apply a surfacing material to a rotating part.   (Wall Colmonoy Corp.)

GAS METAL ARC WELDING

The Lincoln Electric Co.

Gas metal arc welding equipment being used to make a vertical weld. Notice that the arc welding machine and shielding gas cylinder are on a wheeled dolly. (Miller Electric Mfg. Co.)

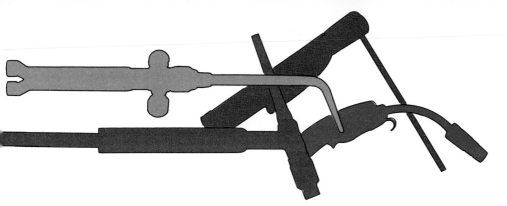

Chapter 17

Gas Metal Arc Welding Equipment and Supplies

After studying this chapter, you will be able to:
- ☐ Define GMAW and FCAW.
- ☐ Identify the correct polarity to use for GMAW and FCAW.
- ☐ List advantages and disadvantages of GMAW as compared to SMAW.
- ☐ Define metal transfer.
- ☐ Describe the three methods of metal transfer.
- ☐ Identify the equipment that makes up a GMAW outfit.
- ☐ Summarize the operation of a wire feeder.
- ☐ List the parts of a welding gun and cables.
- ☐ List four gases used for GMAW and FCAW.
- ☐ Explain the use of a flow meter for GMAW or FCAW.
- ☐ Identify protective clothing and equipment used for GMAW and FCAW.

TECHNICAL TERMS

Active gas, argon, background current, carbon dioxide, constant potential, constant voltage, consumable electrode, contact tube, drive rolls, flow meter, flux-cored arc welding, (FCAW), gas metal arc welding (GMAW), globular transfer, globule, helium, inch switch, inductance, inert gas, inverter, liner, metal inert gas welding (MIG), metal transfer, nozzle, peak current, pinch force, pulsed spray transfer, pulses, purge switch, push-pull setup, self-shielding electrode, short arc, short circuit, short circuiting transfer, slope adjustment, spray transfer, Teflon liner, transformer-rectifier, transition current, welding arc, welding gun, wire feeder, wire tension control knob.

GAS METAL ARC WELDING PRINCIPLES

Gas metal arc welding (GMAW) is a welding process in which metals are joined by heating them with a *welding arc* between the base metal and a continuous consumable electrode. A shielding gas or gas mixture is used to prevent the atmosphere from contaminating the weld. Another term that is often used to describe GMAW is MIG welding. MIG stands for *metal inert gas.* However, because some of the gases used in GMAW are not inert, MIG is an incorrect term. In the welding trade, both GMAW and MIG are used to describe this welding process.

Gas metal arc welding uses a wire as an electrode. A welding arc is struck between the electrode and the base metal. Direct current is always used for GMAW. Direct current electrode positive (DCEP), or direct current reverse polarity (DCRP), is used for GMAW. Direct current electrode negative (DCEN), or direct current straight polarity (DCSP), is seldom used.

A *constant voltage* power supply is used for GMAW. On this type of machine, the welder sets the required voltage. A constant voltage welding machine tries to maintain a constant voltage during the welding operation. Although the voltage is constant, the current varies. Fig. 17-1 illustrates the GMAW process. The electrode is melted and becomes part of the weld. For this reason, the electrode is called a *consumable electrode.* The welding arc, also called simply arc, and the molten base metal are protected from air by a shielding gas. Flux is not used in this process.

Flux-cored arc welding (FCAW) is similar to GMAW. Flux-cored arc welding is a welding process

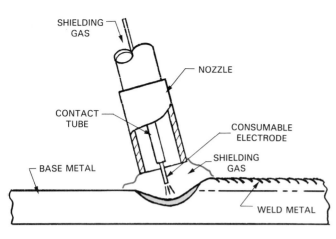

Fig. 17-1. A schematic drawing of a gas metal arc weld in progress.

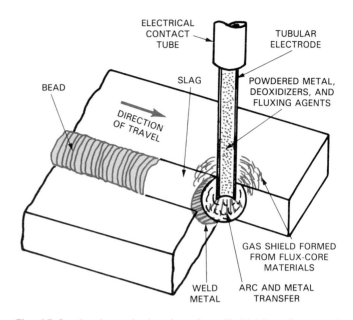

Fig. 17-3. A schematic drawing of a self-shielding, flux-cored arc weld in progress. Neither a nozzle nor shielding gas is required.

in which metals are joined by heating them with an electric arc between the base metal and a continuous consumable electrode. Shielding gas is obtained from the flux inside the tubular electrode. Additional shielding may be provided by a shielding gas. The main difference between FCAW and GMAW is that the electrode used in FCAW is hollow. Flux is contained inside the hollow electrode. The flux creates a shielding gas as it melts. The flux used in FCAW can be used to add alloying elements to the weld.

Some flux-cored electrodes require using a shielding gas from a cylinder. See Fig. 17-2. Although some shielding is created by the flux inside the electrode, additional shielding gas is required. *Self-shielding electrodes* produce their own shielding gas and do not require additional shielding gas. See Fig. 17-3.

Some of the advantages of GMAW and FCAW include:
1. A continuous electrode is used so that longer welds can be made without stopping to change electrodes. This eliminates many of the starts and stops that can be a major cause for weld defects.
2. Welding speeds are faster than for shielded metal arc welding (SMAW).
3. No slag is produced in GMAW. Very little spatter is created. More of the electrode becomes part of the weld.
4. Deeper penetration is possible with GMAW than with SMAW.
5. Less training and skill are required to weld with gas metal arc and flux-cored arc welding than with SMAW.

Some disadvantages of GMAW and FCAW include:
1. GMAW and FCAW equipment costs more than SMAW equipment.
2. Some weld joints are hard to reach with the welding gun.
3. Rapid air movement, such as a strong wind, can blow the shielding gas away, resulting in a weld defect.

METAL TRANSFER

Filler metal from the electrode must leave the electrode and enter the weld. This is called *metal transfer.* Metal transfer can occur in three different ways: short circuiting transfer, globular transfer, and spray transfer. In *short circuiting transfer* the electrode contacts the base metal and some molten metal is transferred. In *globular* and *spray transfer,* the wire does not touch the base metal. In these transfer methods,

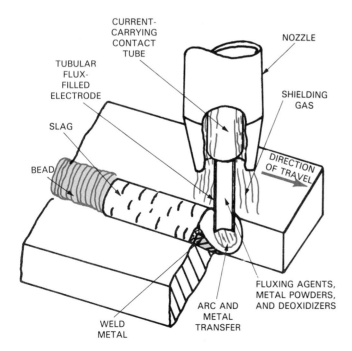

Fig. 17-2. A schematic drawing of a gas-shielded, flux-cored arc weld in progress.

drops of molten metal leave the end of the electrode and enter the weld pool through the arc. These drops travel between the end of the wire and the base metal. The size of the drop in globular transfer is much larger than the size of the drop in spray transfer. Voltage, current, shielding gas, and other factors determine the type of metal transfer you will obtain. Voltage and wire feed speed recommendations are covered in Chapter 18.

SHORT CIRCUITING TRANSFER

Short circuiting transfer is also called *short arc.* Short circuiting transfer uses a low welding current and voltage. A small-diameter electrode, .030″ to .045″ (.76 mm to 1.14 mm), is used. Short circuiting transfer method does not add much heat to the base metal. Therefore, short circuiting transfer is used to join thin sheets of metal. It is also used for out-of-position welding and for welding parts that have large gaps.

In short circuiting transfer, the base metal and the end of the electrode are heated by an arc. Metal is not transferred across the arc. The electrode touches the base metal. There is no arc when the electrode touches the base metal. When the electrode touches the base metal it is called a *short circuit.* While the short circuit exists, metal from the electrode enters the weld. The welding machine tries to maintain the preset voltage, so it increases the current to the electrode. The increased current results in a **pinch force** that causes the end of the electrode to separate from the base metal. After the end of the electrode separates from the weld pool, the arc reignites and the steps are repeated. Fig. 17-4 shows the short circuiting transfer process. The electrode is continuously fed into the weld. The electrode contacts the base metal 20 – 200 times per second. The correct settings to use for short circuiting transfer are covered in Chapter 18.

The end of an electrode explodes if the welding machine increases the welding current too fast. Some welding machines allow the welder to control the inductance. *Inductance* is an electrical means of controlling the rate of current change in a circuit. In a welding machine, inductance controls how fast the welding machine increases the welding current. Inductance is very important when using short circuiting transfer. The inductance should be increased if the end of the electrode appears to explode. Increasing the inductance helps to make the weld pool flatter and more fluid.

GLOBULAR TRANSFER

Globular transfer uses slightly higher voltage and current settings than those used for short circuiting transfer. In globular transfer, the arc melts the end of the electrode and the base metal. The size of the molten metal, *globule* (ball) on the end of the electrode increases until it is larger than the diameter of the elec-

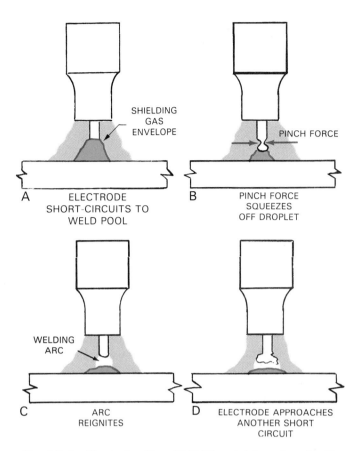

Fig. 17-4. Short circuiting GMAW metal transfer. A—The electrode wire short-circuits to the weld pool. B—A magnetic pinch force squeezes off a droplet of molten electrode metal. C— The welding arc reignites. D—The electrode nears another short circuit condition and the process repeats itself.

trode. It continues to grow in size until the globule falls off the electrode due to its own weight. Fig. 17-5 illustrates globular transfer.

The molten globule does not fall straight into the weld. Forces acting on the arc cause the globule to travel across the arc and land on the base metal in random patterns. A large amount of spatter usually occurs. Globular transfer is difficult to perform for out-of-position welding, because gravity is used to transfer the molten globules to the base metal.

SPRAY TRANSFER

Spray transfer requires higher voltage and current settings than globular transfer. In spray transfer, hundreds of small droplets are formed every second. They travel at a high rate of speed directly into the weld. See Fig. 17-6. Good penetration is obtained with spray transfer. Very little spatter is produced. The weld pool is very fluid; spray transfer is used primarily for welding in the flat position or on horizontal fillet welds. Metal that is 1/8″ (3.2 mm) and thicker can be welded using the spray transfer process.

A shielding gas mixture containing at least 90% argon must be used for spray transfer. Spray transfer

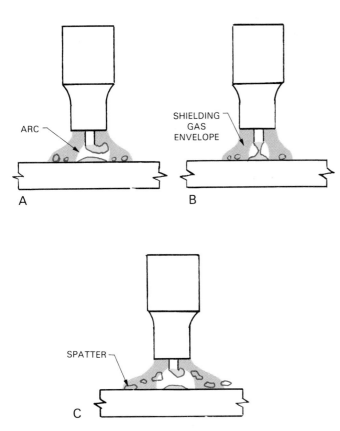

Fig. 17-5. GMAW globular metal transfer processes. A—Large irregular droplets form. B—Droplets may short circuit when they fall. C—Droplets may fall erratically and cause spatter.

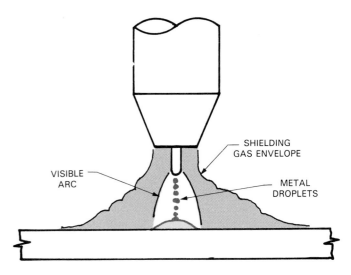

Fig. 17-6. GMAW spray transfer. Hundreds of small droplets leave the end of the electrode every second.

will only occur when the current setting is above the transition current. The **transition current** is the amount of current required to convert from globular transfer to spray transfer. Fig. 17-7 lists the transition currents for various electrodes.

Pulsed spray transfer

Pulsed spray transfer is very similar to standard spray transfer. However, two different currents are used: background current and peak current. The **background current** is an amount of current that is less than the transition current. Background current maintains the arc, but no metal is transferred across the arc during the background current period. **Peak current** is a preset current level that is above the transition current. Spray transfer occurs during periods of peak current. See Fig.

Wire electrode type	Wire electrode diameter		Shielding gas	Minimum spray arc current, (A)
	in.	mm		
Mild steel	0.030	0.76	98% argon-2% oxygen	150
Mild steel	0.035	0.89	98% argon-2% oxygen	165
Mild steel	0.045	1.14	98% argon-2% oxygen	220
Mild steel	0.062	1.59	98% argon-2% oxygen	275
Stainless steel	0.035	0.89	99% argon-1% oxygen	170
Stainless steel	0.045	1.14	99% argon-1% oxygen	225
Stainless steel	0.062	1.59	99% argon-1% oxygen	285
Aluminum	0.030	0.76	Argon	95
Aluminum	0.045	1.14	Argon	135
Aluminum	0.062	1.59	Argon	180
Deoxidized copper	0.035	0.89	Argon	180
Deoxidized copper	0.045	1.14	Argon	210
Deoxidized copper	0.062	1.59	Argon	310
Silicon bronze	0.035	0.89	Argon	165
Silicon bronze	0.045	1.14	Argon	205
Silicon bronze	0.062	1.59	Argon	270

Note: Spray transfer will only occur when high percentages of argon or helium are used.

Fig. 17-7. Approximate transition current levels for various electrodes.   (The American Welding Society)

17-8. The peak current and background current alternate, with current flowing for a short time at each level. Less heat is applied to the base metal in pulsed spray transfer because the weld pool cools during the period of background current.

Peak current *pulses* (turns on and off) approximately sixty to several hundred times per second. Pulsed spray transfer allows for good penetration. It can be performed in all positions, since the weld pool has time to cool during the background current period.

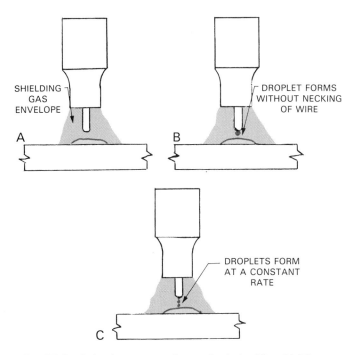

Fig. 17-8. Pulsed spray transfer method. A—The shielding gas envelope surrounds the weld area. B—Droplets form at the tip of the electrode without any necking down of the wire. C—Droplets form at a constant rate and are transferred to the weld crater.

GMAW EQUIPMENT

GMAW requires more equipment than shielded metal arc or gas tungsten arc welding. In addition to the power supply, a wire feeder and a *welding gun* are required. See Fig. 17-9. A complete GMAW setup includes the following:
• A welding machine.
• A wire feeder.
• A welding gun.
• Electrode wire.
• Shielding gas supply and controls.

GMAW POWER SUPPLY

A constant voltage, or *constant potential* power supply is used for gas metal arc welding. This type of power supply is different from one used for shielded metal arc or gas tungsten arc welding. A constant cur-

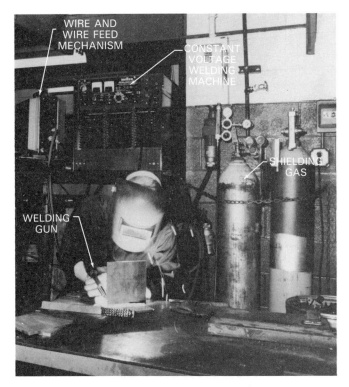

Fig. 17-9. A complete GMAW outfit. The shielding gas cylinders, wire feeder, and welding machine are behind the welder. (Miller Electric Mfg. Co.)

rent welding machine is used for SMAW or GTAW. The power supply can be a *transformer-rectifier,* or an *inverter.* A constant *voltage* power supply has a flat volt-amp curve or slope, Fig. 17-10. The voltage stays about the same, even though the current varies. The relationship between voltage and current is the slope of the power supply. Remember that a constant current power supply has a steep or drooping volt-amp curve.

A constant voltage power supply tries to maintain the preset voltage while welding. The power supply tries to maintain the same arc voltage and arc length,

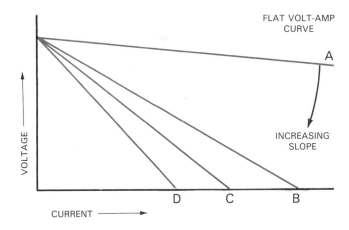

Fig. 17-10. Four power supply curves for a constant voltage welding machine. Increased slope reduces the maximum current during short circuiting transfer. A—A flat volt-amp curve. The short circuiting amperage is almost limitless. As the slope of curves B, C, and D increase, their maximum short circuiting amperage decreases.

even though the welder may try to change it. The welding machine makes the necessary adjustments very quickly to compensate for any changes that the welder makes.

A welding machine for GMAW is considered to be self-adjusting, since it automatically makes adjustments. If the welding gun is *moved closer* to the base metal while welding, the arc length shortens and the voltage decreases. The power supply will automatically increase the current so that the electrode melts faster. This, in turn, increases the arc length. When the electrode burns back to the original arc length and original arc voltage, then the current returns to its original setting or value.

If the welding gun is *pulled away* from the base metal, both the arc length and voltage increase. The power supply automatically reduces the current. Less metal will melt off the electrode. This reduces the arc length. Since the electrode now melts more slowly, the arc becomes shorter and the arc voltage decreases. When the correct arc length and voltage are obtained, the power supply will return the current to the original amount.

A crank or knob is used to adjust the voltage on a GMAW machine. See Fig. 17-11. There is no adjustment for current. Current is controlled by adjusting the wire feed speed. A faster wire feed speed results in higher current.

On some machines, the welder can adjust the amount of inductance. Inductance controls how fast the welding machine increases the welding current. Increasing inductance helps to reduce spatter during short circuiting transfer. A certain amount of inductance provides for a better arc start during spray transfer.

On some machines, the slope of the volt-amp curve can be changed. *Slope adjustment* is used to change the maximum short circuit current. The maximum current decreases as the slope is increased. See Fig. 17-10. An increased slope reduces spatter during short circuiting transfer.

A welding machine also includes connections for the wire feeder, shielding gas, water (when used), and ground cable. Some machines have a built-in wire feeder, so they do not have wire feeder connections. See Fig. 17-12. They do, however, have a place to attach a welding gun. Some machines have a voltmeter and an ammeter. Machines used for pulsed spray transfer have additional adjustments to control the peak current and background current.

GMAW or FCAW machines have a 60% or 100% duty cycle. Most machines used in schools and industry are 220 V single-phase or 220 V three-phase machines. Some small, light-duty welding machines used in auto body repair are 115 V machines.

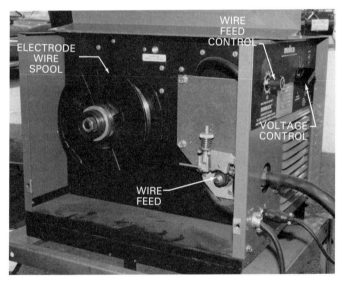

Fig. 17-12. The wire feeder and electrode wire spool holder are built into this GMAW machine.

WIRE FEEDERS

A *wire feeder* is used to feed electrode wire to the welding gun. GMAW and FCAW electrode wire is available in coils or spools. The coils or spools generally contain several hundred feet of wire. The wire feeder smoothly pulls the wire from the spool and pushes it to the welding gun. The gun is attached to the wire feeder.

Wire feed speed is set on the wire feeder by the welder. A DC motor attached to the knob controls the speed of the drive rolls. A wire feeder has two or four drive rolls that contact the wire. The *drive rolls,* or drive wheels, push the wire to the gun at a preset rate. The amount of pressure that each pair of rolls or wheels

Fig. 17-11. Voltage adjustments. A knob or a crank on the face of the welding machine is used to adjust voltage.

applies to the wire is adjustable. A *wire tension control knob* or screw is used to adjust the pressure. Fig. 17-13 shows a wire feeder with two drive rolls.

On a wire feeder that has two drive rolls, only one drive roll is driven. On a wire feeder with four rolls, two rolls are driven. A center drive gear engages two drive rolls, which in turn drive the remaining two rolls. The roll or drive gear is powered by a DC motor.

Two other controls commonly found on a wire feeder are an inch switch and a purge switch. Pressing the *inch switch* causes the electrode to feed out. The electrode will feed slowly as long as the inch switch is pressed. It is often used when loading a new coil or spool of wire. The *purge switch* is used to manually control the flow of shielding gas. When the purge switch is pressed, shielding gas will flow. It is used to set the flow rate of the shielding gas and to purge any air in the hoses. See Fig. 17-14.

WELDING GUNS AND CABLES

A GMAW or FCAW gun must perform the following functions:
1. Make electrical contact with the electrode.
2. Direct the shielding gas to the workpiece (not used in all FCAW).
3. Remove heat from the gun.
4. Have a switch to start and stop the welding operation.

Electrical contact is made between the welding gun and an electrode by a *contact tube.* A contact tube is made of copper alloy. It is screwed into the welding gun. See Fig. 17-15. The electrode wire passes through the contact tube, making electrical contact. The hole in the center of the contact tube becomes worn as the electrode passes through it. When the hole becomes

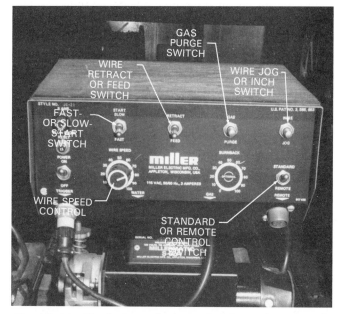

Fig. 17-14. *The inch or wire jog switch is used to move the electrode wire slowly through the cable to the welding gun. The gas purge switch is used to allow shielding gas to flow through the hose to remove any trapped air.*

Fig. 17-15. *Several different sizes of copper contact tubes. The contact tube screws into the contact tube adapter (object at left).*

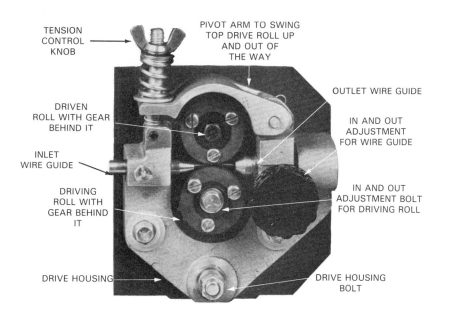

Fig. 17-13. *A wire feeder that uses one driving roll and one driven roll. The upper drive roll can be pivoted out of the way when the pressure adjusting wing nut is loosened.* (Miller Electric Mfg. Co.)

larger it does not make good electrical contact with the electrode. Contact tubes must be replaced when they become worn.

A *nozzle* is used to direct the shielding gas to the weld area. Nozzles are usually made of metal and come in various shapes and diameters. A nozzle is used for FCAW when shielding gas is used. When a shielding gas is not used, the nozzle is not required. A FCAW gun without a nozzle is shown in Fig. 17-16.

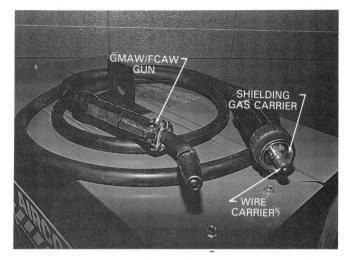

Fig. 17-17. A combination cable used with a gas-cooled GMAW gun. A combination cable carries both the electrode wire and shielding gas to the welding gun.

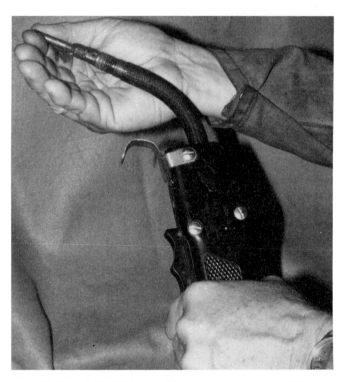

Fig. 17-16. A GMAW gun with the nozzle removed. The nozzle is not required for FCAW when a shielding gas is not used.

GMAW guns can be cooled by using shielding gas or water. Shielding gas-cooled guns are usually limited to 200 A (amps). These guns can be up to 600 A when CO_2 is used as a shielding gas. A gas-cooled gun may have a combination cable assembly or a cable assembly and a hose. A combination cable assembly on a gun carries the electricity, electrode, and shielding gas. See Fig. 17-17. For guns that use a cable assembly and a hose, the hose carries the shielding gas. The cable assembly carries the electricity and the electrode.

A water-cooled gun has two hoses and a cable assembly. One hose carries the water to the gun and the other hose carries the shielding gas. The cable assembly carries the electricity and electrode to the gun, and carries cooling water away from the gun. See Fig. 17-18.

The cable assembly that the electrode wire travels through has a liner in it. See Fig. 17-19. The *liner* keeps the electrode traveling smoothly through the cable. A liner is generally a flexible tube made of coiled wire.

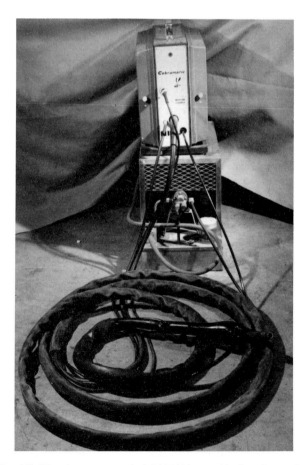

Fig. 17-18. A water-cooled GMAW gun. Notice the various cables and hoses used for electricity, water, and shielding gas.

When welding with soft electrodes, such as aluminum, a *Teflon® liner* is used.

All GMAW and FCAW guns have switches to start and stop the welding operation. When the switch is pressed, the electrode wire begins to feed. Shielding gas, electrical current, and water (when used) are also

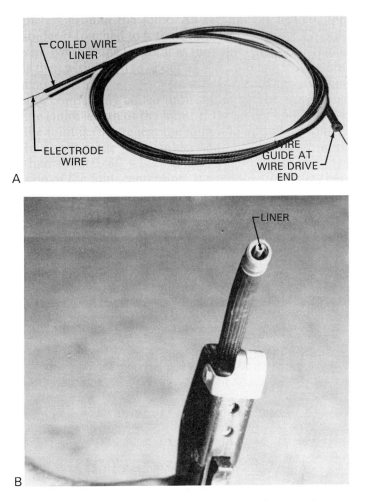

A

B

Fig. 17-19. A cable liner is used to carry the electrode through the combination cable to the welding gun. A— The wire guide used in the wire drive. B— The liner within the welding gun.

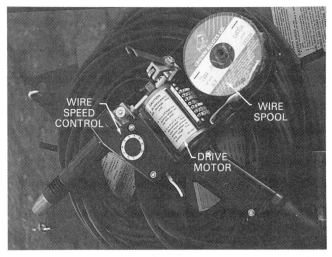

Fig. 17-20. The wire drive motor and a small electrode spool are built into this GMAW gun.

turned on when the switch is pressed. When the switch is released, the wire and current stop immediately. Shielding gas and water may continue to flow for a short time to cool the gun and continue shielding the weld and weld area.

Welding guns are available in a variety of shapes, as shown in Figs. 17-16, 17-17, 17-18, and 17-20. Welding guns that are commonly used for steel electrodes and large diameter electrodes have the electrode pushed through them by the wire feeder. However, the electrode may not travel smoothly through the electrode cable when welding a long distance from the wire feeder, or when using a small soft electrode, such as aluminum. The electrode wire may get bent and jam the machine. A pull-type gun is used to prevent this. A pull-type gun has a motor in it. The motor pulls the wire through the cable and prevents the electrode from being bent. A wire feeder is used to push the electrode while the welding gun pulls the electrode. This is a *push-pull setup.*

Some welding guns allow for a small spool of wire to be placed in the gun. The gun has a motor in it to feed the electrode. See Fig. 17-20. It is almost impossi-

ble to kink or jam the wire with this type of gun. The gun can be easier to handle, but it does weigh more than a standard gun. This is due to the weight of the motor and electrode wire.

SHIELDING GASES

Shielding gases used for GMAW include argon, helium, and carbon dioxide. Mixtures of these gases may also be used. *Argon* and *helium* are inert gases. *Inert gases* do not combine with the weld metal. *Carbon dioxide* and oxygen are active gases. *Active gases* will combine with the weld metal if given the chance. If carbon dioxide or oxygen enters the weld, the weld will become oxidized. Precautions are taken to prevent oxidation. Carbon dioxide and oxygen are only used when welding on ferrous metals (various steels, including stainless). The type of shielding gas selected for a job will affect the following:

1. Type of metal transfer.
2. Penetration and shape of weld bead.
3. Speed of welding.
4. Mechanical properties of the weld.

Argon is the most common shielding gas for GMAW. It is used to weld almost all types of metals with the spray transfer method. Helium or oxygen may be added. When argon is used in spray transfer to weld steel, the weld has very deep penetration in the center. The edges, however, are undercut. Oxygen or CO_2 is added to eliminate undercutting. The penetration is also reduced. There is very little spatter. Typical mixtures consist of argon with $1\% - 5\%$ oxygen or argon with $3\% - 50\%$ carbon dioxide (CO_2). Remember, spray transfer requires at least 90% argon. Fig. 17-21 shows the amount of penetration obtained using various shielding gases. See Chapter 18 (Figs. 18-11 and 18-12) for a list of the shielding gases that are recommended for welding on different base metals.

175

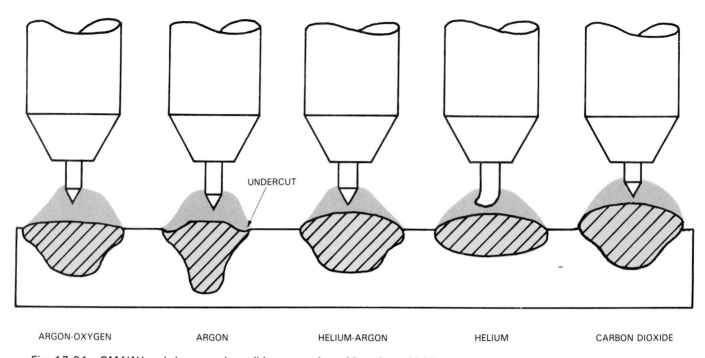

ARGON-OXYGEN   ARGON   HELIUM-ARGON   HELIUM   CARBON DIOXIDE

Fig. 17-21. GMAW bead shapes and possible penetration with various shielding gases and gas combinations. These shapes are typical for welds made with DCEP (DCRP). Notice the deep penetration and undercut when using pure argon.

Argon and carbon dioxide (CO_2) displace oxygen. Excellent ventilation is required to remove the shielding gases and supply fresh air when welding in a closed area. Suffocation can result if adequate ventilation is not provided.

CO_2 is the most common shielding gas used for FCAW. This provides globular transfer of the electrode. A common mixture is 75% argon and 25% CO_2, which allows metal transfer that approaches spray transfer. The shielding gas that is selected is often determined by the type of electrode being used. Remember, however, that not all FCAW electrodes require a shielding gas.

REGULATORS AND FLOW METERS

Shielding gases are available in cylinders similar to those used for fuel gases. The gases in the cylinders are at very high pressures. Care must be taken when handling or moving any gas cylinder. The cylinder valve must be protected by a properly installed safety cap when moving a cylinder or whenever the cylinder is not in use.

Secure a cylinder to a hand truck when moving it. If a hand truck is not available, tilt the cylinder and roll it along on the bottom edge. Use one hand to support the cylinder. Use the other hand to rotate the cylinder. Cylinders should be stored in the upright position. Each cylinder must be secured to a wall, column, or hand truck using chains or straps.

Shielding gases in cylinders have pressures that may exceed 2000 psi (13.79 MPa). A regulator is used to

reduce the cylinder pressure to a much lower working pressure. A regulator used for oxyfuel gas welding often has a high-pressure and a low-pressure gauge. These gauges indicate the cylinder pressure of the gas and the working pressure.

When using a shielding gas it is important to control the *flow* of gas, rather than the *pressure* of the gas. A combination regulator and *flow meter* is used to regulate the pressure and control the gas flow. See Fig. 17-22. The regulator reduces the pressure of the shielding gas to 50 psi (345 kPa). A high-pressure gauge

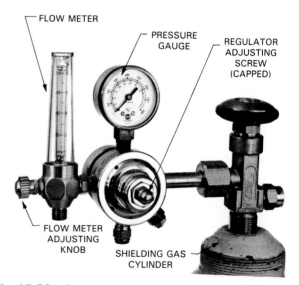

Fig. 17-22. An argon pressure regulator and flow meter. Notice that the regulator adjusting screw is capped and not readily adjustable.

may be used, but is not required. A knob or adjusting screw on the flow meter controls the flow of shielding gas.

One type of flow meter has a ball enclosed in a clear plastic or glass tube. The tube is calibrated to indicate the flow of shielding gas in cubic feet per hour (cfh) or liters per minute (Lpm). Gas flows into the tube from the bottom. As the gas flows, it lifts the ball. A larger gas flow is indicated by the ball rising higher in the tube. Shielding gas flow is indicated by the top of the ball. The flow rate is controlled by adjusting a knob on the flow meter. The correct flow rate for each welding job varies and is covered in Chapter 19. A different type of flow meter must be used for each type of shielding gas. An argon flow meter is calibrated differently from one for helium. In some school and industrial settings, the shielding gas is piped to the welding area using a manifold system. The pressure of this shielding gas has already been reduced to working pressure. A welding station does not need a regulator, but it does require its own flow meter.

PROTECTIVE CLOTHING AND EQUIPMENT

The same type of protective clothing should be worn for GMAW as is worn for SMAW. Fire-resistant or leather clothing must be worn. Your pant legs should not have cuffs. Pockets on your shirt should be buttoned to prevent spatter from getting caught in them. A #12 lens is recommended for your welding helmet. A hat and leathers must be worn when welding out of position. Leather gloves capable of withstanding high temperatures should be worn for GMAW or FCAW.

A small amount of fumes is generated by the GMAW operation. A large amount of fumes is created when performing FCAW. When welding in an enclosed area, some type of fume exhaust should be used. Fig. 17-23

shows an overhead fume exhaust. It should be positioned to capture and remove as much of the fumes as possible. The fume exhaust should be placed over and slightly to the rear of the welding area. The exhaust should pull the fumes away from your face. Fig. 17-24 shows a type of fume remover that is attached to the welding gun.

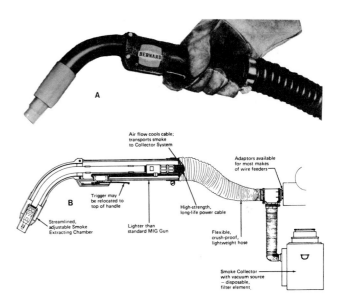

Fig. 17-24. A fume extractor built into a GMAW gun. The extractor is connected to a smoke collector and filter system. A—GMAW gun. B—A complete smoke and fume extracting system.   (Bernard Div./Dover Corp.)

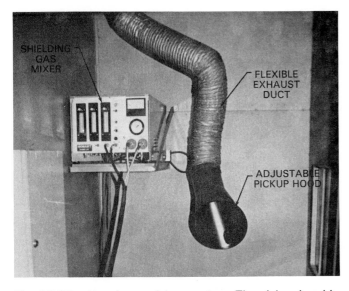

Fig. 17-23. An exhaust pickup system. The pickup hood in this exhaust system can be located near the weld due to the flexible exhaust duct.

REVIEW QUESTIONS

Write your answers on a separate sheet of paper.
1. What does GMAW stand for? Can the terms GMAW and MIG be interchanged? Explain.
2. In gas metal arc welding, the arc is struck between the _____ and the _____.
3. Flux and shielding gas are used to protect the weld from the atmosphere during GMAW. True or False?
4. What type of welding machine is used for GMAW, a constant voltage machine or a constant current machine?
5. List four advantages and two disadvantages of GMAW as compared to SMAW.
6. List the equipment that makes up a complete GMAW outfit.
7. In short circuiting transfer, when does the metal from the electrode enter the weld?
8. What forces the molten metal to leave the electrode in globular transfer?
9. Spray transfer will only occur when at least _____% argon shielding gas is used.
10. During pulsed spray transfer, molten metal is transferred to the base metal during the _____.

current cycle. This current is above the _____ current.

11. Which GMAW metal transfer methods are recommended for out-of-position welding?

12. The volt-amp curve for a constant potential machine is a flat curve. True or false?

13. Will the current of the welding machine increase or decrease if the arc voltage increases while welding?

14. List two adjustments you can make to prevent the metal from exploding off the electrode when using short circuiting transfer.

15. What part of a wire feeder is used to pull the electrode wire off the spool and push it to the gun?

16. Which of the following is not a function of a GMAW gun?
   a. Make electrical contact.
   b. Direct the shielding gas.
   c. Remove heat.
   d. Maintain the correct voltage.
   e. Control the welding with a switch.

17. How many hoses does a water-cooled welding gun have? What does each hose carry?

18. List three shielding gases commonly used in GMAW.

19. List two gases that are added to argon to eliminate the undercutting that occurs when pure argon is used to weld steel.

20. Why is a flow meter used for GMAW and not a low-pressure gauge?

A firetruck's water tank being welded with GMAW equipment. The water cooling system and electrode wire feeder are mounted on top of the welding machine. The welder must set the controls on all of the equipment.   (Miller Electric Mfg. Co.)

Chapter 18

Gas Metal Arc Welding: Assembly and Adjustment

After studying this chapter, you will be able to:
☐ Assemble a GMAW welding outfit.
☐ Adjust the drive mechanism for the proper pressure and alignment.
☐ List the proper sequence for removing a bird's nest.
☐ Adjust the shielding gas flow meter for the proper pressure and flow rate.
☐ Identify the electrode wire designation for GMAW electrodes.
☐ Identify the two adjustments that are made to the welding machine for GMAW.
☐ Identify safety precautions for GMAW and FCAW.

TECHNICAL TERMS

Bird's nest, dash number, open circuit voltage, post flow adjustment, pressure roll, remote contactor control.

ASSEMBLY AND SETUP

GMAW or FCAW outfits must be assembled properly to operate properly. Shielding gas must be connected to the welding machine and gas flow must be set correctly. The wire feeder must be connected to the machine. It should be adjusted so the electrode wire feeds properly to the welding gun. Also, the welding gun and its parts must be set up properly.

CYLINDER, REGULATOR, AND FLOW METER

Shielding gas is stored in cylinders at very high pressures. Care must be taken when handling cylinders so they are not dropped or damaged. A cylinder must always be secured to a wall, column, or hand truck. A safety cap should be secured over the cylinder valve when the cylinder is not in use. See Chapter 4 for additional information regarding the handling of cylinders.

The outlet on the cylinder must be cleaned before attaching a regulator to it. This is done by quickly opening and closing the cylinder valve. The cylinder valve must not be pointed at any workers in the area. The pressure of the gas escaping from the cylinder cleans the outlet. The regulator is then attached to the cylinder. Secure the regulator nut using an open-end wrench, so it is tight and leakproof. However, do not overtighten the nut.

A flow meter is used for GMAW, instead of the low-pressure gauge used in oxyfuel gas welding. A ball-float flow meter must be installed in a vertical position to operate properly. See Fig. 18-1.

Shielding gas can also be supplied through a manifold system. Shielding gas cylinders are stored in a different area from the welding area. Pressure from the cylinders is reduced by a regulator. Delivery pipes carry the low-pressure shielding gas to the welding area. Each welding station requires a flow meter so that each welder can set the desired gas flow rate.

WELDING MACHINE

A constant voltage welding machine is used for GMAW. The machine is either plugged into an electrical outlet or hard-wired (permanently wired) into the electrical source.

Shielding gas is connected to either the welding machine or the wire feeder. Connect the hose coming

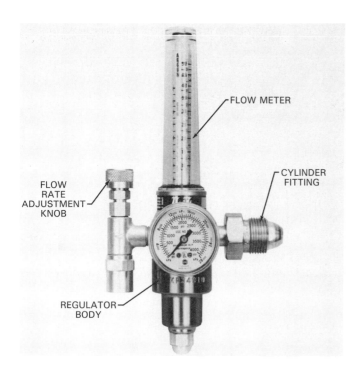

Fig. 18-1. A ball-float flow meter. A ball rises in the inner tube when gas is flowing. (Victor Equipment Co.)

from the flow meter to the proper place on the welding machine or wire feeder.

When a separate wire feeder is used, it is usually placed on top of the welding machine. The wire feeder is then connected to the welding machine. Connect the wire feeder to the *remote contactor control* on the welding machine, using the correct cable. See Fig. 18-2. The wire feeder requires 115 V or 230 V AC, or 24 V

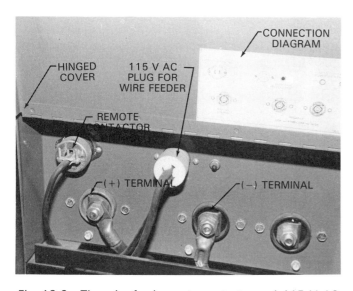

Fig. 18-2. The wire feed remote contactor and 115 V AC power supply cord are plugged into the back of the welding machine. The connections are located under a hinged panel. Notice the wire connection diagram on the panel cover.

DC, for operation. Connect the power cord to the front of the welding machine or to an electrical outlet with the required voltage.

The welding gun is then connected to the wire feeder. The gun may be connected to the welding machine if the wire feeder is inside of the welding machine. If the welding gun has a combination hose, connect it to the welding machine. If there is an additional shielding gas hose, connect it to its proper fitting on the welding machine or wire feeder. Other welding guns have multiple connections. Be careful to connect all hoses and cables properly. See Fig. 18-3.

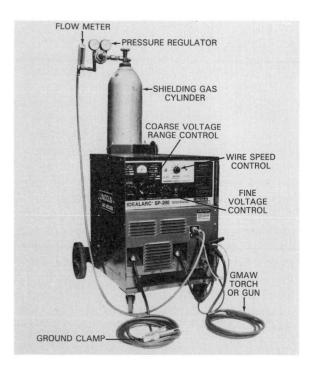

Fig. 18-3. A constant voltage GMAW machine. The welding gun is connected to the wire feeder through the front of the machine. Notice that the shielding gas is fed into the front of the machine. It is carried through the combination hose/cable to the welding gun. (The Lincoln Electric Co.)

WIRE FEEDER

A wire feeder should pull the electrode wire smoothly from the spool and push it to the welding gun. Several adjustments must be made on the wire feeder to ensure proper operation.

First, select the type of electrode wire to be used. Mount the spool of electrode wire on the wire spool hub. Secure the spool on the hub so it does not fall off during operation. A lock pin is generally used to keep the spool in place. See Fig. 18-4A. Next, check that the grooves in the drive rolls are the correct size for the diameter of the electrode wire. Some drive rolls have two grooves. See Fig. 18-4B. Install the proper size drive rolls so that the correct size groove aligns with the electrode wire.

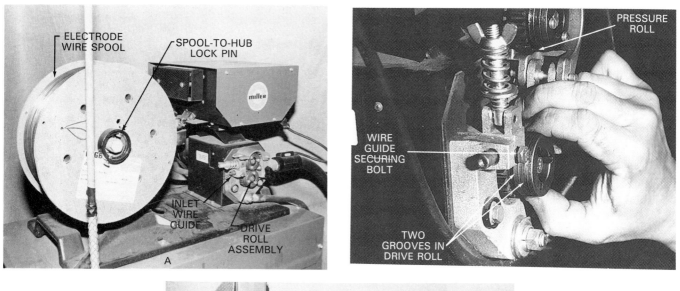

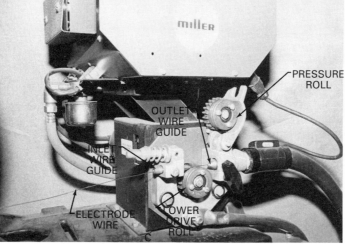

Fig. 18-4. A—The electrode wire spool is held on the wire feed hub by two spring-loaded lock pins. The spool may be replaced by pulling these two pins. B—Two wire grooves are cut in both the pressure and drive rolls. Each groove is a different size. C—Feeding the electrode wire through the wire guides in the wire drive. Swing the pressure roll up and feed the electrode through the inlet and outlet wire guides.

Swing the **pressure roll** (top drive roll) up out of the way. Push the end of the electrode wire through the inlet wire guide directly behind the drive rolls. See Fig. 18-4C. Feed the wire into the grooves of the drive roll, then through the outlet wire guide directly in front of the drive roll. Swing the pressure roll back into position to complete the assembly.

A wire feeder that has four drive rolls is assembled and connected in the same way as a two-roll wire feeder. The only difference is that there is a third guide tube between the two sets of rolls. The electrode wire must pass through this guide tube. See Fig. 18-5.

After the electrode wire has been inserted, look closely at it where it passes between the drive and pressure rolls. The wire should travel in a straight line between the inlet wire guide to the outlet wire guide, Fig. 18-6. If the drive rolls and guide tubes are not properly adjusted, the wire will not feed smoothly. Adjust the drive housing if the wire bends up or down as it passes bet-

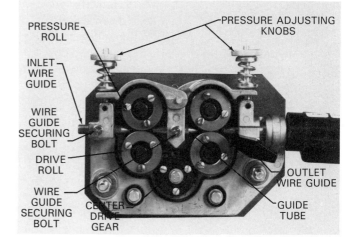

Fig. 18-5. Two drive rolls and two pressure rolls are used in this wire drive mechanism. The center drive gear moves the drive rolls. Note that three wire guides are used.
(Miller Electric Mfg. Co.)

181

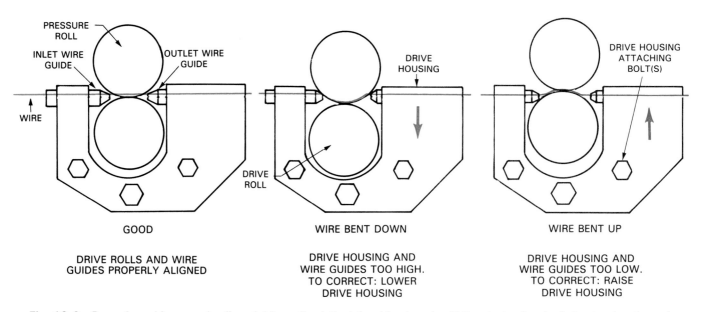

GOOD

DRIVE ROLLS AND WIRE
GUIDES PROPERLY ALIGNED

WIRE BENT DOWN

DRIVE HOUSING AND
WIRE GUIDES TOO HIGH.
TO CORRECT: LOWER
DRIVE HOUSING

WIRE BENT UP

DRIVE HOUSING AND
WIRE GUIDES TOO LOW.
TO CORRECT: RAISE
DRIVE HOUSING

Fig. 18-6. Properly and improperly aligned drive rolls. Adjust the drive housing if the electrode wire is bent going through the pressure and drive rolls. To do so, loosen the drive housing bolts, align the rolls, and retighten the bolts.

ween the drive rolls. The grooves in the drive and pressure rolls must also be properly aligned. The drive roll on most wire feeders is adjustable. See Fig. 18-7.

Once the rolls have been properly adjusted and aligned, the correct amount of pressure must be applied to the electrode wire. The drive and pressure rolls exert this pressure. If not enough pressure is applied,

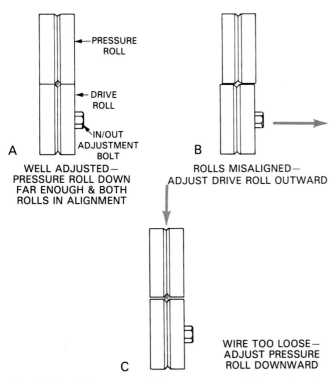

Fig. 18-7. Adjustment of the pressure and drive rolls. The pressure roll is adjusted up and down by means of the pressure adjusting wing nut or knob. The lower drive roll is adjusted in and out by means of an adjustment bolt in the center of the drive roll.

the drive rolls will slip and the electrode wire will not be fed smoothly to the gun. If excessive pressure is applied, the rolls will flatten the electrode wire. The knobs on top of the drive housing are used to adjust the pressure exerted on the electrode wire. Refer to Figs. 18-5 and 18-7.

When all adjustments have been made to the wire feeder, pull the trigger on the welding gun or press the inch switch on the welding machine or wire feeder. Stop feeding electrode wire when it is visible at the end of the gun. Electrode wire will feed from the wire feeder to the gun whenever the trigger is pressed.

Removing a Bird's Nest

Occasionally, the electrode wire in the wire feeder will become tangled, and produce a *bird's nest*. See Fig. 18-8. A bird's nest commonly results from improper feeding as the wire feeder tries to push the wire through the cable to the welding gun. A blockage in the cable,

Fig. 18-8. A bird's nest in the wire feed mechanism.

misaligned guide tubes and rolls, or stubbing the electrode wire on the base metal causes the electrode wire to bunch up in the wire feeder. Stubbing the electrode on the base metal is usually a result of one of the following:

1. Voltage setting on the machine is too low.
2. Wire feed speed is too fast.
3. Welding gun is held too close to the work when starting an arc.

A bird's nest is removed by raising the pressure roll and cutting the wire on both sides of it. See Fig. 18-9. Put your hand over the wire as you cut to prevent it from flying and causing injury. Pull all the electrode wire out of the cable connected to the gun. Check the wire coming from the spool to make sure it is not bent or twisted. Remove any bent or twisted wire.

Feed new electrode wire through the wire guides and into the cable connected to the welding gun. Secure the pressure roll into position. Pull the gun trigger or push the inch switch to feed the electrode wire to the gun. When the wire comes out of the contact tube, you are ready to begin welding again.

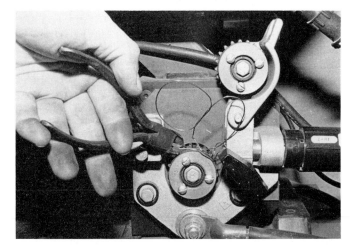

Fig. 18-9. Removing a bird's nest by cutting the electrode wire behind the inlet wire guide and before the outlet wire guide.

WELDING GUN

The welding gun and cable assembly have many parts, as discussed in Chapter 17. Some parts may need to be replaced or cleaned due to wear and build-up of debris. The contact tube will need to be changed most often. Select the correct size contact tube for the diameter electrode you will use. The contact tube is threaded into the end of the welding gun. The hole in the center of the contact tube will get larger due to wear. Eight hours of constant welding with a steel electrode will wear a contact tube so that it will need to be replaced. Fig. 18-10 shows a worn contact tube and several usable contact tubes.

Fig. 18-10. Four electrode contact tubes. The second one from the left has been worn by the wire and must be replaced.

After installing the proper contact tube, place a nozzle on the end of the gun. The position of the nozzle may be adjustable.

Metal may spatter and get into the nozzle and on the contact tube while welding. This spatter must be removed. Special tools are available to remove the spatter, but a small wire brush will usually do the job. Anti-stick compounds that help prevent spatter from sticking to the nozzle and contact tube may also be applied. These compounds make it easier to remove spatter that does stick to the tube and nozzle.

SELECTING THE SHIELDING GAS

Shielding gases are chosen after considering a number of factors. Two important factors to consider are the type of base metal being welded and the metal transfer method to be used. Refer to Chapter 17 for information regarding these two factors. Figs. 18-11 and 18-12 list the recommended shielding gases to use for various base metals.

ADJUSTING THE SHIELDING GAS FLOW METER

The shielding gas flow rate is set after the regulator and flow meter have been attached to the cylinder and the wire feeder and welding gun are connected. An adjusting screw or adjustment knob is used to set the shielding gas flow rate.

Open the cylinder valve before setting or adjusting the shielding gas flow rate. Turn the adjusting screw or knob on the flow meter so no gas flows. The screw or knob is usually turned counterclockwise. Then, slowly open the cylinder valve. If the flow meter is open when the cylinder valve is opened, high pressure from the gas in the cylinder may damage it.

Set the flow rate after opening the cylinder valve. Shielding gas must be flowing when adjusting the flow meter. Press the purge switch if the welding machine or wire feeder has one. This causes the gas to flow. Another way to start gas flowing is to set the *post flow adjustment* for 15—20 seconds and pull the gun trigger for one second or less. Do not strike an arc. Shielding gas will flow for a while so the flow rate can be adjusted.

METAL	SHIELDING GAS	ADVANTAGES
Aluminum, copper, magnesium, nickel, and their alloys	Argon and argon-helium	Argon satisfactory on sheet metal; argon-helium preferred on thicker sheet metal.
Steel, carbon	Argon-20-25% CO_2	Less than 1/8 in. (3.2 mm) thick; high welding speeds without melt-through; minimum distortion and spatter; good penetration.
	Argon-50% CO_2	Greater than 1/8 in. (3.2 mm) thick; minimum spatter; clean weld appearances; good weld pool control in vertical and overhead positions.
	CO_2[a]	Deeper penetration; faster welding speeds; minimum cost.
Steel, low alloy	60-70% Helium-25-35% argon-4-5% CO_2	Minimum reactivity; good toughness; excellent arc stability, wetting characteristics, and bead contour; little spatter.
	Argon-20-25% CO_2	Fair toughness; excellent arc stability; wetting characteristics, and bead contour; little spatter.
Steel, stainless	90% Helium-7.5% argon-2.5% CO_2	No effect on corrosion resistance; small heat-affected zone; no undercutting; minimum distortion; good arc stability.
a - CO_2 is used with globular transfer also.		

Fig. 18-11. Suggested gases and gas mixtures for use in GMAW short circuiting transfer.

METAL	SHIELDING GAS	ADVANTAGES
Aluminum	Argon	0.1 in. (0.25 mm) thick; best metal transfer and arc stability; least spatter.
	75% Helium-25% argon	1-3 in. (25-76 mm) thick; higher heat input than argon.
	90% Helium-10% argon	3 in. (76 mm) thick; highest heat input; minimizes porosity.
Copper, nickel, & their alloys	Argon	Provides good wetting; good control of weld pool for thickness up to 1/8 in. (3.2 mm)
	Helium-argon	Higher heat inputs of 50 and 75% helium mixtures offset high heat conductivity of heavier gages.
Magnesium	Argon	Excellent cleaning action.
Reactive metals (Titanium, Zirconium, Tantalum)	Argon	Good arc stability; minimum weld contamination. Inert gas backing is required to prevent air contamination on back of weld area.
Steel, carbon	Argon-2-5% oxygen	Good arc stability; produces a more fluid and controllable weld pool; good coalescence and bead contour, minimizes undercutting; permits higher speeds, compared with argon.
Steel, low alloy	Argon-2% oxygen	Minimizes undercutting; provides good toughness.
Steel, stainless	Argon-1% oxygen	Good arc stability; produces a more fluid and controllable weld pool, good coalescence and bead contour, minimizes undercutting on heavier stainless steels.
	Argon-2% oxygen	Provides better arc stability, coalescence, and welding speed than 1% oxygen mixture for thinner stainless steel materials.

Fig. 18-12. Suggested gases and gas mixtures for use in GMAW spray transfer.

184

While gas is flowing, turn the adjusting screw or knob until the flow meter indicates the correct flow rate. The flow rate is indicated by the top of the ball on a ball-float flow meter, or by a needle on a gauge flow meter. See Fig. 18-13.

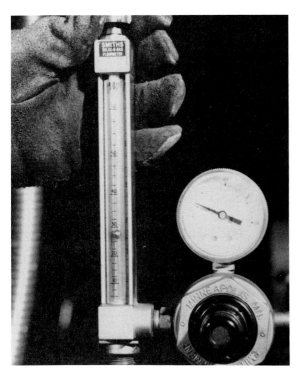

Fig. 18-13. A flow meter with gas flowing through it, raising the ball float in the vertical tube. The correct flow rate is obtained when the top of the ball lines up with the desired flow rate on the gauge.

SELECTING THE ELECTRODE WIRE

Choice of the electrode wire used for GMAW is based on a number of factors. They include:

1. Base metal composition.
2. Base metal properties.
3. Cleanliness of the base metal.
4. Shielding gas.
5. Metal transfer method.
6. Welding position.

Fig. 18-14A lists some GMAW electrodes commonly used for welding different base metals. Fig. 18-14B lists common FCAW electrodes. FCAW electrodes are limited to welding steel and stainless steel base metals. Fig. 18-14C identifies the letters and numbers used in the electrode designations.

Most flux-cored electrodes are designed to be used in the flat position or to make horizontal fillet welds. These electrodes are designated as E70T-X, E80T-X, E90T-X, etc. Electrodes designed to be used in all welding positions are designated as E71T-X, E81T-X, E91T-X, etc. The "1" indicates that the electrode is used for all positions.

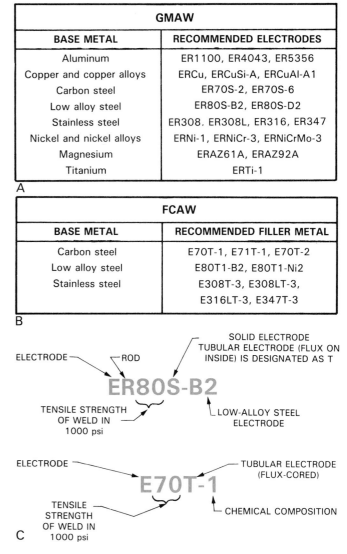

GMAW	
BASE METAL	**RECOMMENDED ELECTRODES**
Aluminum	ER1100, ER4043, ER5356
Copper and copper alloys	ERCu, ERCuSi-A, ERCuAl-A1
Carbon steel	ER70S-2, ER70S-6
Low alloy steel	ER80S-B2, ER80S-D2
Stainless steel	ER308. ER308L, ER316, ER347
Nickel and nickel alloys	ERNi-1, ERNiCr-3, ERNiCrMo-3
Magnesium	ERAZ61A, ERAZ92A
Titanium	ERTi-1

A

FCAW	
BASE METAL	**RECOMMENDED FILLER METAL**
Carbon steel	E70T-1, E71T-1, E70T-2
Low alloy steel	E80T1-B2, E80T1-Ni2
Stainless steel	E308T-3, E308LT-3,
	E316LT-3, E347T-3

B

Fig. 18-14. A—Commonly recommended GMAW electrodes for use with various base metals. There are many more electrodes to choose from. The electrode must be selected for the particular welding job. B—Commonly recommended FCAW filler metals for use with various base metals. C—Explanation of electrode designations.

ELECTRODE WIRE COMPOSITION

Many different alloy electrodes are available. An electrode wire that has a chemical composition similar to the base metal is commonly used, but there are important exceptions. The following AWS publications should be referenced for information on the electrodes listed in Fig. 18-14.

AWS A5.9 — Stainless steel electrodes for GMAW
AWS A5.10 — Aluminum alloy electrodes for GMAW
AWS A5.14 — Nickel alloy electrodes for GMAW
AWS A5.18 — Carbon steel electrodes for GMAW
AWS A5.19 — Magnesium alloy electrodes for GMAW
AWS A5.20 — Carbon steel electrodes for FCAW.
AWS A5.22 — Stainless steel electrodes for FCAW.

AWS A5.28 – Low alloy steel electrodes for GMAW
AWS A5.29 – Low alloy steel electrodes for FCAW

FCAW can be done with or without the addition of shielding gas. Electrodes used for welding with a shielding gas have a different composition than those that do not require shielding gas. Electrodes that have a flux core can produce some gases to protect the welding area. Flux also contains elements to prevent the weld metal from oxidizing.

Fig. 18-15 lists FCAW electrodes used for welding carbon steel. Each FCAW electrode has a *dash number* (numerals or letters at the end of the electrode designation, following a dash). The dash number indicates how the electrode is to be used. It also indicates if a shielding gas is required, and the current and polarity of the electrode. The dash number indicates whether the electrode should be used for single or multiple-pass welding. Electrodes that are used to make multiple-pass welds can also be used to make single pass welds. However, electrodes designed for single pass welds cannot be used to make multiple-pass welds.

AWS ELECTRODE CLASSIFICATION	EXTERNAL SHIELDING MEDIUM	CURRENT AND POLARITY
EXXT-1 (Multiple-pass)	CO_2	DC, electrode positive
EXXT-2 (Single-pass)	CO_2	DC, electrode positive
EXXT-3 (Single-pass)	None	DC, electrode positive
EXXT-4 (Multiple-pass)	None	DC, electrode positive
EXXT-5 (Multiple-pass)	CO_2	DC, electrode positive
EXXT-6 (Multiple-pass)	None	DC, electrode positive
EXXT-7 (Multiple-pass)	None	DC, electrode negative
EXXT-8 (Multiple-pass)	None	DC, electrode negative
EXXT-10 (Single-pass)	None	DC, electrode negative
EXXT-11 (Multiple-pass)	None	DC, electrode negative
EXXT-G (Multiple-pass)	a	a
EXXT-GS (Single-pass)	a	a
a. As agreed upon between supplier and user.		

Fig. 18-15. Recommended welding variables for FCAW electrodes used to weld carbon steel.

ELECTRODE WIRE DIAMETER

The diameter of electrode is selected after the type of electrode is chosen. Common electrode diameters range between .030″ – .0625″ (.76 mm – 1.59 mm). Large-diameter electrodes are not used for out-of-position welding. Small-diameter electrodes deposit large amounts of metal using spray transfer. Soft electrodes like aluminum and magnesium with small diameters cannot be pushed through a welding cable as far as a large-diameter electrode can. A push-pull wire feeder and welding gun must be used with soft, small-diameter electrodes. A Teflon® liner should also be used with soft electrodes.

WELDING MACHINE SETTINGS

Two adjustments must be made on the welding machine: voltage and wire feed speed. The arc length and current are determined by setting these two variables. The metal transfer method is dictated by these two variables, plus the type of shielding gas selected. The electrode diameter, electrode composition, and desired metal transfer method determine the proper settings. Figs. 18-16, 18-17, and 18-18 list the voltage and wire feed speeds required for welding various metals.

WELDING MACHINE SETTINGS FOR MILD AND LOW ALLOY STEEL				
SHORT CIRCUITING TRANSFER[1]				
Electrode Diameter		Wire Feed Speed in./min.	M/min.	Arc Voltage
in.	mm			
.030	.76	150-340	3.81-8.64	15-21
.035	.89	160-380	4.06-9.65	16-22
.045	1.14	100-220	2.54-5.59	17-22
SPRAY TRANSFER[2]				
in.	mm			
.030	.76	390-670	9.91-17.02	24-28
.035	.89	360-520	9.14-13.21	24-28
.045	1.14	210-390	5.33-9.91	24-30
1/16	1.59	150-360	3.81-9.14	24-32
3/32	2.38	75-125	1.91-3.17	24-32
[1]Values are based on using CO_2 for mild steel and argon – CO_2 for low alloy steel.				
[2]Values are based on using argon with 2% to 5% oxygen shielding gas.				

Fig. 18-16. Suggested arc welding machine settings for short circuiting and spray transfer with GMAW on mild and low alloy steel.

The voltage set on the welding machine is the open circuit voltage. The *open circuit voltage* is higher than the actual welding voltage. The voltage on the welding machine will have to be set higher than the voltage required for welding. The actual difference in voltages varies with the slope of each machine. Some machines have controls for slope and inductance. The functions of these controls are covered in Chapter 17.

PREPARING THE BASE METAL

Metal to be welded by GMAW and FCAW should be chemically or mechanically cleaned. A stainless steel brush is commonly used to mechanically clean the metal. Cleaning removes oxides, oil, and other debris.

The edges of some joints may require edge preparation, such as beveling. Refer to Chapter 3 for information on joints and edge preparation. The groove angle required for GMAW is less than the angle re-

WELDING MACHINE SETTINGS FOR SERIES 300 STAINLESS STEEL				
SHORT CIRCUITING TRANSFER[1]				
Electrode Diameter		Wire Feed Speed		Arc Voltage
in.	mm	in./min.	M/min.	
.030	.76	150-430	3.81-10.92	17-22
.035	.89	120-400	3.05-10.16	17-22
.045	1.14	100-240	2.54-6.10	17-22
SPRAY TRANSFER[2]				
in.	mm			
.030	.76	440-650	11.18-16.51	24-28
.035	.89	430-500	10.92-12.70	24-29
.045	1.14	220-400	5.59-15.16	24-30
1/16	1.59	110-210	2.79-5.33	24-32
3/32	2.38	50-80	1.27-2.03	24-32

[1]Values are based on shielding gas mixture of 90% helium, 7 1/2% argon, and 2 1/2% CO_2 with a flow rate of 20 cfh (566 Lpm).

[2]Values are based on a shielding gas mixture of argon and 1%—5% oxygen.

Fig. 18-17. Suggested arc welding machine settings for short circuiting and spray transfer GMAW on series 300 stainless steel.

WELDING MACHINE SETTINGS FOR ALUMINUM AND ALUMINUM ALLOYS				
SHORT CIRCUITING TRANSFER[1]				
Electrode Diameter		Wire Feed Speed		Arc Voltage
in.	mm	in./min.	M/min.	
.030	.76	300-580	7.61-14.73	15-18
.035	.89	250-450	6.35-11.43	17-19
3/64	1.19	200-350	5.08-8.89	16-20
SPRAY TRANSFER[1]				
in.	mm			
.030	.76	470-680	11.94-17.27	22-28
.035	.89	350-475	8.89-12.06	22-28
3/64	1.19	235-375	5.97-9.52	22-28
1/16	1.59	180-300	4.57-7.63	24-30
3/32	2.38	100-210	2.54-5.33	24-32

[1]Values based on the use of argon shielding gas.

Fig. 18-18. Suggested arc welding machine settings for short circuiting and spray transfer GMAW on aluminum and aluminum alloys.

quired for SMAW. See Fig. 18-19. A smaller groove angle requires less weld metal and less of the welder's time to fill the joint.

SAFETY PRECAUTIONS

GMAW and FCAW can result in spatter flying through the air. They also produce intense light. It is

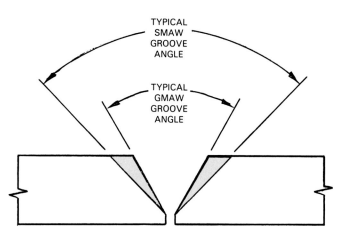

Fig. 18-19. Comparison of typical SMAW and GMAW groove angles. The red area represents the amount of weld metal that is saved. The smaller angle used in GMAW also saves a great deal of welding time.

important to wear proper clothing and eye protection. Refer to Chapter 17 for a discussion of proper protective clothing. It is also important to check that the welding equipment is assembled and operating properly. Each day prior to welding, check the following:

1. The welding gun, wire feeder, and welding machine should be properly connected.
2. The cables are in good condition and are properly connected to the machine and welding gun.
3. The regulator and flow meter are correctly installed on the cylinder.
4. The contact tube is in good condition.
5. An excessive amount of spatter is not built-up on the inside of the nozzle.
6. A #12 or darker filter lens is installed in your helmet.
7. Appropriate protective clothing and gloves are being worn.
8. Adequate ventilation for removing fumes and shielding gases is present.

SHUTTING DOWN A GMAW OR FCAW STATION

The proper sequence must be followed to properly shut down a GMAW or FCAW welding station. Damage to equipment and supplies can be avoided by following the proper sequence.

1. Adjust the flow meter so that shielding gas does not flow.
2. Close the cylinder valve by turning it clockwise or close the valve on the manifold.
3. Readjust the flow meter to allow shielding gas to flow.
4. Press the purge switch on the wire feeder to remove all shielding gas from the hose.
5. Turn the wire feeder off.
6. Turn the welding machine off.
7. Coil the cable connected to the welding gun, then the ground cable. Place cables where they will not get damaged.

REVIEW QUESTIONS

Write your answers on a separate sheet of paper.

1. When shielding gas is supplied to a welding station through a manifold system, what piece of equipment is used at the welding station to control the shielding gas?
2. When a separate wire feeder and welding machine are used, the cable from the wire feeder is plugged into the _____.
   a. Remote voltage receptacle.
   b. Remote contactor receptacle.
   c. Remote control receptacle.
   d. Flow control.
3. How is the pressure that the drive rolls exert on the electrode wire adjusted?
4. What part of the welding gun needs to be replaced most often?
   a. Nozzle.
   b. Liner.
   c. Switch.
   d. Contact tube.
5. During welding, spatter gets into the _____ and onto the _____ _____.
6. How is the correct shielding gas flow rate set on the flow meter?
7. Select the factors below that should be considered when selecting an electrode.
   a. Base metal composition.
   b. Manufacturer of the power supply.
   c. Base metal properties.
   d. Shielding gas used.
   e. Welding position.
   f. Flow rate of shielding gas.
   g. Cleanliness of the base metal.
   h. Type of metal transfer.
8. What do the letters and numbers in the electrode designation ER70S-2 represent?
9. The type of shielding gas recommended for welding low alloy steel using spray transfer is _____.
10. The type of shielding gas recommended for welding stainless steel using short circuiting transfer is _____.

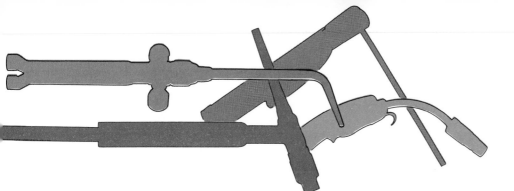

Chapter 19

GMAW: Flat Welding Position

After studying this chapter, you will be able to:
☐ Describe the GMAW process.
☐ Determine the appropriate electrode to use when GMAW or FCAW in the flat welding position.
☐ Identify the correct electrode extension to use when GMAW or FCAW using different metal transfer methods.
☐ Lay a bead on a plate using GMAW or FCAW.
☐ Make a fillet weld on a lap joint in the flat welding position.
☐ Make a fillet weld on a T-joint in the flat welding position.
☐ Weld a butt joint in the flat welding position.
☐ Describe how to weld aluminum using GMAW or FCAW.
☐ Identify various weld defects.

TECHNICAL TERMS

Backing, electrode extension, keyhole, pull gun, push-pull system, root pass.

GAS METAL ARC WELDING PRINCIPLES

In GMAW and FCAW, a welding arc is struck between the electrode and the base metal. The welder presses the trigger on the welding gun to start the process. Shielding gas begins to flow. Electrode wire feeds out of the gun and a welding arc is struck. The arc melts the base metal and forms a weld pool. Metal from the electrode melts and enters the weld pool. The gun is moved forward to keep the weld pool the correct size.

Welding continues until the welder releases the trigger. Electrode wire stops feeding; the arc also stops. Shielding gas flows for a short time after the welding stops to protect the weld and base metal as they cool.

PREPARING TO WELD

Prepare to begin GMAW or FCAW by setting up the equipment. Connect the welding machine, wire feeder, welding gun, and shielding gas as described in Chapter 18. Check that all electrical, shielding gas, and water connections are tight and leakproof. See Fig. 19-1. Also, make sure that you are wearing the proper protective clothing for welding.

Decide on the type of welding you will be doing after checking the equipment. Answer the following questions:
1. What type of base metal is to be welded?
2. How thick is the base metal?

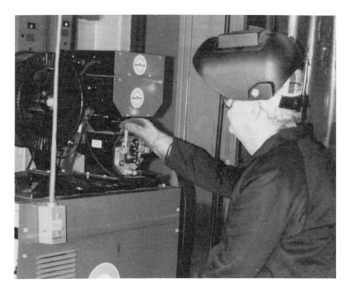

Fig. 19-1. A welder checking the welding machine and adjusting the wire feed tension.

3. In what position will the welding be done?

4. What type of metal transfer will be used?

Select an electrode recommended for the type of base metal you will be welding. Refer to Chapter 18 for information regarding the type and diameter of GMAW or FCAW electrodes to be used with the base metal. Install a spool of electrode wire on the hub of the wire feeder. Select a contact tube for the diameter of electrode wire you have chosen. Examine the contact tube. The hole in the center should not be worn or contain spatter. Install the contact tube into the gun.

Next, select the shielding gas and the flow rate recommended for the base metal you will be welding and the type of metal transfer to be used. Chapter 18 describes the type of shielding gas that should be used for various base metals and types of metal transfer. Fig. 19-2 lists recommended flow rates for shielding gases. Set the recommended flow rate on the flow meter as discussed in Chapter 18. Finally, set the voltage and wire feed speed recommended for the type of metal transfer you have selected. Chapter 18 lists recommended voltages and wire feed speeds. Remember that the voltage setting on the welding machine is the open circuit voltage. The open circuit voltage is always higher than the voltage used when welding. Allow for the difference in voltages by setting the voltage on the machine a few volts higher than the recommended welding voltage.

ELECTRODE EXTENSION

In GMAW and FCAW, the voltage set on the welding machine determines the arc length. The wire feed speed determines the current. The welder must control the electrode extension. *Electrode extension* or electrode stick out is the distance from the end of the contact tube to the end of the electrode. See Fig. 19-3. The electrode must extend beyond the contact tube to preheat the electrode wire. Some preheating is desirable. Shallow penetration of the electrode results from too much preheating. Proper electrode extension is different for GMAW than for FCAW, Fig. 19-4.

The contact tube usually cannot be seen when welding because it is inside the nozzle. Therefore, to control the electrode extension, you must control the distance from the nozzle-to-work distance. Hold the nozzle off the base metal to obtain the correct electrode extension.

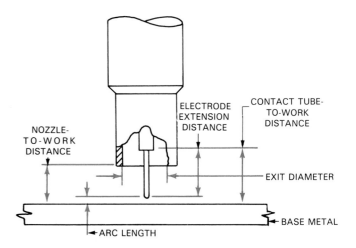

Fig. 19-3   Note the various distances when GMAW or FCAW. Notice the electrode extension distance.

METAL	TYPE JOINT	THICKNESS		WELD POSITION	ARGON FLOW	
		in.	mm		cfh	Lpm
Aluminum and Aluminum Alloys	All	1/16	1.6	F	25	11.8
		3/32	2.4	F,H,V,O	30	14.2
		1/8	3.1	F,H,V,O	30	14.2
		3/16	4.8	F,H,V,O	30	14.2
		1/4	6.4	F	40	18.9
				H,V	45	21.2
				O	60	28.3
		3/8	9.5	F	50	23.6
				H,V	55	26.0
				O	80	37.8
		3/4	19.0	F	60	28.3
				H,V,O	80	37.8
Steel and Stainless Steel	Butt	1/16	1.6		30	14.2
	Butt	1/8-3/16	3.2-4.8		(98Ar-2O₂) 35	16.5
	60° Bevel	1/4-1/2	6.4-12.7		35	16.5
	60° Double Bevel	1/2-5/8	12.7-15.9		35	16.5
	Lap, 90° Fillet	1/8-5/16	3.2-7.9		35	16.5
Nickel and Nickel Alloys	All	Up to 3/8	Up to 9.5		25	11.8
Magnesium	Butt	.025-.190	.6-4.8		40-60	18.9-28.3
		.250-1.000	6.4-25.4		50-80	23.6-37.8

Fig. 19-2. Suggested shielding gas flow rates for use with various metals and thicknesses.

PROCESS	METHOD	ELECTRODE EXTENSION
GMAW	Short circuiting	1/4″ – 1/2″ (6 mm – 13 mm)
GMAW	Spray transfer	1/2″ – 1″ (13 mm – 25mm)
FCAW	Gas-shielding	3/4″ – 1 1/2″ (19 mm – 38 mm)
FCAW	Self-shielding	3/4″ – 3 3/4″ (19 mm – 95 mm)

Fig. 19-4. Proper electrode extension for GMAW and FCAW, using different metal transfer methods.

HOLDING THE WELDING GUN

Pick up and hold the welding gun. Use a grip that is comfortable, so you can weld for long periods of time without fatigue. Hold the gun so your finger or thumb (depending on the type of gun) is on the switch. Find a position that is comfortable. You may want to place the cable across your arm. See Fig. 19-5.

Press the inch switch on the wire feeder or press the trigger on the welding gun until the electrode sticks out about 2″. Using a pair of wire cutters, cut the electrode wire to obtain the proper electrode extension. Caution: Place one hand over the electrode wire as it is cut to prevent the excess piece from flying through the air. The piece of electrode wire could hit someone in the eye resulting in temporary or permanent damage.

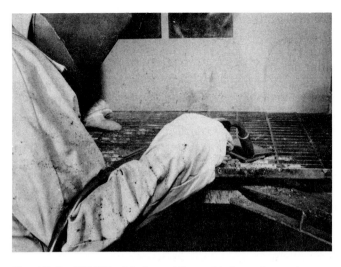

Fig. 19-5. GMAW with the welding cable looped over the arm. This position reduces the weight and makes it more comfortable to hold the gun.

LAYING A BEAD

Welding is performed using a forehand or backhand method, or with the torch held vertically. Better penetration is obtained with the backhand method. When using the backhand method, hold the gun at approximately a 25° angle. See Fig. 19-6. The backhand method also has a more stable arc and produces less spatter than the other two methods.

Position the gun over the area where you will begin welding. Tilt the gun about 20° – 25° in the backhand

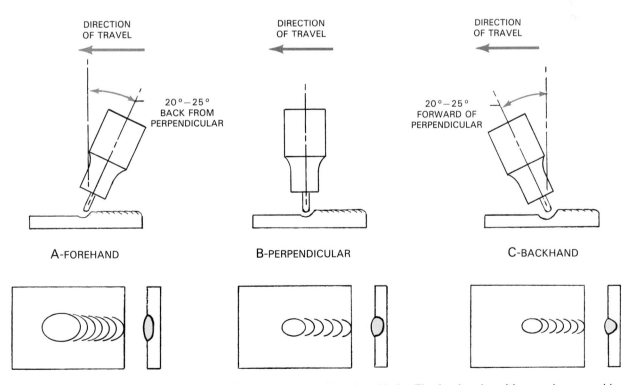

A-FOREHAND   B-PERPENDICULAR   C-BACKHAND

Fig. 19-6. The effects of the torch position on the coompleted weld. A – The forehand position produces a wide bead with shallow penetration. B – The perpendicular position produces a medium-width bead and deeper penetration. C – The backhand position produces a narrow bead with deep penetration.

position. Touch the electrode wire to the base metal, but do not press the switch. Raise the gun about 1/16″ (1.6 mm) or less. Lower your helmet and press the switch. See Fig. 19-7. A welding arc will then form. The arc melts the base metal and forms a *weld pool* (depression). The depth of this weld pool is the amount of penetration of the weld. Fig. 19-8 shows these parts of the weld.

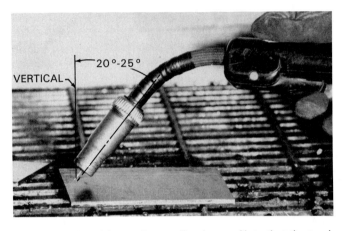

Fig. 19-7. A welder ready to strike the arc. Note that the torch is held at 20°−25° from vertical. This weld will be made in the backhand position.

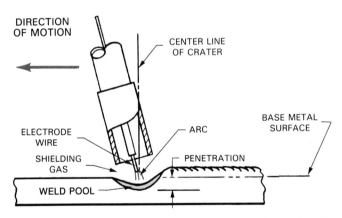

Fig. 19-8. GMAW in progress. Note that the electrode wire is kept ahead of the center line of the weld crater.

Move the welding gun in the direction of travel. Metal from the electrode will constantly be added to the pool. The weld pool shape is round to oval. A wider weld bead is obtained by moving the welding gun from side to side. The electrode should be in front of the center line of the pool. See Fig. 19-8. Continue moving the arc and pool until you reach the end of the weld. Keep the nozzle-to-work distance and the angle of the welding gun as constant as possible.

The weld pool must be filled at the end of the joint or when you are ready to stop. Move the gun backwards over the weld pool 1/4″−1/2″ (6 mm to 13 mm), then release the switch. The electrode wire will stop feeding

and the welding arc will also stop. Hold the gun in the same position until the shielding gas stops flowing.

For the same diameter electrode, more metal is deposited using spray transfer than globular or short circuiting transfer. The welder must move the welding gun much faster when using spray transfer than with the other two transfer methods. Regardless of the transfer method, watch the weld pool size. It determines the travel speed. If the pool gets too large, move the gun faster. If you are moving the arc in front of the pool, slow the travel speed down. Keep the electrode wire directly in front of the center of the pool.

A good weld bead should have evenly spaced ripples. The edges of the weld should be the same width. There should be about 1/16″−1/8″ (1.6 mm to 3.2 mm) of reinforcement. Fig. 19-9 shows an example of a good fillet weld bead on a T-joint. Practice moving the gun at a constant speed if the ripples are not evenly spaced. If the edges are not the same width along the entire weld, practice holding the gun steady. Try to avoid moving the gun from side to side.

Fig. 19-9. A well-formed fillet weld on a T-joint. Note that the bead is slightly convex.

The amount of reinforcement is controlled by the travel speed. A wide, shallow bead is produced by too slow a travel speed. A narrow bead with undercutting along the edges results from a too-fast travel speed. At the correct travel speed, a weld bead with good penetration and no undercutting is produced.

Spatter is reduced by changing the angle of the welding gun or the electrode extension. A longer electrode extension reduces spatter. However, an electrode that extends too far results in poor penetration. Too long an extension may cause porosity because the shielding gas is not able to cover the welding area. A low flow rate of shielding gas may also cause weld porosity.

EXERCISE 19-1 − LAYING A BEAD ON A PLATE

1. Obtain a piece of mild steel measuring 3/16″ x 3″ x 6″ (4.8 mm x 76 mm x 152 mm).
2. Clean both surfaces with a wire brush.
3. Lay out five straight lines, using a ruler and a

piece of soapstone.

4. Set up the welding machine for short circuiting transfer. Refer to Chapter 18 for details. Select the appropriate shielding gas and set the flow meter. Figs. 18-11 and 19-2 list the correct gases and flow rates.

5. Use a backhand method. Fig. 19-7 shows the correct angle between the welding gun and the plate. This angle is the same for most GMAW and FCAW in any position.

6. Watch the metal as the welding arc starts. A molten weld pool develops quickly. Move the arc toward the forward edge of the pool. Note the width of the pool and the amount of penetration. Keep the same width and penetration as the weld progresses.

7. Fill the weld pool at the end of the bead.

8. Examine the bead. Change the voltage, wire feed speed, or shielding gas flow rate as required.

9. Weld beads on the other four lines on the plate.

Inspection:

Each weld bead should be straight with evenly spaced ripples and continuous reinforcement. The crater at the end of the weld should be filled. No evidence of porosity should be seen.

Turn the plate over and repeat this exercise. Set the welding machine for spray transfer. Change the shielding gas to argon or argon with oxygen. See Figs. 18-12 and 18-16.

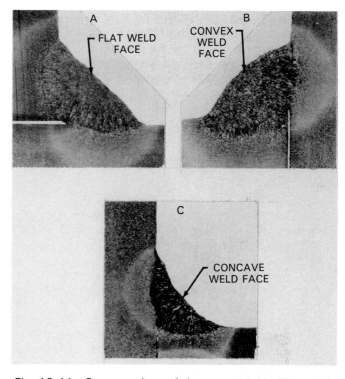

Fig. 19-11. Cross sections of three acceptable fillet welds made on an inside corner joint. These macrographs were etched and magnified four times. (U.S. Steel)

MAKING A FILLET WELD

A fillet weld made with GMAW or FCAW looks similar to one made with other welding processes. The two legs should be equal in size. The face of the bead should be convex or flat with evenly spaced ripples. Undercutting, porosity, or cracks should not be visible. Fig. 19-10 shows a good fillet weld made with GMAW. Fig. 19-11 shows cross sections of three good fillet welds.

A fillet weld is started in the same manner as starting to lay a bead on a plate. The electrode and the arc are centered over the root of the weld. The welding gun should be held at about a 45° angle to the base metal. The gun can be held in a forehand, perpendicular, or backhand position. See Fig. 19-12. Be sure

Fig. 19-10. A well-formed fillet weld on a lap joint. The bead is slightly convex.

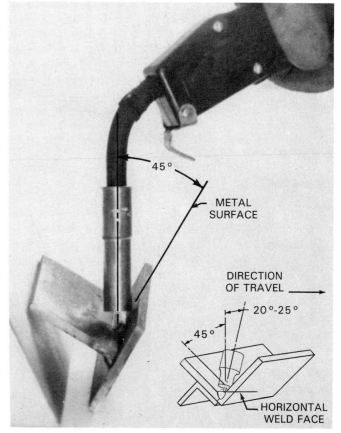

Fig. 19-12. The correct angle of the GMAW gun for making a fillet weld on a joint in the flat welding position. The gun is held at 45° to the metal surfaces and at 20°—25° from vertical. This weld is being made in the backhand position.

193

to use the proper electrode extension.

Once the welding arc is started, a pool will develop under it. Some side-to-side movement of the gun may be required to form a C-shaped weld pool. The arc should be centered over the root of the weld and kept toward the front of the pool. Watch the pool as you move along the weld joint. The pool should flow into both pieces of the base metal. Keep the pool the same width as the weld progresses. Weld defects and ways to prevent them are discussed later in this chapter.

EXERCISE 19-2 — MAKING A FILLET WELD ON A LAP JOINT

1. Obtain two pieces of mild steel measuring 1/8" x 1 1/2" x 6" (3.2 mm x 38 mm x 152 mm).
2. Clean both surfaces and edges with a wire brush.
3. Select the correct electrode wire and metal transfer method.
4. Set the shielding gas flow rate and the machine adjustments.
5. Place the two pieces together to form a lap joint.
6. Tack weld the pieces in three places on each side.
7. Position the gun over one end of the joint using the correct angles. It may be necessary to point the gun more toward the surface than the edge when welding an edge to a surface.
8. Press the switch and begin welding.
9. Make a fillet weld along the seam. Fill the crater at the end of the bead.
10. Make a second fillet weld on the other side.

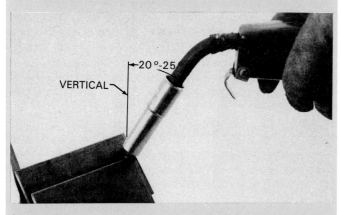

Inspection:
The completed welds should be convex with evenly spaced ripples. The edges of the bead should be straight and should not overlap.

EXERCISE 19-3 — PLACING A FILLET WELD ON A T-JOINT

1. Obtain two pieces of mild steel measuring 1/8" x 1 1/2" x 6" (3.2 mm x 38 mm x 152 mm).
2. Clean one edge of one piece and one surface of the other piece.

3. Select the correct welding electrode and shielding gas. Set up the machine for the metal transfer method you will be using.
4. Tack weld the clean edge to the clean surface in three places on each side.
5. Place the weldment so the weld is in the flat position, Fig. 19-12.
6. Make a fillet weld on each side of the T-joint. Watch the weld pool as it forms. Keep a C-shaped pool.

Inspection:
The weld should be straight with evenly spaced ripples. Undercutting or overlap should not be visible.

WELDING A BUTT JOINT

A butt joint can be formed with a square groove, or the edges can be prepared with a groove, such as a V-groove. Steel that is 3/16" (4.8 mm) or thicker usually has a prepared groove. Grooves used in GMAW and FCAW have a smaller angle than grooves used in SMAW.

Full penetration of butt joints is usually required. A *root opening* (gap) is left between the two pieces of base metal when tack welding them. The edges of both pieces must be melted by the arc when welding. The root of the weld pool looks like a *keyhole* when welding. The round portion of the keyhole is smaller than the keyhole in oxyfuel gas welding or SMAW. The keyhole is only seen on the *root pass* (first pass) of a multiple-pass weld. A keyhole at the root opening indicates that both pieces of base metal are being melted. It also indicates that the molten metal is flowing through the joint, resulting in full penetration. See Fig. 19-13. The size of the pool and the keyhole must be kept constant as the weld progresses, to produce consistent penetration.

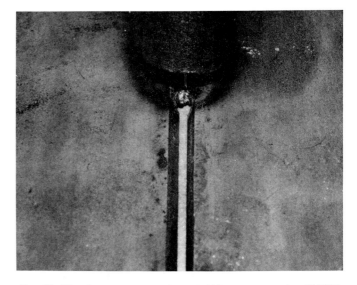

Fig. 19-13. A square-groove butt weld in progress using GMAW. Note the keyhole at the leading edge of the weld pool.

A backing may be used to produce consistent penetration. **Backing** is attached to the root side of the joint to control penetration. Backings can be used in all welding positions. Some backings become part of the completed joint while others are removed after welding. Removable metal backings can be flat or have a radius machined in them. See Fig. 19-14. Removable metal backings are not melted while welding. A copper backing is used when welding mild or stainless steel. A stainless steel backing is used when welding aluminum.

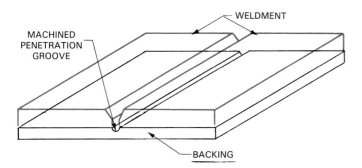

Fig. 19-14. Backing with a round groove milled into it. The groove ensures uniform penetration. A copper plate is used for welding steel. The metal to be welded is shown in red.

When welding a butt joint, the electrode should be directly over the center line of the weld. Both pieces being welded should be melted equally by the welding arc. If one piece is melting more than the other, adjust the location or angle of the gun, so both pieces melt equally. Use either a forehand, perpendicular, or backhand angle from the axis of the weld. Fig. 19-15 illustrates the angles used to weld a butt joint.

Multiple passes or weld beads are required to fill the grooves for thick base metal. The keyhole method is used only on the root pass. After the root pass, each weld pass is similar to laying a bead on a plate. Good penetration into the base metal and the previous weld bead is required. Fig. 19-16 shows a multiple pass groove weld. Each bead must be thoroughly cleaned before another pass is made. When FCAW, all slag must be removed after each weld pass is made.

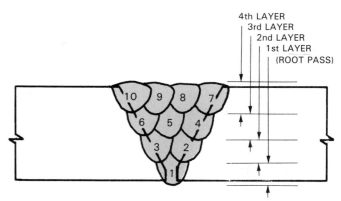

Fig. 19-16. A multiple-pass weld. This weld has four layers and 10 passes in it.

EXERCISE 19-4 — WELDING A SQUARE-GROOVE BUTT JOINT

1. Obtain two pieces of mild steel measuring 1/8'' x 1 1/2'' x 6'' (3.2 mm x 38 mm x 152 mm).
2. Clean one edge on each piece.
3. Select the correct welding electrode and shielding gas. Set up the machine for the metal transfer method you will be using.

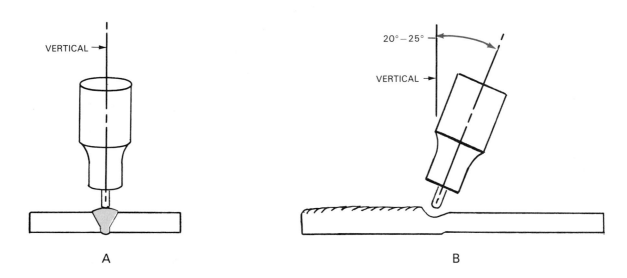

Fig. 19-15. The suggested gun angles for welding a butt joint. A—The gun is aligned vertically with the weld axis, as shown in the end view. B—It is tipped forward or back from vertical at 20°−25° for forehand or backhand welding, as shown in the side view.

4. Position the two pieces so there is a 3/32'' (2.38 mm) gap between them. Tack weld the pieces together in three places.
5. Support the pieces so a full penetration weld can be made. A backing is optional.
6. Align the gun and electrode with the center of the joint before welding. Press the switch on the gun. Move the gun forward when the pool forms. Keep the keyhole and the pool as consistent as possible.
7. Continue welding the butt joint to the end of the plate. Fill the weld pool at the end of the weld.

Inspection:

A butt weld should have complete penetration. The root of the weld should indicate complete penetration. This penetration and the face of the weld should have evenly spaced ripples. The bead should be consistent in width. Signs of defects should not be visible. The weld bead should be built up slightly higher than the base metal. Excessive spatter on the surface is not acceptable.

WELDING ALUMINUM

A few problems are encountered when welding aluminum that are not encountered when welding mild steel. These include frequent bird's nests and the forming of aluminum oxide. The electrode wire in the wire feeder may become tangled, producing a bird's nest. The number of bird's nests may be reduced by:

• Selecting the correct electrode wire.
• Adjusting the equipment properly.

Aluminum electrode wire is very soft. Bird's nests can be avoided by selecting the largest possible diameter electrode wire when welding aluminum. A large-diameter wire is stiffer than a smaller-diameter wire. Use an aluminum alloy electrode that is stiffer. This allows the electrode to be pushed through the cable to the welding gun without kinking. Two common aluminum electrode wires are ER5356 and ER4043. ER5356 is stiffer than ER4043 and will result in fewer bird's nests.

Proper adjustment of the equipment is an important consideration in reducing the number of bird's nests. One necessary adjustment is to position the guide tubes in the wire feeder as close to the drive rolls as possible. The guide tubes should also be aligned with each other.

Aluminum electrode wire is soft and cannot be easily pushed through a long cable to the welding gun. Use the shortest cable assembly possible from the wire feeder to the welding gun. Also use a Teflon® liner in the cable. A *pull gun* which pulls the electrode wire through the cable may be used. A *push-pull system* is formed when a pull gun pulls the electrode wire through the cable and the wire feeder pushes the wire. A push-pull system helps to prevent a bird's nest from forming.

Aluminum oxidizes very easily, forming aluminum oxide. Aluminum oxide has a higher melting temperature than aluminum. The base metal should be chemically or mechanically cleaned before welding. A stainless steel wire brush is recommended to mechanically clean the joint. The tip of the electrode wire may also oxidize. Use a pair of wire cutters to cut the tip off the electrode wire. This should be done each time before starting a new weld. Place your hand over the area to be cut to prevent the piece of wire from flying through the air and injuring someone.

The forehand welding method helps to remove oxides on unwelded areas. However, penetration may be less than when metal is welded using the backhand method. Aluminum conducts heat very well. The metal you are welding is already preheated as you move along the weld joint. Therefore, it may be necessary to move the welding gun faster to prevent melt-through.

WELD DEFECTS

Defects produced in GMAW and FCAW are similar to those found in work done using other welding processes. These defects include incomplete penetration, lack of fusion, slag inclusions, undercut, overlap, and porosity. Defects cause a weld to be weaker than its design requirements. Fig. 19-17 shows a test sample of a fillet weld on a T-joint. The sample was tested until it failed. It had no apparent inclusions.

Fig. 19-17. A fillet weld test sample. This weld was bent until it failed. There are no apparent inclusions in the weld.

Two common defects in a fillet weld are undercut and overlap. It is important to use the correct gun angles and gun motion to avoid these defects. Make sure both pieces of metal are being melted. You may need to move the gun toward the toe of the fillet weld to melt the base metal.

Another common defect is slag in the weld. This often happens on a multiple-pass weld. Be sure to completely remove all slag from one weld pass before starting another pass.

Most defects can be avoided by:
- Selecting the correct shielding gas and electrode.
- Setting the correct shielding gas flow rate, voltage, and wire feed speed.
- Maintaining the proper electrode extension (contact tube-to-workpiece distance).
- Using the proper gun angles.
- Moving the gun properly.
- Removing slag between weld passes.

REVIEW QUESTIONS

Write your answers on a separate sheet of paper.
1. The molten metal under the welding arc is called the _____.
2. List four factors that a welder must consider before selecting the electrode and shielding gas.
3. Refer to Figs. 18-13, 18-17, and 18-20. What electrode and shielding gas are recommended for gas metal arc welding of 308 stainless steel, using short circuiting transfer?
   Electrode: _____
   Shielding gas: _____
4. Refer to Figs. 18-23 and 19-2. What voltage, wire feed, and flow rate should be used to weld 1/8 aluminum using .047 diameter electrode wire and spray transfer?
   Voltage: _____
   Wire feed: _____
   Flow rate: _____

5. Which of the following does the welder actually observe and control to obtain a good-quality weld bead?
   a. Nozzle-to-work distance.
   b. Electrode extension.
   c. Contact tube-to-work distance.
   d. Arc length.
6. The electrode should extend _____ beyond the end of the contact tube for spray transfer with GMAW.
   a. 1/4″−1/2″ (6 mm−13 mm).
   b. 1/2″−1″ (13 mm−25 mm).
   c. 3/4″−1 1/2″ (19 mm−38 mm).
   d. 3/4″−3 3/4″ (19 mm−95 mm).
7. List two advantages of backhand welding over forehand welding.
8. _____ transfer deposits more metal for a given size electrode wire.
   a. Short circuiting.
   b. Globular.
   c. Spray.
   d. Pulsed-spray.
9. List two ways to reduce spatter.
10. The weld pool shape for a fillet weld is a _____ shape. The opening at the root pass of a butt weld is a _____ shape.

GMAW being used to fabricate a truck cab. The welder is welding out of position. (Hobart Brothers Co.)

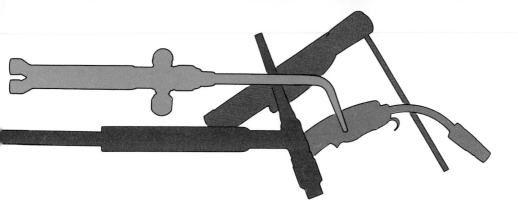

Chapter 20

GMAW: Horizontal, Vertical, and Overhead Welding Positions

After studying this chapter, you will be able to:
□ Explain why flat position welding is preferred over out-of-position welding.
□ Identify the correct welding gun and electrode angle for out-of-position welding.
□ Weld in the horizontal welding position using GMAW or FCAW.
□ Weld in the vertical welding position using GMAW or FCAW.
□ Weld in the overhead welding position using FCAW or GMAW.

TECHNICAL TERMS

Downhill, drag angle, overlap, underfill, uphill.

OUT-OF-POSITION GMAW or FCAW

Both GMAW and FCAW can be done out of position. When GMAW, only short circuiting and pulsed spray transfer methods can be used. Pulsed spray transfer requires a welding machine with pulse controls. Since not all machines have pulse spray capabilities, short circuiting transfer is more often used for out-of-position welding.

The flat welding position should be used whenever possible. Welding in the flat position is preferred over out-of-position welding for the following reasons:
1. The pool will not sag or fall from the weld joint.
2. All metal transfer methods can be used.
3. The flat welding position is more comfortable for a welder.

4. Better welds may be produced because welders are usually better at welding in the flat position.

However, some industrial weldments require out-of-position welding to be performed. Weldments such as pipelines and structural welding are too large to move. See Fig. 20-1. Most welding qualification tests are performed on out-of-position weld joints. A welder who can make out-of-position welds successfully can also produce good welds in the flat welding position. Welds made out of position should look as good and should be as strong as welds made in the flat welding position.

Fig. 20-1. This welder is using GMAW in the overhead position to assemble a coal mining rail car. The wire feeder is supported overhead by a small crane. (Miller Electric Mfg. Co.)

WELDING IN THE HORIZONTAL WELDING POSITION

The weld axis and weld face are horizontal or close to horizontal when making a weld in the horizontal position. The weld axis can vary from 0° − 15°. The weld face can vary from 80° − 150° or from 210° − 280°.

Welding in the horizontal welding position can be more difficult than welding in the flat welding position. Gravity pushes the molten weld pool down. This causes the weld bead to sag and not properly fill the joint. It is important to use proper gun angles and techniques to obtain high-quality welds.

Fillet welding

A fillet weld in the horizontal welding position is the most common out-of-position weld. A horizontal fillet weld is made on inside corner joints and T-joints. Spray and globular transfer can be used.

The process for making a horizontal fillet weld is similar to making a fillet weld in the flat position. Notice the gun angles in Fig. 20-2. The gun points toward the root of the weld. A C-shaped weld pool must be formed when welding. The best penetration is produced using the backhand technique. The same electrode extension is used for making horizontal welds and welding in the flat position. Chapter 19 describes the proper electrode extension. The recommended electrode extension for short circuiting transfer with GMAW is 1/4″ − 1/2″ (6 mm − 13 mm).

A common weld defect encountered when making a fillet weld in the horizontal welding position is undercutting on the vertical plate. To prevent undercutting, use a slight weaving bead by moving or pointing the

arc at the vertical piece for a short period of time (less than a half second). More electrode metal is directed to the vertical piece. The gun angle can also be changed to 40° from the horizontal, instead of 45°, to help eliminate undercutting.

Welding a butt joint

Short circuiting or pulsed spray transfer is used when welding a butt joint. The weld pool remains fairly small and not too fluid in the horizontal position. When using these methods, a pool will sag downward if it gets too large. A slight sag is acceptable. However, overlap and underfill are not acceptable. *Overlap* is a condition in which the metal sags downward and overlaps the lower piece. *Underfill* is a condition in which the molten metal sags downward and does not fill the top part of the weld joint. See Fig. 20-3.

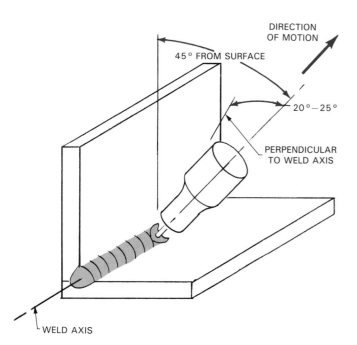

Fig. 20-2. Suggested electrode and gun angles for a fillet weld on an inside corner joint in the horizontal welding position.

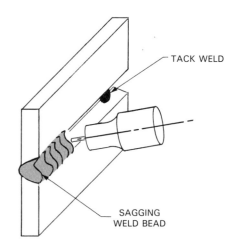

Fig. 20-3. The bead on this horizontal butt joint was too hot and sagged downward. This resulted in an underfilling of the bead on the upper plate. The weld becomes weaker at this point due to the thinning.

The opening at the bottom of the weld bead should look like a keyhole to obtain a full penetration weld. This indicates that both pieces are melting and flowing together. It also shows that the metal is flowing through the joint to the backside for proper penetration. See Fig. 20-4. The round part of the keyhole should not be allowed to get too large. Backing may be used to control the penetration. Refer to Chapter 19 for more information on backing.

Fig. 20-4. A square-groove butt joint being welded in the horizontal welding position.

The angle of the welding gun and welding arc should be considered when making a horizontal butt weld. The arc must melt both pieces. Excessive melting and undercutting may result if the electrode is pointed too much toward the upper piece. If the electrode is not directed to the upper piece, then it will not melt enough and a lack of fusion will result. Watch the pool carefully. Adjust the gun angle as required to maintain a pool that is equally melted on both pieces. Fig. 20-5 shows the angles to be used when making a horizontal butt weld. The weld pool should remain molten. However, it

should not sag and cause overlap or underfill. Point the welding gun upward at a 5° – 10° angle to help control any sagging of the bead. Using a backing helps to control the penetration.

Higher shielding gas flow rates are used on horizontal and vertical position welds than on flat position welds. Argon, CO_2, and other shielding gases tend to fall away from the weld area. Therefore, more gas is used to protect the weld.

EXERCISE 20-2 — WELDING A BUTT JOINT IN THE HORIZONTAL WELDING POSITION

1. Obtain two pieces of mild steel measuring 1/8″–3/16″ x 1 1/2″ x 6″ (3.2 mm–4.8 mm x 38 mm x 152 mm).
2. Clean one edge on each piece.
3. Set the proper voltage and wire feed speed for short circuiting transfer. Set the shielding gas flow.
4. Position the pieces to form a butt joint. Tack weld the joint in three places along the joint. There should be a 1/8″ (3.2 mm) root opening.
5. Position the pieces for welding in the horizontal welding position.
6. Make a horizontal butt weld. Watch and control the weld pool. Fill the weld pool at the end of the weld. It may be necessary to weld more than one pass to fill the joint.

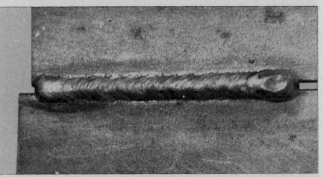

FRONT OF PLATE

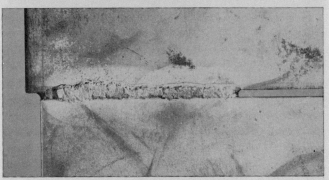

BACK OF PLATE

Inspection:
The completed weld should have a flat-to-convex bead shape. The ripples should be even. There should not be evidence of undercutting or other defects. Even penetration should be achieved on the back of the plates.

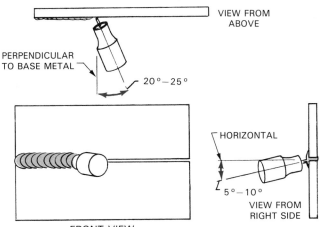

PERPENDICULAR TO BASE METAL

VIEW FROM ABOVE

20°–25°

HORIZONTAL

5°–10°

VIEW FROM RIGHT SIDE

FRONT VIEW

Fig. 20-5. The suggested gun and electrode angles for welding a butt joint in the horizontal welding position.

WELDING IN THE VERTICAL POSITION

Vertical position welding can be done traveling up or down (welding *uphill* or *downhill*). The forehand welding method is used when welding uphill. The backhand method is used when welding downhill. When traveling up or down, the gun is usually pointed up. When welding in the vertical position, the weld pool should not become too large. The proper gun angle and control of the weld pool size will help prevent the pool from sagging.

The electrode is directed toward the root of the weld when welding in the vertical position. The electrode should be centered over the root of the weld. This heats both pieces of metal equally. A side-to-side weaving motion is used to obtain a wider bead. Hesitate at the edges of the weld to prevent undercutting when using a weaving motion. The gun angles used to make a vertical weld are the same as those used for flat welding. Gun angles for fillet and butt welds are discussed in Chapter 19.

A C-shaped weld pool is used when making a fillet weld. This indicates that both surfaces are melting. If one plate is thicker than the other, the electrode may have to be directed more toward the thicker plate. When welding a lap joint, the electrode may need to be pointed more toward the surface than the edge.

A fillet weld made in the vertical welding position should appear similar and have the same strength as a weld made in the flat welding position. More skill is required to make the vertical weld, however. Fig. 20-6 shows a fillet weld being made in the vertical position.

More than one weld pass or weld bead may be required to complete a weld. The weld is cleaned after making the first or additional passes. Remove all slag from the weld and wire-brush the weld. The second pass uses a slightly different angle of the gun. The *drag angle* remains the same ($20° - 25°$), but the gun is pointed between the first pass and the base metal. A third pass requires the same angle but in the opposite direction. Each pass must melt the base metal and the previous pass. All slag must be removed from the weld before making an additional weld pass. See Fig. 20-7.

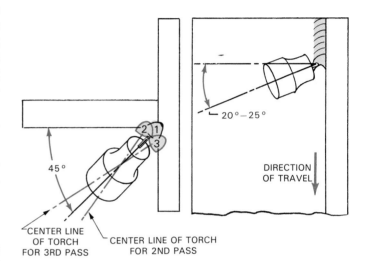

Fig. 20-7. The suggested gun and electrode angles for welding a multiple-pass fillet weld on a lap joint in the vertical position. Notice that the torch angle must change from 45° for the second and third passes.

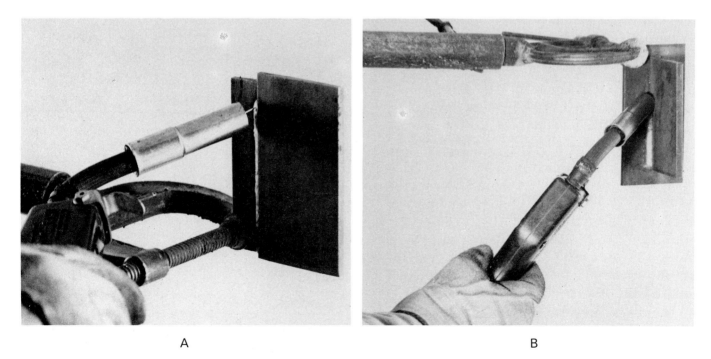

A   B

Fig. 20-6. A—A fillet weld being made on a lap joint in the vertical welding position. This weld is being made uphill using the forehand welding method. B—A fillet weld being made on a T-joint in the vertical welding position. The weld is being made uphill using the forehand welding method.

The same gun angles are used for making butt welds in the vertical and flat welding positions. A keyhole-shaped opening should be seen on the first pass on a butt joint to obtain complete penetration. Baking can be used to control the penetration.

When making a multiple-pass weld, each weld must be cleaned prior to making the next pass. Each pass must melt into the previous pass and the base metal. Either stringer beads or weaving beads can be used to fill and build up the weld.

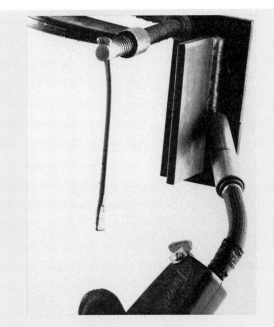

EXERCISE 20-3 — MAKING A FILLET WELD ON A LAP JOINT IN THE VERTICAL WELDING POSITION

1. Obtain two pieces of mild steel measuring 1/8''—3/16'' x 1 1/2'' x 6'' (3.2 mm—4.8 mm x 38 mm x 152 mm).
2. Clean an edge and one surface on each piece.
3. Set the proper voltage and wire feed speed for short circuiting transfer. Set the shielding gas flow.
4. Clamp the pieces together to form a lap joint.
5. Tack weld the joint in three places along the joint.
6. Position the pieces in a welding positioner for welding in the vertical position.
7. Make a vertical fillet weld traveling downhill. Maintain the proper electrode extension. Make sure the weld penetrates into the surface of the lower piece of metal.
8. Weld the joint on the other side of the metal. This weld can be made uphill or downhill. See Fig. 20-6A.

Inspection:
The weld should have evenly spaced ripples on a convex or flat face. The width of the bead should be consistent. There should not be signs of defects.

EXERCISE 20-4 — MAKING A FILLET WELD ON A T-JOINT IN THE VERTICAL WELDING POSITION

1. Obtain two pieces of mild steel or stainless steel measuring 3/16''—1/4'' × 1 1/2'' × 6'' (4.8 mm— 6.4 mm × 38 mm × 152 mm).
2. Clean an edge on one piece and a surface on the other piece.
3. Set the voltage, wire feed speed, and shielding gas flow rate for short circuiting transfer.
4. Place the clean edge on the clean surface to form a T-joint. Tack weld the joint in three places on each side of the joint.
5. Place the pieces in a welding positioner for welding in the vertical position.
6. Make a multiple-pass vertical fillet weld with three passes on each side of the T-joint. Weld one side uphill and the other downhill. Do not allow the weld pool to get too large when welding downhill. Reduce the wire feed speed or increase the travel speed if it gets too large.

Inspection:
The completed welds should not penetrate through the base metal. The beads should be convex, even in width, and have evenly spaced ripples. Spatter on the plates should be kept to a minimum. Refer to Fig. 20-7.

EXERCISE 20-5 — WELDING A BUTT JOINT IN THE VERTICAL WELDING POSITION

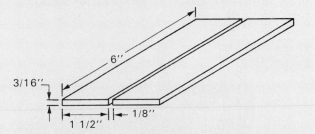

1. Obtain two pieces of mild steel measuring 3/16'' x 1 1/2 '' x 6 '' (4.8 mm x 38 mm x 152 mm).
2. Set the proper voltage, wire feed speed, and

shielding gas flow for short circuiting transfer.

3. Tack weld the pieces in three places along the joint to form a square-groove butt joint. There should be a root opening of 1/8" (3.2 mm), as shown in the drawing.

4. Mount the tack-welded pieces in a vertical position for welding.

5. Make a vertical butt weld using the keyhole method. Watch and control the weld pool. Do not let it sag.

Inspection:

The completed weld should have a flat-to-convex bead shape. The ripples should be even. There should not be evidence of undercutting or other defects. There should be even penetration on the backside of the plates.

WELDING IN THE OVERHEAD POSITION

The overhead position is considered to be the most difficult of all welding positions. The voltages and wire feed speeds used for overhead welding are lower than those used in the flat position. Refer to Chapter 18 for the proper voltage and wire feed speed. Use the values that are on the lower end of the charts. Short circuiting transfer and pulsed spray transfer are the only types of metal transfer that can be used. The weld pool must be kept to a small and controllable size.

Shielding gas flow rates are higher for welding in the overhead position than for other welding positions. Argon and carbon dioxide fall away from the weld area, thus more shielding gas must be used.

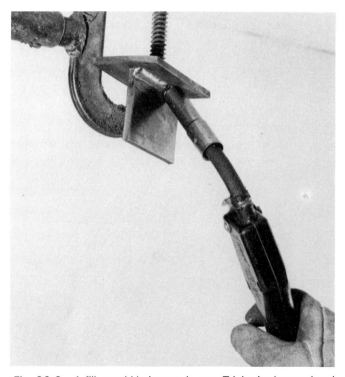

Fig. 20-8. A fillet weld being made on a T-joint in the overhead welding position. The gun angles are the same as those used for the flat or vertical welding positions.

Welding gun angles used in the overhead welding position are the same as those used in the flat position. The gun should be pointed at the root of a fillet weld. See Fig. 20-8. The gun is held at a 90° angle to the surface when welding a butt joint. A 5°–15° drag angle or backhand angle should be used. See Fig. 20-9. Hold the gun in a comfortable position. Sit or stand so that you can see the joint while welding.

Be sure to wear the proper protective clothing when welding in the overhead welding position. Cover all pockets and button your shirt collar. A cap and leathers are strongly recommended. Both GMAW and FCAW produce a lot of spatter. Very little spatter sticks to the base metal; instead, it falls downward.

Fig. 20-9. A square-groove butt joint being welded in the overhead welding position. The gun angles are the same as those used for the flat or vertical welding positions.

EXERCISE 20-6 — MAKING A FILLET WELD ON A T-JOINT IN THE OVERHEAD WELDING POSITION

1. Obtain two pieces of mild steel or stainless steel measuring 1/8"–3/16" × 1 1/2" × 6" (3.2 mm–4.8 mm × 38 mm × 152 mm).

2. Clean an edge on one piece and a surface on the other piece.

3. Set the voltage, wire feed speed, and shielding gas flow rate.

4. Place the clean edge on the clean surface to form a T-joint. Tack weld the joint in three places on each side of the joint.

5. Position the pieces in a welding positioner for overhead welding.

6. Make a fillet weld in the overhead welding position. Keep the electrode pointed at the root of the weld. Some weaving motion can be used. Watch the weld pool so that it does not get too large. Increase the

travel speed or reduce the wire feed speed if the weld sags. Maintain the correct electrode extension.

Inspection:

The completed welds should not penetrate through the base metal. They should not show evidence of a large weld pool that sags. The weld should be convex, even in width, and have evenly spaced ripples.

EXERCISE 20-7 — WELDING A BUTT JOINT IN THE OVERHEAD WELDING POSITION

1. Obtain two pieces of mild steel or stainless steel, measuring 1/8''—3/16'' × 1 1/2'' × 6'' (3.2 mm—4.8 mm × 38 mm × 152 mm).
2. Clean one edge on each piece.
3. Set the voltage, wire feed speed, and shielding gas flow.
4. Align the pieces to form a butt joint. Tack weld the pieces in three places along the joint. There should be a 3/32''—1/8'' (2.4 mm—3.2 mm) root opening.
5. Position the pieces for welding in the overhead position.
6. Make an overhead butt weld. Weld the root pass and watch for the keyhole. A second pass may be required to fill the joint.

Inspection:

The completed weld should have a flat-to-convex bead shape. There should not be evidence of any defects.

REVIEW QUESTIONS

Write your answers on a separate sheet of paper.

1. Which out-of-position weld can be made with all metal transfer methods?
   a. Horizontal fillet weld.
   b. Horizontal butt weld.
   c. Downhill butt weld.
   d. Downhill fillet weld.
2. Where should you point the gun and electrode when making a fillet weld?
3. What can be done to prevent undercutting when making a horizontal fillet weld?
4. Select three of the following defects that are more common in a horizontal butt weld than in a flat butt weld.
   a. Slag inclusions.
   b. Weld pool sag.
   c. Overlap.
   d. Rough surface on weld bead.
   e. Underfill.
5. What weld pool shape is used when welding a fillet weld in the vertical and overhead welding positions?
6. When using the backhand method for making a vertical weld, in which direction will you be traveling?
   a. Uphill.
   b. Downhill.
7. What two types of weld beads can be used to fill or build up a weld?
8. Are out-of-position welds stronger, weaker, or the same strength as those welded in the flat welding position.
9. What shielding gas and flow rate should be used to weld 1/4″ (6.4 mm) aluminum in the overhead welding position?
10. What type of protective clothing is strongly recommended when welding in the overhead welding position?

205

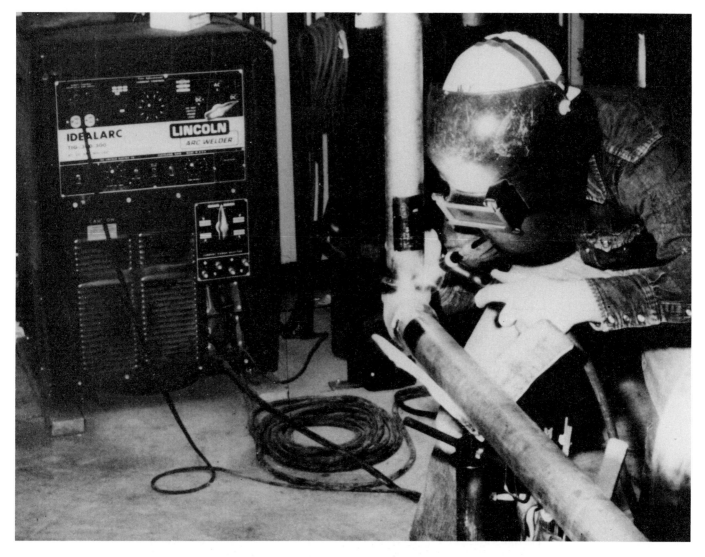

GTAW being used to make a fillet weld on a stainless steel pipe and fitting.
(The Lincoln Electric Co.)

GAS TUNGSTEN ARC WELDING

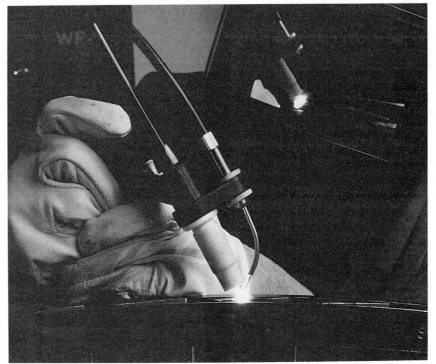

CK Worldwide, Inc.

A welder using GTAW equipment to fabricate a large part for a papermaking machine.   (Miller Electric Mfg. Co.)

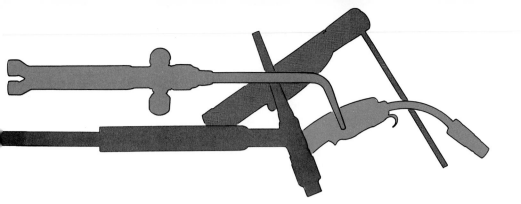

Chapter 21

Gas Tungsten Arc Welding Equipment and Supplies

After studying this chapter, you will be able to:
☐ Define gas tungsten arc welding.
☐ Describe the principles of gas tungsten arc welding.
☐ Identify the equipment and supplies involved with GTAW.
☐ List the three different types of GTAW machines.
☐ Identify the parts of a GTAW torch.
☐ Describe the functions of the cables and hoses.
☐ Identify safety considerations when GTAW.

TECHNICAL TERMS

Background current, collet, collet body, foot pedal, gas-cooled torch, gas lens, gas tungsten arc welding (GTAW), heliarc welding, inert gas, nonconsumable electrode, nozzle, partially rectified current, peak current, post flow, pulsed current, rectified current, thumb switch, torch body, tungsten inert gas (TIG), water-cooled torch.

The American Welding Society defines *gas tungsten arc welding (GTAW)* as: "A process that joins metals together by heating them with an arc between a nonconsumable tungsten electrode and the work. The electrode and weld area are shielded with an inert gas or gas mixture. Filler metal may or may not be used." In industry, gas tungsten arc welding is commonly referred to as TIG. *TIG* stands for "tungsten inert gas." Some people also refer to this process as *heliarc welding.*

GAS TUNGSTEN ARC WELDING PRINCIPLES

Gas tungsten arc welding is a welding process in which an arc is struck between a tungsten electrode and the base metal. See Fig. 21-1. The tungsten electrode does not melt and become part of the weld. It is, therefore, referred to as a *nonconsumable electrode.*

The tungsten electrode and the molten weld pool must be protected from the atmosphere while welding. Oxygen in the atmosphere causes the electrode and base metal to oxidize. An inert gas is used to shield the arc area from the atmosphere. *Inert gases* do not react with the electrode, arc, or molten weld metal. GTAW is considered to be a very clean process, since oxygen is kept away from the electrode and the molten metal.

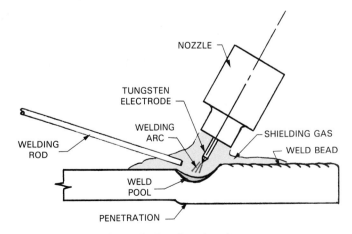

Fig. 21-1. A schematic drawing showing a gas tungsten arc weld in progress. The tungsten electrode is not consumed. A welding rod is often used to supply the filler metal.

GTAW EQUIPMENT AND SUPPLIES

The gas tungsten arc welding process requires use of equipment designed for this type of welding. The equipment must be able to start and maintain a welding arc, and must be able to withstand the heat generated by welding. The equipment must also provide shielding gas to protect the electrode and the base metal. The basic parts of a GTAW setup include:
- A welding machine.
- A torch.
- An electrode.
- An inert gas supply and controls.

Other equipment that is regularly used includes:
- Filler metal.
- Remote control (usually a foot pedal).
- Water cooling equipment for the torch.

A person welding using GTAW must wear proper protective clothing, a welding helmet with a number 10 filter (minimum), and gloves. Fig. 21-2 shows a properly dressed welder working at a GTAW station.

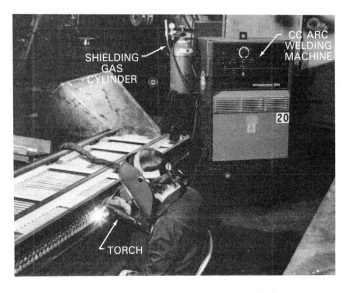

Fig. 21-2. A GTAW outfit being used to weld the ends of a cluster of tubes. Note that the welder is wearing the proper clothing and equipment. (Miller Electric Mfg. Co.)

GTAW MACHINE

A constant current welding machine is used for GTAW. This is the same type of machine used in shielded metal arc welding (SMAW). Constant current machines allow the welder to slightly vary the welding current by changing the arc length. Chapter 11 provides detailed information about constant current machines. Most GTAW welding machines provide for the following.
- Selection of current type (AC, DCEP, or DCEN).
- Selection of current range and setting.
- Selection and adjustment of the high frequency.

- Connection of welding torch.
- Connection of ground lead.
- Connection of shielding gas.
- Connection of water, on water-cooled machines.
- Controls to allow shielding gas to flow after the current has stopped.
- Current control on the machine panel or from a remote control.

Most of the controls that a welder uses are on the front panel of a welding machine. See Fig. 21-3. Electrical, water, and shielding gas connections are usually on the bottom front or the back of a machine. A GTAW machine can be an alternating current (AC) machine, a direct current (DC) machine, or a machine that can provide either AC or DC current.

Direct current (DC) welding

Two types of direct current are used for GTAW; direct current electrode negative (DCEN) and direct current electrode positive DCEP). DCEN was also known as direct current straight polarity (DCSP). DCEP was also referred to as direct current reverse polarity (DCRP).

DCEN provides the deepest penetration. In DCEN, the electrons leave the negative electrode and flow to the base metal. Most of the heat, approximately 70 percent, is developed at the base metal. Only 30 percent of the heat is developed in the electrode. See Fig. 21-4.

When DCEP is used, the electrode is positive and the base metal is negative. Electrons leave the base metal and flow toward the positive electrode. In this case, approximately 70 percent of the heat is developed in the electrode and only 30 percent is developed at the base metal. The weld bead is wide and shallow. See Fig. 21-4. The electrode used during DCEP welding is larger than one used during DCEN for the same current. This allows the electrode to handle the higher heat of DCEP. A small-diameter electrode will melt using DCEP if there is too much current.

One major advantage of DCEP is the cleaning action it has on certain metals, like aluminum and magnesium. Oxides on the base metal surface are removed. This results in better quality welds, Fig. 21-5.

Alternating current (AC) welding

Alternating current (AC) welding combines the characteristics of DCEN and DCEP. The electrode alternates between negative and positive. This change occurs 120 times a second, so the arc stops and restarts 120 times a second. There is an equal amount of heat on the base metal and the electrode. Penetration in AC welding is less than it is in DCEN, but greater than DCEP. A cleaning action occurs, as in DCEP. See Fig. 21-5.

Alternating current should flow equally in both directions. However, current in the electrode *positive* half cycle is less than in the electrode *negative* half cycle. In some cases, current does not flow at all in the elec-

Fig. 21-3. The control panel on one type of GTAW machine.   (Miller Electric Mfg. Co.)

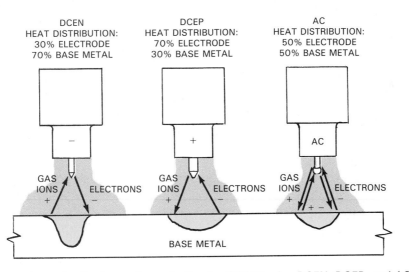

Fig. 21-4. The approximate heat distribution and penetration for GTAW using DCEN, DCEP, and AC. Notice the direction of travel for the gas ions and the electrons in each case. In AC, the gas ions and electrons are constantly reversing direction. Notice the larger diameter electrode needed for DCEP.

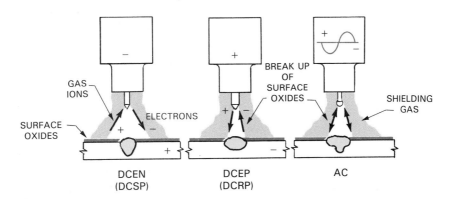

Fig. 21-5. A diagram showing how surface oxides are broken up when DCEP or AC current are used. This is called a cleaning action. The heavy gas ions strike the surface and cause the oxides to break up near the weld bead. No cleaning action takes place when DCEN is used.

trode positive half cycle, Fig. 21-6. When current does not flow in the electrode positive half cycle, it is said to be *rectified* (stopped). When the current in the electrode positive half cycle is less than desired, it is referred to as *partially rectified current.*

DCEP current may be kept from being totally rectified by using a high-frequency, high-voltage supply. This several-thousand-volt supply jumps the gap each time the arc stops. The arc stops 120 times per second when AC welding; this high frequency arc creates a path. It allows the arc to restart, especially during the DCEP half cycle. The amount of the high frequency that a welding machine supplies is adjustable.

A balanced AC cycle occurs when the current during the electrode positive and electrode negative half cycles are equal. Capacitors in the welding machine are used to obtain a balanced AC cycle.

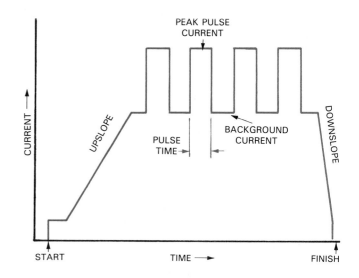

Fig. 21-7. Pulsed GTAW. The peak current is used for welding. The lower background current maintains the arc and allows the weld to cool.

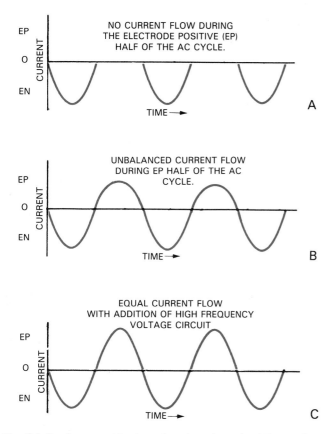

Fig. 21-6. As current is plotted against time, the AC waveform can be seen. A—In this curve, the electrode positive half cycle is rectified (stopped). B—An unbalanced curve where partial rectification is occuring. C—A balanced curve that results from the use of capacitors in the welding circuit.

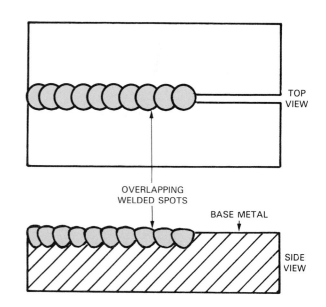

Fig. 21-8. Two views of the weld bead created by a pulsed current GTAW. A series of overlapping welded spots makes up the weld bead.

Pulsed current is excellent for out-of-position welding. The metal becomes molten during the peak current and forms a spot. The weld then has a chance to cool as the welder moves the torch a short distance during the background current period. Another spot forms during the following peak current period. This pulsed current produces a weld that is actually a series of overlapping welded spots. See Fig. 21-8.

TORCHES

GTAW torches perform many functions. The four major functions are:

1. Hold the electrode.

Pulsed GTAW

Certain GTAW welding machines produce a pulsed current. *Pulsed current* is like direct current except that it has two current settings. There is a high current period followed by a low current period. The high current is the *peak current;* the low current is the *background current.* See Fig. 21-7.

2. Make good electrical contact with the electrode.
3. Direct shielding gas to the weld area.
4. Provide a means to cool the electrode.

Fig. 21-9 shows an exploded view of a GTAW torch. A *collet* is used to hold the electrode and make good electrical contact. Collets are made in different sizes; each size for one specific size of electrode. A *collet body* is used to support the collet. Collet bodies and collets are usually available as a set. The set is used for one specific electrode size only. Whenever the electrode size is changed, the collet and collet body must also be changed. Collets have a slot in them. When the *end cap* is tightened, the collet tightens around the electrode. A tight fit holds the electrode in its proper place for welding. It also provides a good electrical contact.

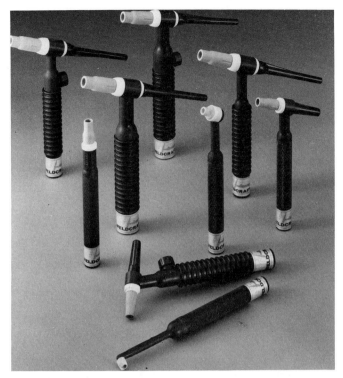

Fig. 21-10. Some of the many torch styles, nozzle sizes, and end cap lengths used for GTAW. (Weldcraft Products, Inc.)

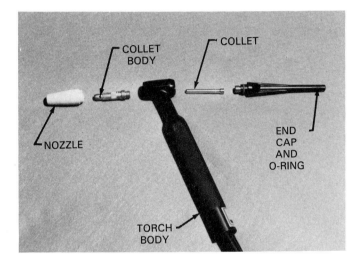

Fig. 21-9. An exploded view of the parts of a GTAW torch.

The *torch body* contains connections for shielding gas, water, and electricity. An O-ring and an end cap are used to prevent shielding gas from escaping. They also prevent the atmosphere from getting into the torch. End caps are available in different lengths.

The *nozzle,* or cup, is used to direct the shielding gas to the weld area. A number of nozzle sizes are available. See Fig. 21-10. If the nozzle that is selected is too small, there will not be enough shielding gas to cover the weld area. If a too-large nozzle is used, shielding gas will be wasted. The nozzle must be able to withstand the high temperatures produced by the arc. Most nozzles are made of ceramic or metal.

A gas lens may be used in a GTAW torch. A *gas lens* is a series of very fine stainless steel wire screens. It makes the shielding gas exit the nozzle in a column. When using a gas lens, the electrode can stick out of the cup farther. A welder can see the weld better. Hard-to-reach areas, like inside corner joints, can be welded more easily because the electrode sticks out farther.

The heat that develops in the electrode, cables, and nozzle must be removed or the torch will overheat.

GTAW torches are either gas- or water-cooled. See Fig. 21-11. *Gas-cooled torches* are used for light duty work and are not recommended for use with over 200 A. *Water-cooled torches* are used to carry currents over 200 A. They have a continuous flow of water passing through the torch body to remove heat.

CABLES AND HOSES

Cables convey the electrical current to the torch. A hose carries the shielding gas to the torch. The cables also carry the water to and from the torch when a water-cooled torch is used. Checking the cables and hoses for wear or leaks once a day is a good idea. When cables are installed properly, little or no adjustment is needed.

In Fig. 21-11B, the shielding gas and water supply to the torch have their own hoses. The power cable, however, has a dual purpose. It carries the electrical current to the torch and is also used to take water away from the torch. The water cools the power cable. Fig. 21-12 shows a cable assembly for a water-cooled torch. Each fitting shown is attached to the welding machine.

SHIELDING GASES

Inert shielding gases are used in GTAW. An inert gas does not react with the weld. Two types of shielding gas are used: argon and helium. Each gas can be used by itself, or a combination of the two can be used. Each has advantages and disadvantages. It is easier to

213

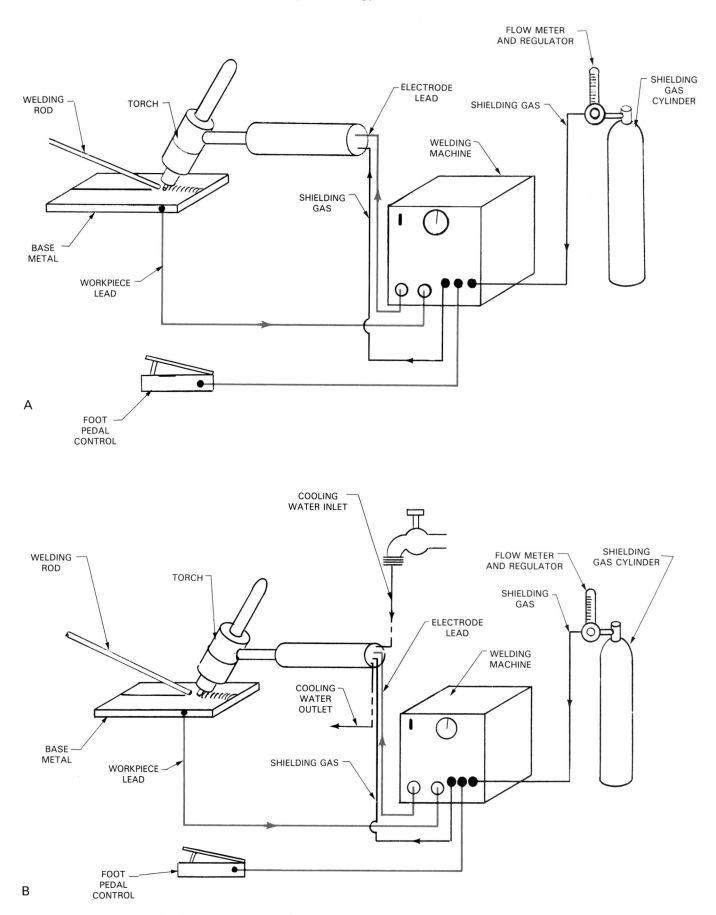

Fig. 21-11. A—A drawing of a gas-cooled GTAW outfit. Notice that the shielding gas hose and electrode lead enter the torch. B—A drawing of a water-cooled GTAW outfit. Note that the electrode lead and hoses for cooling water and shielding gas enter the torch. A cooling water hose exits the torch.

Fig. 21-12. A water-cooled GTAW torch with its hoses and fittings. The water outlet hose also carries the current to the torch and electrode.

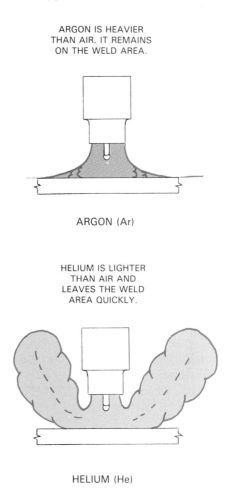

ARGON IS HEAVIER THAN AIR. IT REMAINS ON THE WELD AREA.

ARGON (Ar)

HELIUM IS LIGHTER THAN AIR AND LEAVES THE WELD AREA QUICKLY.

HELIUM (He)

Fig. 21-13. The efficiency of argon and helium as shielding gases. Argon is heavier and stays on or near the weld area for a relatively long period of time. Helium is a lighter gas. It rises rapidly and leaves the weld area.

start and maintain an arc with argon. Helium provides a hotter arc and provides for faster travel speeds. It is commonly used in automatic welding. Helium, however, is more expensive than argon. Less shielding gas is required when using argon because it is heavier than air. Argon covers the weld area better when welding in the flat position. See Fig. 21-13.

Caution: Argon will displace oxygen. When welding in a closed area, there must be good ventilation to remove the shielding gas and supply fresh air. Suffocation can occur if good ventilation is not provided.

REGULATORS AND FLOW METERS

Argon and helium are available in cylinders. Care must be taken when handling, moving, and storing gas cylinders. Chapter 4 provides detailed information regarding cylinders.

A regulator is used to reduce the cylinder pressure to a usable pressure. A flow meter controls the amount of shielding gas that goes to the welding torch. Flow meters measure gas flow in cubic feet per hour (cfh) or liters per minute (Lpm). Argon is heavier than helium, so an argon flow meter is calibrated differently from a helium flow meter. Sometimes, there are different scales on opposite sides of a single flow meter. Each scale is calibrated for a different shielding gas. Fig. 21-14 shows a typical flow meter.

The gas flow is adjusted using a screw or knob on the flow meter. See Chapters 4 and 17 for additional information on regulators and flow meters.

As discussed earlier, a shielding gas covers and protects the electrode and the base metal during GTAW. If you stopped welding and the shielding gas also stopped, oxygen and other gases could contaminate the weld. The tungsten electrode would then become oxidized. To prevent this, shielding gas flow continues

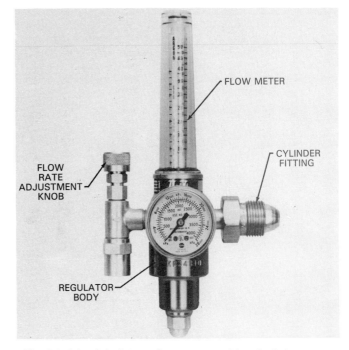

Fig. 21-14. A ball-type flow meter with a built-in pressure regulator. (Victor Equipment Co.)

215

for a short time after the weld current stops. This gas flow is called *post flow.* The length of time that the gas flows is set on the welding machine. Refer to Fig. 21-3 for position of the post flow timer.

REMOTE CURRENT CONTROL

A foot pedal or thumb switch is commonly used in GTAW. The *foot pedal* and *thumb switch* are remote control devices. They control the current and turn shielding gas on and off. These devices can be used to turn the current on or off and also to control the current within the range on the welding machine. See Fig. 21-15.

A button or switch on the welding machine controls the foot pedal or thumb switch. The button or switch has two positions; one labeled *control panel* and the other labeled *remote.* With the switch or button in the control panel position, the foot pedal or thumb control is used as an ON-OFF switch. The current that is preset on the welding machine is delivered to the torch.

In the remote position, the foot pedal or thumb switch is used to vary the current within the range preset on the welding machine. The foot pedal allows more current to flow as it is pressed down. This gives a welder additional control of the welding operation.

PROTECTIVE EQUIPMENT

Proper protective clothing must be worn for GTAW, as for other welding processes. Proper clothing includes a shirt with long sleeves, pants, and gloves. Clothing should be fire-resistant. Flammable objects like matches and lighters should not be carried in your pockets. Gloves worn for GTAW are usually thinner than those used for SMAW or GMAW. GTAW requires good skill in handling the torch and filler rod. Thin gloves are used so the welder can get a good feel for the torch and filler metal.

A GTAW welding arc is very intense. It is not shielded by flux or fumes, so a No. 10, 12, or 14 lens is recommended in your welding helmet. A clear cover lens should be placed in front of the shade lens. Flash goggles with a No. 2 lens should be worn under your helmet.

REVIEW QUESTIONS

Write your answers on a separate sheet of paper.

1. List two names other than GTAW that are used to refer to the gas tungsten arc welding process.
2. A tungsten electrode is called a _____, since it does not become part of the weld.
3. A _____ _____ welding machine is used in GTAW. This type of machine allows the welder to vary the _____ slightly by varying the arc length.
4. Which type of welding current applies most of the heat to the base metal and gives the best penetration?
5. What is used to prevent the current from stopping during the electrode positive half cycle when AC welding?
6. Which of the following part(s) of the torch must also be changed when a different size tungsten electrode is used?
   a. Torch body.
   b. Collet.
   c. End cap.
   d. Collet body.
7. Which of the cables on a water-cooled torch serves two purposes? What are those two purposes?
8. A _____ is used to reduce the cylinder pressure of shielding gas to a usable level. A _____ is used to control the amount of shielding gas used.
9. Why is post flow of shielding gas used?
10. Gloves worn for GTAW are thick leather to protect the welder from the intense heat and rays of the arc. True or False?

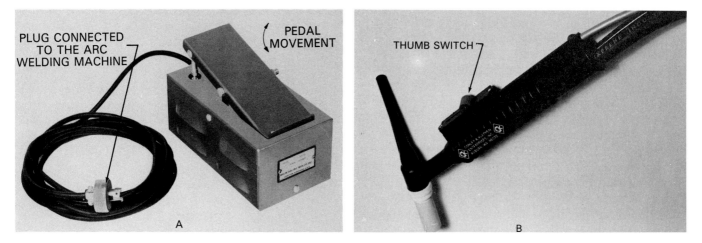

Fig. 21-15. A—A foot pedal for remote current control. Pressing down the foot switch also turns on the water and shielding gas. (Miller Electric Mfg. Co.) B—A thumb-operated current control switch mounted on a GTAW torch. The amperage is varied by moving the sliding thumb switch. (CK Worldwide, Inc.)

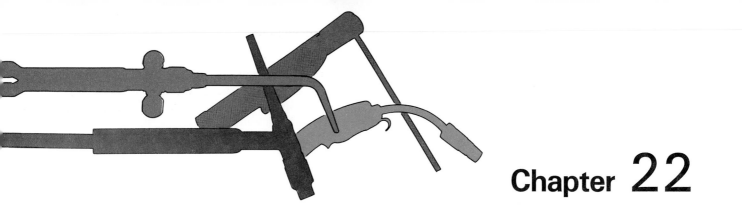

Chapter 22

Gas Tungsten Arc Welding: Assembly and Adjustment

After studying this chapter, you will be able to:
- ☐ Assemble a GTAW welding outfit.
- ☐ Adjust the drive mechanism for the proper pressure and alignment.
- ☐ Prepare an electrode for GTAW.
- ☐ Adjust the shielding gas flow meter for the proper flow rate.
- ☐ Identify electrode type designations for GTAW electrodes.
- ☐ Select the proper current amount and type for the metal to be welded.
- ☐ Know and be able to use the processes used to clean metal for GTAW.

TECHNICAL TERMS

Ceria, circuit breaker, contactor switch, current switch, electrode tip, lanthana, thoria, zirconia.

EQUIPMENT ASSEMBLY

Equipment used in a GTAW outfit must be assembled properly. The main piece of equipment in a welding station is the welding machine. All cables and hoses, the shielding gas supply, and the welding torch are connected to the welding machine. Equipment used to control the flow of shielding gas must be assembled and adjusted. The torch must be assembled correctly, as well.

ELECTRICAL CONNECTIONS

GTAW welding machines are connected to 115 V or 220 V electrical circuits. Machines are often wired into a *circuit breaker* panel or a fuse box. Only a qualified electrician should connect, inspect, or repair the wiring inside a circuit breaker panel or fuse box. A welder should not try to repair wiring inside a welding machine.

Know where the circuit breaker panel for the welding shop is located. In case of an emergency, know how to turn off power at the circuit breaker. In addition to turning off the welding machine at the end of each shift, some welding shops turn off the power at the circuit breaker at the end of the day. A circuit breaker is shown in Fig. 22-1.

CABLE AND HOSE CONNECTIONS

A welding machine has places to connect all needed cables and hoses. Electrical connections for the electrode and workpiece lead are mechanical. They must be tight to make good contact. Water and shielding gas connections are threaded to help prevent leaks. Use the proper size wrench to tighten these connections. Fig. 22-2 shows the cable and hose connections on a welding machine. See Fig. 21-13 for connections on a torch.

Most machines have a remote control connection. This is where the foot pedal, thumb switch, or other control feature is connected.

CYLINDER, REGULATOR, AND FLOW METER

Cylinders containing compressed gases must be handled with great care. They must be secured to a wall, column, or welding hand truck with chains or steel

Fig. 22-1. A disconnect switch or circuit breaker for an arc welding machine. This switch may be turned off to make the welding machine power switch inoperable.

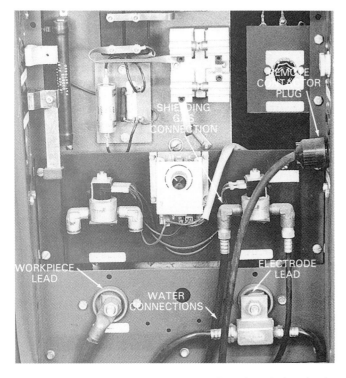

Fig. 22-2. Connections for the electrode and workpiece leads, water, and shielding gas are made under the hinged panel of this welding machine.

bands. This keeps the cylinder from falling over and damaging the cyinder valve. See Chapter 4 for more information.

Once the cylinder is secure, a regulator and flow meter can be attached to it. First, clean the outlet of the cylinder valve. Make sure the outlet is not pointing toward anyone, including yourself. Quickly open, then close, the cylinder valve. After the outlet is clean, a regulator and flow meter can be attached. Thread the regulator nut onto the cylinder valve. If a ball float-type flow meter is used, the flow meter must be in a vertical position. See Fig. 22-3. Tighten the regulator nut with an open-end wrench of the correct size. Do not overtighten or you may damage the threads.

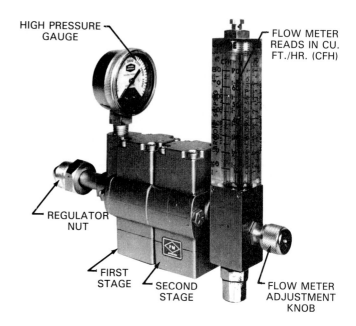

Fig. 22-3. A two-stage regulator with a flow meter. This flow meter is calibrated for four different shielding gases. (ESAB Welding and Cutting Products)

ASSEMBLING THE TORCH

A separate collet and collet body are needed for each diameter of electrode. Choose the correct-size collet and collet body for the electrode's diameter. Assemble the torch as follows:

1. Thread the collet body into the torch body.
2. Slide the collet into the collet body from the top of the torch, as shown in Fig. 22-4.
3. Thread an end cap with an O-ring loosely into the torch body. Do not tighten the end cap yet.
4. Select the gas nozzle you will be using. Thread it into the torch body. A nozzle is also called a cup.
5. Insert the electrode into the collet. When correctly inserted, the electrode will extend past the end of the gas cup only a short distance. The distance depends on the type of joint being welded, as shown in Fig. 22-5.

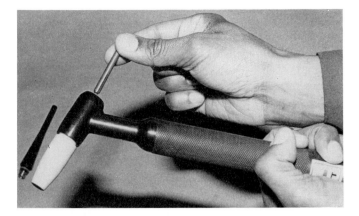

Fig. 22-4. A collet being inserted into the body of a GTAW torch. Notice the end cap in the background.

Fig. 22-6. The electrode is clamped tightly by the collet when the end cap is tightened.

6. When the electrode extends the right amount, tighten the end cap. See Fig. 22-6. This will make the collet clamp the electrode and secure it. To remove or adjust the electrode, loosen the end cap.

ADJUSTING THE SHIELDING GAS FLOW METER

After the regulator and flow meter are attached to the cylinder, the correct rate of gas flow can be set. Flow meters use a knob or adjusting screw to control gas flow. Some flow meters use a gauge to indicate the gas flow. The most common type of flow meter, however, uses a ball float in a clear tube. The gas flow is read at the *top* of the ball. Fig. 22-7 shows how to read this type of flow meter properly.

To set the gas flow rate, shielding gas must be flowing. Open the cylinder valve, then use the foot pedal or thumb switch to make the shielding gas flow. Do not strike an arc. Another way to make the shielding gas flow is to set the post flow timer for about 10 seconds. Then, press the foot pedal or thumb switch for a second or less. While the shielding gas is flowing, adjust the flow meter knob to obtain the proper rate for the welding job you will be doing. Figs. 22-8 through 22-12 list gas flow rates for use on different base metals.

Do not change the regulator pressure setting. Shielding gas flow meters are calibrated for a given pressure. If this pressure is changed, flow rate readings will not be accurate.

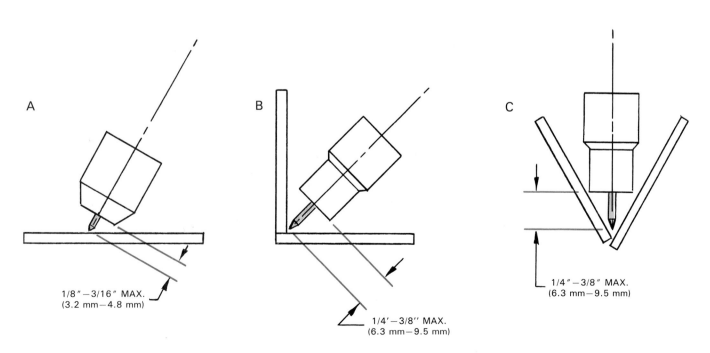

Fig. 22-5. The electrode extension distance varies for different types of joints. It should never be greater than the nozzle diameter.

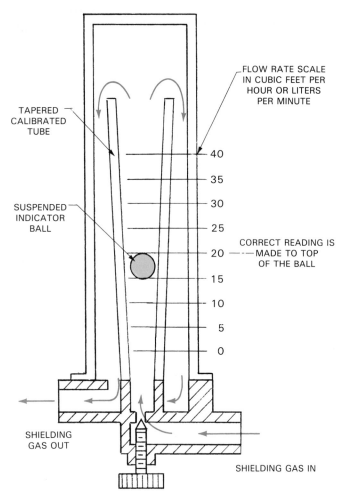

TAPERED CALIBRATED TUBE

FLOW RATE SCALE IN CUBIC FEET PER HOUR OR LITERS PER MINUTE

— 40

— 35

— 30

— 25

SUSPENDED INDICATOR BALL

— 20 --- CORRECT READING IS MADE TO TOP OF THE BALL

— 15

— 10

— 5

— 0

SHIELDING GAS OUT

SHIELDING GAS IN

Fig. 22-7. A schematic drawing of a flow meter. The calibration tube is tapered so that a greater flow of gas is required to hold the ball at a higher level. The flow reading is made at the top of the ball.

SELECTING AND PREPARING THE ELECTRODE

The process of selecting and preparing the electrode for GTAW includes:
- Selecting the type of electrode.
- Selecting the diameter of electrode.
- Selecting the shape for the tip.

The electrode used in GTAW may be one of the following types:
- Pure tungsten.
- Tungsten with one or two percent *thoria* (thorium dioxide).
- Tungsten with .15 to .40 percent *zirconia* (zirconium oxide).
- Tungsten with two percent *ceria* (cerium oxide).
- Tungsten with one percent *lanthana* (lanthanum oxide).

Selecting the correct type of tungsten electrode to use is very important. The advantages of each electrode type are described below.

Pure tungsten - Forms a ball on the end when heated.
    - Recommended for AC welding.
    - Can be used for DC welding.

Thoria tungsten - Makes starting the arc easier.
    - Allows electrons to flow easier.
    - Electrode can carry more current than a pure tungsten electrode of the same size.
    Recommended for DC welding.

Zirconia tungsten - Similar to pure tungsten, but with easier arc-starting abilities.
    - Carries the same current as thoria tungsten.
    - Used for AC welding.

Metal Thickness	Joint Type	Tungsten Electrode Diameter	Filler Rod Diameter (If req'd.)	Amperage	Gas		
					Type	Flow cfh	Flow Lpm
1/16 in. (1.59 mm)	Butt	1/16 in. (1.59 mm)	1/16 in. (1.59 mm)	60-70	Argon	15	7.08
	Lap	1/16 in.	1/16 in.	70-90	Argon	15	
	Corner	1/16 in.	1/16 in.	60-70	Argon	15	
	Fillet	1/16 in.	1/16 in.	70-90	Argon	15	
1/8 in. (3.18 mm)	Butt	1/16-3/32 in. (1.59-2.38 mm)	3/32 in. (2.38 mm)	80-100	Argon	15	7.08
	Lap	1/16-3/32 in.	3/32 in.	90-115	Argon	15	
	Corner	1/16-3/32 in.	3/32 in.	80-100	Argon	15	
	Fillet	1/16-3/32 in.	3/32 in.	90-115	Argon	15	
3/16 in. (4.76 mm)	Butt	3/32 in. (2.38 mm)	1/8 in. (3.18 mm)	115-135	Argon	20	9.44
	Lap	3/32 in.	1/8 in.	140-165	Argon	20	
	Corner	3/32 in.	1/8 in.	115-135	Argon	20	
	Fillet	3/32 in.	1/8 in.	140-170	Argon	20	
1/4 in. (6.35 mm)	Butt	1/8 in. (3.18 mm)	5/32 in. (4.0 mm)	160-175	Argon	20	9.44
	Lap	1/8 in.	5/32 in.	170-200	Argon	20	
	Corner	1/8 in.	5/32 in.	160-175	Argon	20	
	Fillet	1/8 in.	5/32 in.	175-210	Argon	20	

Fig. 22-8. Variables for manually welding mild steel using GTAW and DCEN (DCSP) polarity.

Metal Thickness	Joint Type	Tungsten Electrode Diameter	Filler Rod Diameter (If req'd.)	Amperage	Gas		
					Type	Flow cfh	Flow Lpm
1/16 in. (1.59 mm)	Butt	1/16 in. (1.59 mm)	1/16 in. (1.59 mm)	60-85	Argon	15	7.08
	Lap	1/16 in.	1/16 in.	70-90	Argon	15	
	Corner	1/16 in.	1/16 in.	60-85	Argon	15	
	Fillet	1/16 in.	1/16 in.	75-100	Argon	15	
1/8 in. (3.18 mm)	Butt	3/32-1/8 in. (2.38-3.18 mm)	3/32 in. (2.38 mm)	125-150	Argon	20	9.44
	Lap	3/32-1/8 in.	3/32 in.	130-160	Argon	20	
	Corner	3/32-1/8 in.	3/32 in.	120-140	Argon	20	
	Fillet	3/32-1/8 in.	3/32 in.	130-160	Argon	20	
3/16 in. (4.76 mm)	Butt	1/8-5/32 in. (3.18-4.0 mm)	1/8 in. (3.18 mm)	180-225	Argon	20	11.80
	Lap	1/8-5/32 in.	1/8 in.	190-240	Argon	20	
	Corner	1/8-5/32 in.	1/8 in.	180-225	Argon	20	
	Fillet	1/8-5/32 in.	1/8 in.	190-240	Argon	20	
1/4 in. (6.35 mm)	Butt	5/32-3/16 in. (4.0-4.76 mm)	3/16 in. (4.76 mm)	240-280	Argon	25	14.16
	Lap	5/32-3/16 in.	3/16 in.	250-320	Argon	25	
	Corner	5/32-3/16 in.	3/16 in.	240-280	Argon	25	
	Fillet	5/32-3/16 in.	3/16 in.	250-320	Argon	25	

Fig. 22-9. Variables for manually welding aluminum and aluminum alloys using GTAW with AC and high frequency.

Metal Thickness	Joint Type	Tungsten Electrode Diameter	Filler Rod Diameter (If req'd.)	Amperage	Gas		
					Type	Flow cfh	Flow Lpm
1/16 in. (1.59 mm)	Butt	1/16 in. (1.59 mm)	1/16 in. (1.59 mm)	40-60	Argon	15	7.08
	Lap	1/16 in.	1/16 in.	50-70	Argon	15	
	Corner	1/16 in.	1/16 in.	40-60	Argon	15	
	Fillet	1/16 in.	1/16 in.	50-70	Argon	15	
1/8 in. (3.18 mm)	Butt	3/32 in. (2.38 mm)	3/32 in. (2.38 mm)	65-85	Argon	15	7.08
	Lap	3/32 in.	3/32 in.	90-110	Argon	15	
	Corner	3/32 in.	3/32 in.	65-85	Argon	15	
	Fillet	3/32 in.	3/32 in.	90-110	Argon	15	
3/16 in. (4.76 mm)	Butt	3/32 in. (2.38 mm)	1/8 in. (3.18 mm)	100-125	Argon	20	9.44
	Lap	3/32 in.	1/8 in.	125-150	Argon	20	
	Corner	3/32 in.	1/8 in.	100-125	Argon	20	
	Fillet	3/32 in.	1/8 in.	125-150	Argon	20	
1/4 in. (6.35 mm)	Butt	1/8 in. (3.18 mm)	5/32 in. (4.0 mm)	135-160	Argon	20	9.44
	Lap	1/8 in.	5/32 in.	160-180	Argon	20	
	Corner	1/8 in.	5/32 in.	135-160	Argon	20	
	Fillet	1/8 in.	5/32 in.	160-180	Argon	20	

Fig. 22-10. Variables for manually welding stainless steel with GTAW and DCEN (DCSP).

Note that zirconia tungsten electrodes are recommended only for AC welding. Thoria tungsten is best for DC welding, but can be used for either AC or DC welding. When welding steel or stainless steel, a tungsten electrode with thoria is normally used. When AC-welding aluminum, use either a pure tungsten electrode or an electrode with zirconium.

WELDING ELECTRODE CLASSIFICATIONS

Each tungsten electrode type is given a classification by the American Welding Society. Fig. 22-13 lists the different classifications. Each letter and number in the classification has a meaning. The letters can be interpreted as follows:

Metal Thickness	Joint Type	Tungsten Electrode Diameter	Filler Rod Diameter (If req'd. Red)	Amperage[1] With Backup	W/O Backup	Gas		
						Type	Flow chf	Flow Lpm
1/16 in. (1.59 mm)	All	1/16 in. (1.59 mm)	3/32 in. (2.38 mm)	60	35	Argon	13	6.14
3/32 in. (2.38 mm)	All	1/16 in. (1.59 mm)	1/8 in. (3.18 mm)	90	60	Argon	15	7.08
1/8 in. (3.18 mm)	All	1/16 in. (1.59 mm)	1/8 in. (3.18 mm)	115	85	Argon	20	9.44
3/16 in. (4.76 mm)	All	1/16 in. (1.59 mm)	5/32 in. (4.0 mm)	120	75	Argon	20	9.44
1/4 in. (6.35 mm)	All	3/32 in. (2.38 mm)	5/32 in. (4.0 mm)	130	85	Argon	20	9.44
3/8 in. (9.53 mm)	All	3/32 in. (2.38 mm)	3/16 in. (4.76 mm)	180	100	Argon	25	11.80
1/2 in. (12.7 mm)	All	5/32 in. (4.0 mm)	3/16 in. (4.76 mm)	—	250	Argon	25	11.80
3/4 in. (19.05 mm)	All	3/16 in. (4.76 mm)	1/4 in. (6.35 mm)	—	370	Argon	35	16.52

1 - Use alternating current with a constant high frequency (AC-HF)

Fig. 22-11. Variables for manually welding magnesium using GTAW with AC and high frequency.

Metal Thickness	Joint Type	Tungsten Electrode Diameter	Filler Rod Diameter (If req'd. Red)	Amperage[1]	Gas		
					Type	Flow cfh	Flow Lpm
1/16 in. (1.59 mm)	All	1/16 in. (1.59 mm)	1/16 in. (1.59 mm)	110-150	Argon	15	7.08
1/8 in. (3.18 mm)	All	3/32 in. (2.38 mm)	3/32 in. (2.38 mm)	175-250	Argon	15	7.08
3/16 in. (4.76 mm)	All	1/8 in. (3.18 mm)	1/8 in. (3.18 mm)	250-325	Argon	18	9.50
1/4 in. (6.35 mm)	All	1/8 in. (3.18 mm)	1/8 in. (3.18 mm)	300-375	Argon	22	10.38
3/8 in. (9.53 mm)	All	3/16 in. (4.76 mm)	3/16 in. (4.76 mm)	375-450	Argon	25	11.80
1/2 in. (12.7 mm)	All	3/16 in. (4.76 mm)	1/4 in. (6.35 mm)	525-700	Argon	30	14.16

1 - Use DCEN (DCSP)

Fig. 22-12. Variables for manually welding deoxidized copper using GTAW with DCEN (DCSP).

AWS Classification	Defined	Color
EWP	Pure Tungsten	Green
EWTh-1	1% Thoria	Yellow
EWTh-2	2% Thoria	Red
EWZr-1	0.15-0.40% Zirconia	Brown
EWCe-2	2% Ceria	Orange
EWLa-1	1% Lanthana	Black
EWG	Other	Gray

Fig. 22-13. The AWS tungsten electrode classifications and the color codes painted on the ends of the electrodes. (AWS A5.12)

E- electrode
W - tungsten
P - pure
Th - electrode contains thoria
Zr - electrode contains zirconia
Ce - electrode contains ceria
La - electrode contains lanthana

The diameter of an electrode is determined by the amount of current used to weld. As the amount of current increases, so must the diameter of the electrode. Tungsten electrodes are available in diameters from .010″ − .250″ (.25 mm − 6.4 mm). The most common sizes and the current ranges of each electrode type are listed in Fig. 22-14. Select an electrode with the ability to carry the current needed for the job you will be welding.

Electrodes come in 3″ − 24″ (76.2 mm − 609.6 mm) lengths. The most common lengths for manual gas tungsten arc welding electrodes are 3″, 6″, and 7″ (76.2 mm, 152.4 mm, and 177.6 mm). Electrodes are purchased with either a ground or a chemically cleaned surface.

PREPARING THE TIP OF THE ELECTRODE

Finally, the *electrode tip* must be prepared for welding. The tip of an electrode can be a blunted point or a ball end. For DC welding, grind the electrode to a point. After forming the point, blunt it slightly. The grinding marks on the electrode should run lengthwise. See Fig. 22-15.

Most welding shops will have one or more grinding wheels that are used strictly for tungsten electrodes. Aluminum and other metals should never be ground on these wheels. Metal particles left on these wheels may become embedded in a tungsten electrode. A contaminated electrode will give poor results when welding. Caution: Always wear safety glasses when you use a grinding wheel.

CURRENT RANGE—AMPERES				
Pure Tungsten Diameter (mm)	DCEN or DCSP Argon	DCEP or DCRP Argon	AC-HF Argon	AC Balanced Wave-Argon
.010'' (.254)	Up to 15	*	Up to 15	Up to 10
.020'' (.508)	5-20	*	5-20	10-20
.040'' (1.02)	15-80	*	10-60	20-30
1/16'' (1.59)	70-150	10-20	50-100	30-80
3/32'' (2.38)	125-225	15-30	100-160	60-130
1/8'' (3.18)	225-360	25-40	150-210	100-180
5/32'' (3.97)	360-450	40-55	200-275	160-240
3/16'' (4.76)	450-720	55-80	250-350	190-300
1/4'' (6.35)	720-950	80-125	325-450	250-400
2% Thorium Alloyed Tungsten Diameter (mm)				
.010'' (.254)	Up to 25	*	Up to 20	Up to 15
.020'' (.508)	15-40	*	15-35	5-20
.040'' (1.02)	25-85	*	20-80	20-60
1/16'' (1.59)	50-160	10-20	50-150	60-120
3/32'' (2.38)	135-235	15-30	130-250	100-180
1/8'' (3.18)	250-400	25-40	225-360	160-250
5/32'' (3.97)	400-500	40-55	300-450	200-320
3/16'' (4.76)	500-750	55-80	400-550	290-390
1/4'' (6.35)	750-1000	80-125	600-800	340-525
Zirconium Alloyed Tungsten Diameter (mm)				
.010'' (.254)	*	*	Up to 20	Up to 15
.020'' (.508)	*	*	15-35	5-20
.040'' (1.02)	*	*	20-80	20-60
1/16'' (1.59)	*	*	50-150	60-120
3/32'' (2.38)	*	*	130-250	100-180
1/8'' (3.18)	*	*	225-360	160-250
5/32'' (3.97)	*	*	300-450	200-320
3/16'' (4.76)	*	*	400-550	290-390
1/4'' (6.35)	*	*	600-800	340-525
*NOT RECOMMENDED				
The figures listed are intended as a guide, and are a composite of recommendations from American Welding Society and electrode manufacturers.				

Fig. 22-14. Suggested current ranges for various types and sizes of tungsten electrodes.

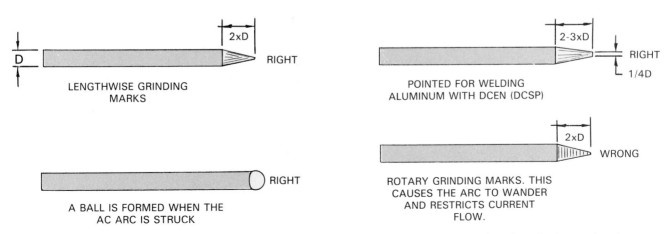

Fig. 22-15. Electrode tip shapes for welding with DC and AC. Grinding marks should run lengthwise on the electrode for better arc control.

For AC welding, the electrode tip is formed into a ball. This is done by striking an arc on a piece of steel or copper. When an arc is struck, the electrode tip will melt and form a ball shape. Use AC or DCEP to form the ball. The ball should be no larger than the diameter of the electrode. For a ball slightly smaller in diameter than the electrode, taper the electrode first on a grinding wheel, then form the ball. Fig. 22-16 shows an electrode with a correctly formed taper and one with a correctly formed ball.

Whenever you use a new electrode, you must first grind or form it into the correct shape. Also, whenever you dip an electrode into the molten weld pool, it must be reground before being used again. This removes any contamination from the electrode.

Never allow a tungsten electrode to become so hot that part of it falls into the weld pool. If this happens, change to a larger size electrode or use less welding current.

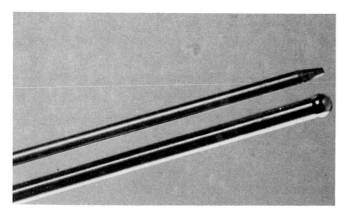

Fig. 22-16. An electrode tapered and blunted for DC welding and a properly balled electrode for use with AC.

FILLER METAL

Filler metal is needed to make most welding joints. Filler metal used in GTAW is solid wire. It comes in precut lengths or in large spools or coils. Precut lengths, or *rods,* are usually 36″ (914 mm) long. Spools or coils are used for automatic GTAW, but also can be used for manual welding. The welder cuts the wire to the length needed. The wire diameter can be as small as .015″ (.38 mm). The most common *precut* wire diameters are from 1/16″ −3/16″ (1.6 mm−4.8 mm).

For a given welding job, filler wire with a composition similar to the base metal is usually used. There are many special alloy filler metals available for specific types of welds. The most common filler metals used in GTAW are shown in Fig. 22-17. Note that each type begins with the letters ER. The E stands for electrode and the R stands for rod.

NOTE: The filler metals used when GTAW carbon

BASE METAL	RECOMMENDED FILLER METAL
Aluminum	ER1100, ER4043, ER5356
Copper and copper alloys	ERCu, ERCuSi-A, ERCuAl-A1
Magnesium	ERAZ61A, ERAZ92A
Nickel and nickel alloys	ERNi-1, ERNiCr-3, ERNiCrMo-3
Carbon steel	ER70S-2, ER70S-6
Low alloy steel	ER80S-B2, ER80S-D2
Stainless steel	ER308, ER308L, ER316, ER347
Titanium	ERTi-1
Zirconium	ERZr2

Fig. 22-17. The most-often-recommended filler metals for use when GTAW different base metals.

steel are not copper-coated, as are the filler metals used for oxyfuel gas welding of carbon steel.

WELDING MACHINE SETTINGS

Each GTAW welding machine has controls on its front panel, Fig. 22-18. You must be able to properly adjust these controls to obtain a good weld. The first step is to select the type of current and the amount of current you will use.

Fig. 22-18. Front panel of a typical gas tungsten arc welding machine. Controls are provided somewhere on each machine for current selection, shielding gas post flow time, polarity selection, and remote or machine control of current.

SELECTING THE WELD CURRENT

Before selecting the weld current, these questions must be answered:
• What type of metal will be welded?
• How thick is the metal?
• What type joint will be used?
• In what position will the welding be done?

After these questions are answered, you can choose the type of welding current. Chapter 21 discussed the three different types of current used in GTAW. The types are: direct current electrode positive (DCEP), direct current electrode negative (DCEN), and alternating current (AC). Fig. 22-19 lists which current is preferred for welding various metals.

Base Material	Direct Current		Alternating Current
	DCEN	DCEP	
	DCSP	DCRP	AC
Aluminum up to 3/32	P	G	E
Aluminum over 3/32	P	P	E
Aluminum bronze	P	G	E
Aluminum castings	P	P	E
Beryllium copper	P	G	E
Brass alloys	E	P	G
Copper base alloys	E	P	G
Cast Iron	E	P	G
Deoxidized copper	E	P	P
Dissimilar metals	E	P	G
Hard facing	G	P	E
High alloy steels	E	P	G
High carbon steels	E	P	G
Low alloy steels	E	P	G
Low carbon steels	E	P	G
Magnesium up to 1/8"	P	G	E
Magnesium over 1/8"	P	P	E
Magnesium castings	P	G	E
Nickel & Ni-alloys	E	P	G
Stainless steel	E	P	G
Silicon bronze	E	P	P
Titanium	E	P	G

E-Excellent   G-Good   P-Poor

Fig. 22-19. Suggested current and polarity choices for use when GTAW on various base metals. (Welding and Fabrication Data Book)

Next, you can select the *amount* of current you will use for welding. Figs. 22-8 — 22-12 list the recommended currents for welding different base metals.

To choose the desired current type, set the selector switch to AC, DCEP, or DCEN. Set the current range selector to the range (low, medium, high) you need for the amount of current you will be using. Then set the desired current, using the current control knob.

When AC welding, you must use continuous high fre-quency. When DC welding, you can use high frequency to start the arc, but it does not have to be used at all. Set the high frequency control to the desired position.

The *contactor switch* has two positions, standard and remote. In the standard position, open circuit voltage is applied to the welding torch at all times. To strike an arc, the electrode is touched to the work. This is similar to shielded metal arc welding. In GTAW, however, the contactor switch is usually placed in the remote position. A foot pedal or thumb switch is used for on/off control of the current and on/off control of the shielding gas.

The current can be set to a specific amount, or can be varied by the welder. To set a specific amount, the *current switch* is placed in the *panel* position. In this position, the current setting on the machine will be the current used to weld with; it cannot be changed while welding. If the current switch is in the *remote* position, the current can be changed while welding. This is done by pressing a foot pedal or a thumb switch. Place the contactor and current switches in the positions you have selected.

After the arc stops, shielding gas will continue to flow. This is called post flow. The post flow timer on the machine controls how long shielding gas will flow. A general rule is one second of gas flow for each 10 amps of current. Even when welding with less than 100 amps, a minimum of 10 seconds of post flow should be used.

Some machines have a voltmeter and an ammeter. These show the voltage and current used when welding. All welding machines have a power switch. This is used to turn the machine on and off.

When you set up a GTAW machine, use this checklist:
1. Set the type of current.
2. Set the current range.
3. Set the desired current amount.
4. Set the high frequency switch to the desired position.
5. Set the contactor switch to the desired position.
6. Set the current switch to the desired position.
7. Set the post flow timer for the desired post flow.

PREPARING METAL FOR WELDING

Gas tungsten arc welding is a very clean process. Prior to welding, the base metal should be cleaned. In industry, chemicals are used to remove grease and other dirt from the surface of metals. Chemicals are also used to remove the oxides from aluminum and magnesium. If chemical cleaning is not used, then mechanical cleaning is done.

Mechanical cleaning is done by scrubbing a metal surface with an abrasive cloth or a wire brush. A grinding wheel can also be used. All rust, grease, oil, and oxides must be removed before welding. If not removed, they could cause a weak or defective weld.

Cleaning is done after all edge preparations (chamfers, grooves, beveled edges) are completed.

REVIEW QUESTIONS

Write your answers on a separate sheet of paper.

1. What is a welding machine often wired into?
2. What type of connection is used for shielding gas and water hoses?
3. In what position must a ball float-type flow meter be mounted?
4. In order to read the flow in a ball float-type flow meter, where should the ball be with respect to the gauge lines?
5. List three of the major types of electrodes used in GTAW.
6. When grinding an electrode, in what direction should the grind marks go?
7. What kind of a tip should an electrode used for DCEN welding have?
   a. Tapered with a sharp point.
   b. Tapered with a blunted point.
   c. Balled on the end.
8. How is the electrode held in place?
9. What type of current requires the high frequency to be used continuously?
10. If you want to set the current to a certain amount and not vary it during welding, what position should the current switch be in?

A titanium air duct for a turbine engine being welded, using GTAW. The welding is being done in the flat welding position.
(Miller Electric Mfg. Co.)

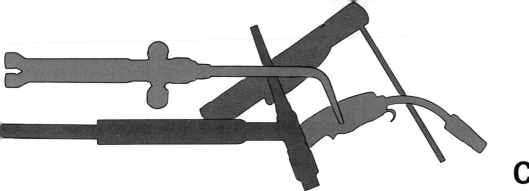

Chapter 23

GTAW: Flat Welding Position

After studying this chapter, you will be able to:
- ☐ Describe the GTAW process.
- ☐ Determine the appropriate filler rod to use when GTAW.
- ☐ Lay a bead on a plate using GTAW.
- ☐ Make a fillet weld on a lap joint in the flat welding position.
- ☐ Make a fillet weld on a T-joint in the flat welding position.
- ☐ Weld a butt joint in the flat welding position.
- ☐ Describe the use of a backing when welding aluminum using GTAW.
- ☐ Identify various weld defects.

TECHNICAL TERMS

Backing, hot shortness, oxidized, porosity, tungsten inclusions, undercutting, unstable arc, weld pool.

GAS TUNGSTEN ARC WELDING PRINCIPLES

Gas tungsten arc welding is done by striking an arc between the tungsten electrode and the base metal. The welding arc melts a small spot on the base metal. This small melted area of base metal is called the *weld pool.* *Filler metal* is often used, but not always required. Filler metal is added to the molten weld pool to fill the weld joint. Fig. 23-1 shows a welder using the GTAW process.

In GTAW, the torch is moved along the weld joint and filler metal added. This continues to the end of the joint. The speed of GTAW is similar to that of oxy-fuel gas welding.

The electrode, arc, and base metal are protected during welding by a shielding gas. Even after the welding arc stops, shielding gas continues to flow. This post flow protects the electrode and base metal from contamination as they cool.

PREPARING TO WELD

Before beginning to weld, you must assemble the equipment. Connect the torch cables to the welding machine. Connect a source of shielding gas. Check that all electrical, shielding gas, and water connections are

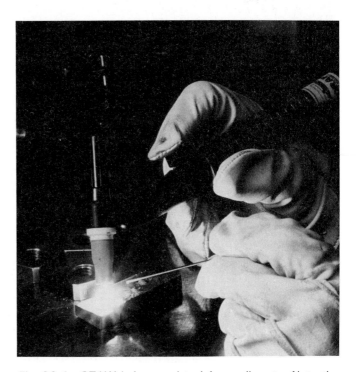

Fig. 23-1. GTAW being used to join small parts. Note the welding rod.

tight. Refer to Chapter 22 for further information on assembling equipment.

Select and set the current you will use on the machine, as described in Chapter 22. Fig. 23-2 shows the controls you may need to adjust on a welding machine. Next, choose the correct type and diameter of tungsten electrode for the welding you will be doing. Prepare the tip of the electrode as described in Chapter 22 and install it in the torch. Finally, adjust the shielding gas flow rate. See Chapter 22.

Make sure that the filter lens in your welding helmet is a #10, #12, or #14. Check that you are wearing the proper clothes for welding. Make sure all parts of your body are protected from the intense rays of the welding arc. Remove all combustible materials from the area where you will be welding.

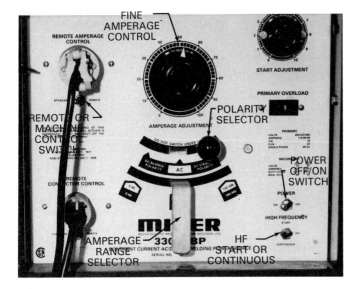

Fig. 23-2. Typical controls on a GTA welding machine.

SELECTING A FILLER ROD

Welding (filler) rods are available in many different alloys. Usually a filler rod with a composition similar to the base metal is used. Fig. 22-17 lists filler rods recommended for use with different base metals.

The diameter of the rod is determined by the size of the weld. Filler metal should have to be added to the weld only about once every two seconds to obtain the correct weld size.

If the rod diameter is too small, metal will have to be added very often. The rod will be used up quickly. If the rod is too large, metal will be added in too-large amounts. The large amount of filler metal will cool the weld pool, causing large bumps in the weld. These bumps will remain unless you take time to smooth them out by reheating the area.

When using filler metal, keep the end of the rod close to the arc. This protects the end of the rod with shielding gas.

USING THE GTAW TORCH

Hold the torch as shown in Fig. 23-3. This should be a comfortable grip. It is similar to holding a pencil. You can also hold the torch like a hammer, Fig. 23-4. This is used on larger torches. When welding out of position, this grip allows for greater flexibility and control of the torch.

Fig. 23-3. Light duty torches may be held like a pencil.

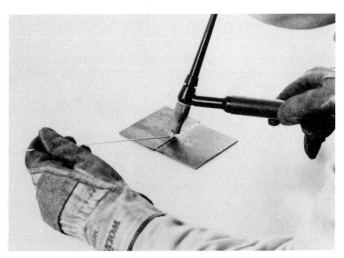

Fig. 23-4. This GTAW torch is being held like a hammer. Note the torch and welding rod angles.

If you use a foot pedal, place it on the floor where your foot can reach it comfortably. Sit or stand at the welding station. Place the metal to be welded in front of you on the table or in a welding fixture. You must be able to move the torch and filler metal along the joint without hitting anything. See Fig. 23-5.

Make sure the ground cable is securely attached to the table or to the workpiece. Now you are ready to begin welding.

Starting the arc

There are two different ways to start or strike an arc. One is to use high frequency. The other is to touch the electrode to the base metal.

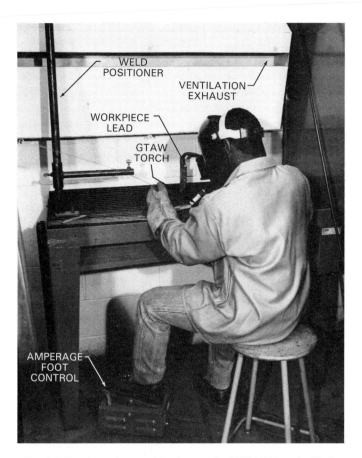

Fig. 23-5. A student welder in a typical GTAW booth. Notice the foot control, workpiece lead, and ventilation.

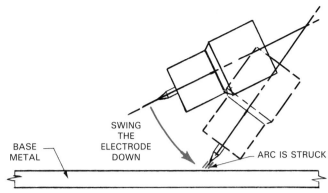

Fig. 23-6. Starting the GTAW arc with a swinging motion. The foot control is depressed and the electrode is swung down toward the base metal until the arc is struck.

The most common way to start the arc is to use high frequency. When you weld with AC, you use high frequency continuously. When high frequency is used to start the arc for DC welding, it will shut off automatically a few seconds after the arc stabilizes.

One way to start the arc with high frequency is to hold the electrode about 1/8″ (3.2 mm) above the base metal. Press the foot pedal or thumb switch. The high frequency arc will jump the gap. Wait about one or two seconds and the arc will stabilize. High frequency can actually jump a gap of more than 1/2″ (13 mm), but the main current will not flow with a gap this large.

You can also start the arc with high frequency by holding the electrode about 2″ (51 mm) above the base metal. Press the foot pedal or thumb switch, then quickly move the torch down toward the work. This can be a swinging motion, as shown in Fig. 23-6. When the electrode gets close enough to the work, the high frequency will jump the gap and start the arc.

The second way to start an arc is to touch the electrode to the work. Touch-starting is used only when DC welding. When using this method, very quickly touch the electrode to the work, then pull it back to about 3/32″ − 1/8″ (2.4 mm − 3.2 mm). Wait for the arc to become stable. Touch-starting the arc can contaminate the electrode. A contaminated electrode will produce an unstable arc. An unstable arc makes welding difficult and can be the cause of a poor quality weld. Refer to the WELD DEFECTS heading in this chapter for additional information.

After you start the arc, hold the torch so the electrode is about 1/16″ − 1/8″ (1.6 mm − 3.2 mm) above the base metal. Wait as a weld pool forms, then begin welding.

Stopping the arc

When the weld is complete, you must stop the arc. Stop the current by removing your foot from the pedal or turning off the thumb switch. Keep the torch over the area where you stopped welding. The continued shielding gas flow will protect the weld metal as it cools.

If the current switch on the welding machine is in the remote position, you can use the foot pedal or the thumb switch to control the amount of current. You can slowly reduce the current near the end of the weld. This gives you time to fill the weld pool with filler metal, if necessary.

WELDING A BEAD ON PLATE

Start the welding arc as described in the preceding section. After the arc stabilizes, hold the electrode about 1/16″ − 1/8″ (1.6 mm − 3.2 mm) above the work. A molten weld pool will form from the heat of the arc. Hold the torch in line with the weld axis and at an angle of 60° − 75° from the work (15° − 30° from the vertical). See Fig. 23-7.

Hold the torch in one place until a weld pool forms. When welding on thin metals, the molten pool will sag slightly. This will occur when the amount of penetration is equal to the thickness of the metal. On thicker metals, no sag will occur. Wait until the pool is the proper size. Usually, the weld pool diameter should be two to three times the diameter of the electrode. Then, begin to move forward. Keep the pool the same size as you move along the length of the weld joint.

You can change the current slightly by moving the electrode toward or away from the base metal. If you

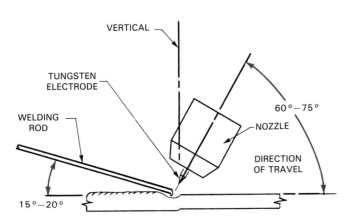

Fig. 23-7. The suggested electrode and welding rod angles for welding a bead on plate. The same angles are used when making a butt weld. The torch is held 60°—75° from the metal surface. This is the same as holding the torch 15°—30° from the vertical.

need more current (for more penetration or more heat), move the electrode closer to the base metal. See Fig. 23-8. If the weld pool is too large and you want to reduce the current, pull the electrode away from the base metal. Do not pull the electrode so far away that the arc becomes unstable. A long arc length, and especially an unstable arc, will cause *porosity* (trapped gas bubbles or holes) in the weld.

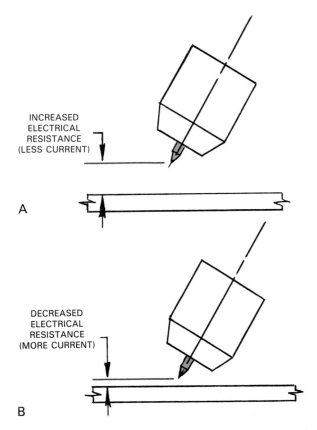

Fig. 23-8. The current can be changed slightly by moving the electrode toward or away from the base metal. A larger gap will decrease the current, heat, and penetration. A smaller-than-usual gap will increase these factors.

Another way to control current is to set the current switch on the machine to the remote position. This allows the current to be changed by using a foot pedal or thumb switch. These changes can be large, compared to moving the electrode, which can cause only small changes of up to 10 amps.

Once you have formed a good weld pool, continue moving the electrode and the pool until you reach the end of the plate. GTAW is a fairly slow process, but produces high-quality welds.

EXERCISE 23-1 — LAYING A BEAD ON PLATE WITHOUT FILLER ROD

1. Obtain a piece of mild steel measuring 1/16″ x 3″ x 6″ (1.6 mm x 76 mm x 152 mm).
2. Clean both surfaces with a wire brush.
3. Mark five straight lines lengthwise on each side of the metal. Start the lines 1/2″ (12.7 mm) in from one edge, and end 1/2″ (12.7 mm) from the opposite edge. Place the metal in the flat position.
4. Set up the welding machine for use with mild steel. Use DCEN (DCSP) and 70 amps current. Place the current switch in the panel position, so the foot pedal or thumb switch can act as an on/off switch. Set the high frequency switch to the start position.
5. Set the shielding gas flow. Refer to Fig. 22-9 for shielding gas flow rates. The shielding gas should be argon.
6. Obtain a 2 percent thoria tungsten electrode 1/16″ (1.6 mm) in diameter. Form the tip into a blunted point. Make sure the grinding marks are lengthwise, as shown in Fig. 22-15.
7. Select the correct size collet and collet body. Install the collet, collet body, and electrode in the torch. The electrode should extend 1/8″—3/16″ (3.2 mm—4.8 mm).
8. Hold the torch at the correct angle about 1/8″ (3.2 mm) above the base metal.
9. Press the foot pedal or thumb switch. The welding arc will start.
10. Move the electrode closer to the base metal, so it is about 3/32″ (2.4 mm) above the surface. Watch the weld pool as it forms. When it reaches about 3/16″ (4.8 mm) in width, it will begin to sag. Move the torch forward along the line marked on the plate.
11. Maintain the same size weld pool as you move along the plate. Keep the electrode a constant distance above the surface.
12. Stop welding at the end of the line.
13. Repeat the process until you have welded five beads on each side of the plate.

Inspection:
Each weld bead should be straight and have even ripples. There should be signs of slight penetration on the back side of the plate. There should not be any porosity in the weld.

WELDING A BEAD ON PLATE WITH FILLER METAL

When welding a butt joint or a bead on plate, hold the torch so it is in line with the weld axis and at an angle of 60° − 75° from the base metal. When welding a fillet weld, the torch should be held at 45° from the base metal and 60° − 75° from the weld axis. The position of the welding torch is the same, whether or not filler metal is used. As shown in Fig. 23-4, the filler rod is held in the opposite hand from the torch, and is grasped like a pencil. The filler rod is held at an angle of 15° − 20° to the base metal. See Fig. 23-7.

Start the arc, as described earlier in this chapter. When the weld pool reaches the correct size (two to three times the diameter of the electrode), add the filler metal at the front of the weld pool. At the same time that you add the filler metal, move the torch toward the back of the pool. This will prevent the filler metal from touching and contaminating the electrode. Before the filler metal is added, the weld pool has a slight depression. After it is added, the pool is filled and slightly crowned (convex).

After adding the filler metal, remove the rod from the weld pool. Keep the end of the filler rod close to the arc and the pool. The rod end should be about 1/8″ − 3/16″ (3.2 mm − 4.8 mm) away from the pool. This will keep the filler metal in the area protected by the shielding gas.

Move the torch forward. As the weld pool moves, it will again form a slight depression. Watch the weld pool. When filler metal is needed, add the rod to the pool's leading edge. Fig. 23-9 illustrates the steps used to add filler metal. Keep the amount of reinforcement or build-up equal for the full length of the bead.

To form a wider bead, move the torch from side to side. Filler metal is still added to the weld pool's leading or front edge. Instead of being added to the center of the pool, though, it is added to one side or the other. When the torch moves to the left, filler metal is added on the right front side of the weld pool. The next time filler metal is needed, it is added to the left front side of the pool while the torch is on the right side.

As you finish a weld, add filler metal to the weld pool to fill the crater. Move the electrode to the rear of the pool, and add filler metal to the front edge. Remove the filler metal, and move the torch over the center of the pool again. This will smooth out the weld pool. Then release the foot pedal to stop the arc.

EXERCISE 23-2 − LAYING A BEAD ON PLATE WITH FILLER ROD

1. Obtain a piece of mild steel measuring 1/16″ x 3″ x 6″ (1.6 mm x 76 mm x 152 mm).
2. Clean both surfaces with a wire brush.
3. Mark five straight lines lengthwise on each side of the metal. Start the lines 1/2″ (12.7 mm) in from one edge, and end 1/2″ (12.7 mm) from the opposite edge. Place the metal in the flat position.
4. Set up the welding machine for use with mild steel. Use DCEN (DCSP) and 70 amps current. Place the current switch in the panel position, so foot pedal or thumb switch can act as an on/off switch. Set the high frequency switch to the start position.
5. Set the shielding gas flow. Refer to Fig. 22-9 for shielding gas flow rates. The shielding gas should be argon.
6. Obtain a 2 percent thoria tungsten electrode 1/16″ (1.6 mm) in diameter. Form the tip into a blunted point. Make sure the grinding marks are lengthwise, as shown in Fig. 22-15.

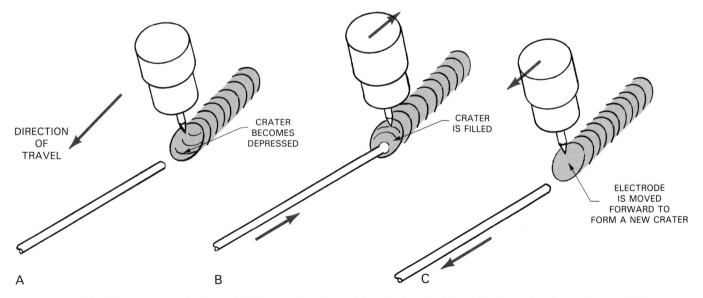

DIRECTION OF TRAVEL

CRATER BECOMES DEPRESSED

CRATER IS FILLED

ELECTRODE IS MOVED FORWARD TO FORM A NEW CRATER

A   B   C

Fig. 23-9. The steps required to add filler metal to the weld pool. A—A weld pool is formed and sags down slightly. B—The torch is moved to the rear of the pool as the rod is added to the front of the pool. C—The rod is withdrawn and the electrode moved to the front of the weld pool again. This process is repeated to the end of the weld.

7. Select the correct size collet and collet body. Install the collet, collet body, and electrode in the torch. The electrode should extend 1/8″−3/16″ (3.2 mm−4.8 mm).

8. Obtain a mild steel filler rod 1/16″ (1.6 mm) in diameter.

9. Hold the torch at the correct angle about 1/8″ (3.2 mm) above the base metal.

10. Press the foot pedal or thumb switch. The arc will start.

11. Move the electrode closer to the base metal, so it is about 3/32″ (2.4 mm) above the surface. Watch the weld pool as it forms.

12. When the weld pool reaches about 3/16″ (4.8 mm) in width, it will begin to sag. Add the filler rod to the front edge of the pool. Begin moving the torch forward along the line marked on the plate.

13. Keep the same size weld pool as you move along the plate. When the pool begins to sag, add filler rod. Keep the electrode a constant distance above the surface. Keep the filler rod close to the weld pool.

14. Stop welding at the end of the line. Fill the weld pool at the end of the weld.

15. Repeat the process until you have welded five beads on each side of the plate.

Inspection:

Each weld bead should be straight and have even ripples. The reinforcement should be about 1/16″ (1.6 mm) or less. There should be signs of slight penetration on the back side of the plate. There should not be any porosity along the weld.

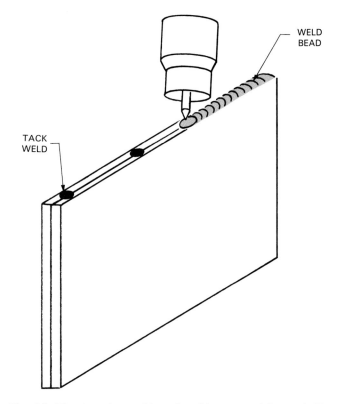

Fig. 23-10. An edge weld made without a welding rod. The base metal is melted by the arc to form a bead.

WELDING WITHOUT FILLER METALS

Some joints do not require filler metal. These include the edge joint, the lap joint, and the outside corner joint. When welding these joints, the base metal is melted until the two metals flow together. Figs. 23-10−23-12 show GTAW on these joints.

The processes of starting the arc and controlling the weld pool are like those used to weld a bead on plate. When welding an edge joint, the torch is held in line with the weld axis. When welding a lap joint or outside corner joint, the torch angle is 45° to the surface of the base metal. The torch is held at 60°−75° angle from the weld axis for all of these weld joints.

A side-to-side motion may be needed to obtain a bead that will flow and join the two metals. When you come to a tack weld, continue to move forward at a constant speed. Do not skip over it. You must have a continuous weld bead from the start to the end of a weld joint.

The welding current, electrode diameter, and shielding gas flow rates are selected on the basis of the thickness of one piece of metal. Refer to Figs. 22-8−22-12 for correct settings.

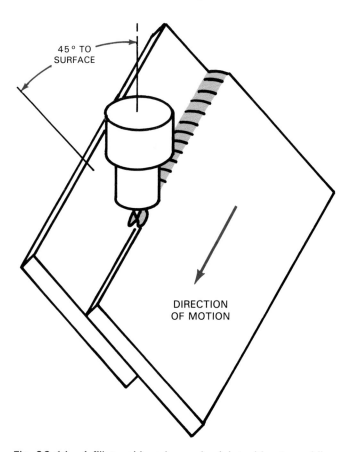

Fig. 23-11. A fillet weld made on a lap joint without a welding rod. The edge is melted into the surface piece to form a bead. The weld face should be horizontal.

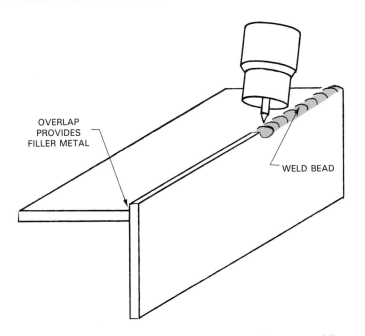

OVERLAP PROVIDES FILLER METAL

WELD BEAD

Fig. 23-12. An outside corner weld made without a welding rod. One edge is overlapped one metal thickness. This edge is then melted down to form the bead.

of 60° − 75° to the axis of the weld and 45° to the surface of the base metal. The weld pool has a "C" shape, as shown in Fig. 23-13. This C-shaped weld pool shows that both pieces of metal are being melted by heat from the arc.

Often, the edge will melt more than the surface. Only half a C-shaped weld pool will form. To correct this, point the torch more toward the surface to reduce the melting of the edge. Change the angle of the torch as needed so both pieces melt evenly and the C-shaped weld pool forms.

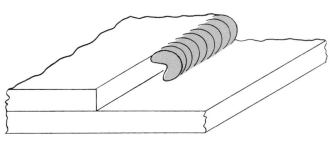

Fig. 23-13. A C-shaped weld pool must form whenever a fillet weld is made on a lap, T, or inside corner joint. The "C" shape indicates that both pieces of metal have melted and the filler metal may be added.

EXERCISE 23-3 — MAKING AN EDGE WELD WITHOUT FILLER ROD

1. Obtain two pieces of mild steel measuring 3/32'' x 3'' x 6'' (2.4 mm x 76 mm x 152 mm).
2. Clean the edges with a wire brush.
3. Set up the welding machine for welding mild steel.
4. Set the shielding gas flow.
5. Select an electrode of the proper type and correct diameter. Prepare the electrode for welding.
6. Select the correct size collet and collet body. Install the collet, collet body, and electrode in the torch. Make sure the electrode extends the proper amount.
7. Tack weld the two pieces in three places to form an edge weld.
8. Strike an arc and weld the edge joint. Make the weld bead as wide as the two pieces of metal are thick. No filler metal is required. A weaving motion may be necessary, depending on the diameter of electrode you are using.
9. Keep the electrode a constant distance above the work.
10. Stop welding at the end of the joint.
11. Repeat the process and weld the other side of the joint.

Inspection:
Each weld bead should be as wide as the two pieces are thick.

FILLET WELDING A LAP JOINT

A lap joint can be made with or without filler metal. When welding a lap joint, hold the torch at an angle

A filler rod may not be required when welding thin metals. When filler metal is not used, the metal from the upper piece flows down and forms a fillet. When using filler metal, add it to the upper edge of the weld pool, and hold the filler rod at an angle of 15° − 20° to the base metal. A completed weld will have evenly spaced ripples. When filler metal is added, the bead should have a slight convex surface. See Fig. 23-14. If filler metal is not used, the bead usually will have a flat surface. Remember to weld with a constant for-

Fig. 23-14. A well-formed fillet weld on a lap joint. The bead has an even width, evenly spaced ripples, and is convex (curved outward). There is no undercutting or underfilling.

ward motion, even when welding over a tack weld. Proper technique is needed to obtain a weld that is free of defects. Weld defects are discussed near the end of this chapter.

EXERCISE 23-4 — MAKING A LAP JOINT WITHOUT FILLER ROD

1. Obtain two pieces of mild steel measuring 3/32'' x 3'' x 6'' (2.4 mm x 76 mm x 152 mm).
2. Clean one edge and one surface on each piece with a wire brush.
3. Set up the welding machine for welding mild steel.
4. Set the shielding gas flow.
5. Select and prepare the proper type and diameter electrode. Select the correct size collet and collet body.
6. Install the electrode in the torch with the proper amount extending.
7. Tack weld the two pieces together to form a lap joint. Make sure there is no gap between them. Place the pieces at an angle so the weld is made in the flat welding position.
8. Strike an arc and weld the joint. Use a C-shaped weld pool. As the upper piece melts, it will form a fillet. No filler metal is needed.
9. Hold the torch steady and complete the weld.
10. Repeat the process and weld the other side of the joint.

Inspection:
Each weld should be even in width. It should have a flat face with evenly spaced ripples. The weld should be free of defects. Refer to the heading WELD DEFECTS at the end of this chapter.

EXERCISE 23-5 — MAKING A LAP JOINT WITH FILLER ROD

1. Obtain two pieces of mild steel measuring 3/32'' x 3'' x 6'' (2.4 mm x 76 mm x 152 mm).
2. Clean one edge and one surface on each piece with a wire brush.
3. Set up the welding machine for welding mild steel. Place the current control switch in the remote position.
4. Set the shielding gas flow.
5. Select and prepare the proper type and diameter electrode. Select the correct size collet and collet body.
6. Install the electrode in the torch with the proper amount extending.
7. Tack weld the two pieces together to form a lap joint. Make sure there is no gap between them. Place the pieces at an angle so the weld is made in the flat welding position.
8. Select the correct filler rod type and diameter.
9. Strike an arc and weld the joint. Use a C-shaped

weld pool. Add the rod as needed to form a convex bead. The upper plate should not be melted as much when adding filler metal as it is when no filler metal is used.
10. Hold the torch steady and complete the weld. When you reach the end of the weld, slowly reduce the welding current. Add filler metal to fill the weld pool.
11. Repeat the process and weld the other side of the joint.

Inspection:
Each weld should be even in width. It should have a convex face with evenly spaced ripples. Compare your weld beads to Fig. 23-14.

WELDING INSIDE CORNER AND T-JOINTS

Fillet welds are made on inside corner joints and T-joints. The arc must heat both surfaces to be welded. A C-shaped weld pool indicates that the surfaces are being melted equally. Remember that the electrode must extend about 1/4'' (6.4 mm) from the nozzle. Refer to Fig. 22-5.

The torch is held at an angle of 45° to the surface of the metal and 60° − 75° to the weld axis. The filler rod is held at an angle of 15° − 20° to the metal. See Fig. 23-15.

To hold the metal in position, tack weld about every 3'' (76 mm) along the joint. When welding, start the arc about 1/8'' − 1/4'' (3.2 mm − 6.4 mm) from the edge of the joint. Once the arc is started, move it back to the beginning of the joint and begin welding.

To make the fillet weld, filler metal is added. When welding in the flat position, the rod is added to the center of the weld pool. When welding in the horizontal position, it is added to the top edge of the weld. This can help reduce undercutting. The proper torch

Fig. 23-15. A fillet weld in progress on a T-joint, using GTAW. Notice the electrode extension and the electrode and welding rod angles.

angle is most important in preventing undercutting. Fig. 23-16 shows the correct angles for the torch and filler metal.

Keep the fillet the same size as you move along the joint to the end of the weld. Fill the weld pool before stopping the arc. After the arc stops, hold the torch over the end of the weld to protect it with shielding gas while it cools.

To obtain the proper size weld, add filler metal as needed. As a rule of thumb, a fillet weld should be as large as the metal being welded is thick. The size of a fillet weld is measured by the lengths of its legs. See Fig. 23-17. Thick metals will require more than one pass to complete a fillet weld of the proper size.

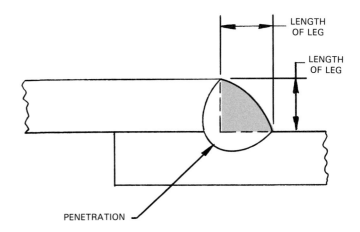

Fig. 23-17. *This drawing shows how the legs of a fillet weld are measured. The size of a fillet weld is usually about the same as the metal is thick.*

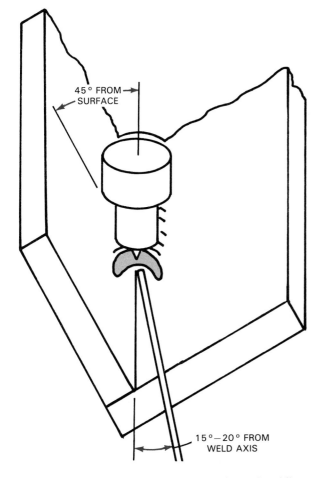

Fig. 23-16. *The suggested torch, electrode, and welding rod angles for making a fillet weld using GTAW in the flat welding position.*

EXERCISE 23-6 — WELDING THE T-JOINT WITH A FILLET WELD

1. Obtain two pieces of mild steel measuring 1/8″ x 3″ x 6″ (3.2 mm x 76 mm x 152 mm).
2. Clean one edge on one piece and one surface on the other.
3. Set up the welding machine for welding mild steel. Place the current control in the remote position.

4. Set the shielding gas flow. Select, prepare, and install the correct electrode. Select the correct filler rod.
5. Tack weld the two pieces together to form a T-joint. Place the pieces at an angle so the weld can be made in the flat welding position. The position is similar to that shown in Fig. 23-15.
6. Weld the joint, using a C-shaped weld pool. Add the filler metal to the front of the pool. Use the foot pedal or thumb switch to control the current and the size of the weld pool.
7. Repeat the process and weld the other side of the joint.

Inspection:

The weld beads should be of high quality and not show any defects.

WELDING BUTT JOINTS

Full-penetration butt welds can be made on a square-groove joint up to 3/16″ (4.8 mm). Thicker metals are beveled to allow for complete penetration. Even 3/16″ (4.8 mm) metal is sometimes beveled for this reason. Full penetration is obtained by using the keyhole method. It is important to tack weld the metals to hold them in position for welding.

The torch and electrode are centered over the root of the joint. The electrode is pointed forward at 60°−75° from the weld axis (15°−30° from the vertical). It is held at 90° to the surface of the base metal. Filler rod is held at 15°−20° above the plate, as shown in Fig. 23-18.

Start the welding arc as described earlier. When the weld pool is properly formed, add filler metal to the front of the pool. While the filler metal is being added, the torch is moved to the rear of the weld pool. After the rod is removed, the torch is moved forward again. Keep the rod close to the weld pool, adding it as needed. Fig. 23-19 shows a welder making a butt weld.

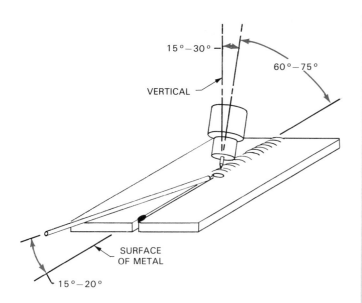

Fig. 23-18. Suggested electrode and welding rod angles for welding a butt joint in the flat welding position.

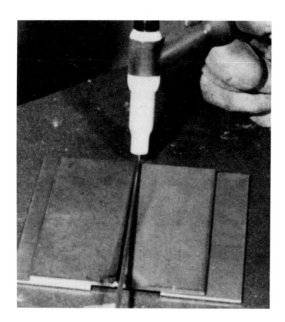

Fig. 23-19. A butt weld in progress in the flat welding position. Note that the welding rod is held in line with the weld axis and at a low angle.

On thick metals, more than one pass is needed to fill the groove. After the root pass, either stringer (straight) beads or weaving beads can be used to fill the groove. These beads are made the same way as welding a bead on plate. You must be sure to get good penetration into the previous bead and into the sides of the joint.

A completed butt weld should show even ripples on the face (top surface) of the metal. There should be even ripples of penetration on the back side, as well. No defects should be seen on either side.

EXERCISE 23-7 — WELDING A BUTT JOINT

1. Obtain two pieces of mild steel measuring 3/32'' x 3'' x 6'' (2.4 mm x 76 mm x 152 mm).
2. Clean one edge on each piece.
3. Set up the welding machine for welding mild steel. Place the current control in the remote position.
4. Set the shielding gas flow.
5. Select and prepare the correct electrode.
6. Select and install the correct size of collet. Install the electrode and nozzle.
7. Select the correct diameter and type of filler rod.
8. Align the two pieces to form a butt weld with a 1/32''–1/16'' (0.8 mm–1.6 mm) root opening. Tack weld the two pieces together.
9. Start the arc. Move the torch until both pieces of metal begin to melt. Push on the foot pedal or adjust the thumb switch to adjust the current as required. When both pieces are molten, add the welding rod. Move the torch ahead, making sure that both pieces melt equally. Also watch that the size and shape of the keyhole remains constant. Complete the butt joint.

Inspection:
The weld beads should be of high quality and not show any weld defects. If you do not get enough penetration, the size of the keyhole must be made larger.

WELDING OTHER METALS

Base metals other than mild steel are often welded with the gas tungsten arc welding process. The most commonly welded metals include stainless steel and aluminum. Magnesium, titanium, and other types of metals require the high-quality welds that GTAW produces. The inert gas shielding prevents oxidation and contamination of the weld.

Stainless steel is welded much like mild steel. Aluminum and magnesium, however, are somewhat different. In these metals, the molten weld pool tends to fall through if it gets too large. If the weld pool falls through the metal, a large hole results. These metals do not have much strength when they are very hot. This is called *hot shortness.* If the pool begins to sag more than it should when welding these metals, either add welding rod to cool it, or reduce the current.

One way to prevent the weld pool from falling through the plate in hot-short metals is to use a backing. A *backing,* often referred to as a "back-up," is a piece of metal that will not melt and will not join to the metal being welded. A stainless steel backing is used when welding aluminum, magnesium, or copper. On metals that do not have hot shortness, a backing is used to control penetration. For example, copper is used when welding steel or stainless steel.

When welding other base metals, refer to Fig. 22-19 for the recommended current type. Figs. 22-9 through

22-12 give the recommended current setting, electrode diameter, filler rod diameter, and shielding gas flow for aluminum, stainless steel, magnesium, and copper.

EXERCISE 23-8 — WELDING OTHER BASE METALS

Each of the welding exercises in Chapter 23 can be repeated using different base metals. You should learn to weld stainless steel and aluminum with GTAW. The techniques of starting the arc, forming the weld pool, and ending a weld are the same as those used for mild steel!

1. Clean the metal surfaces to be joined.
2. Set up the welding machine for the type of metal to be welded.
3. Set the shielding gas flow.
4. Select an electrode of the proper type and correct diameter. Prepare the electrode for welding.
5. Select the correct size collet and collet body. Install the collet, collet body, and electrode in the torch. Make sure the electrode extends the proper amount.
6. Make the weld.

Inspection:
Completed welds in stainless steel, aluminum, and other metals should all exhibit high quality weld beads and not have any defects.

WELD DEFECTS

Common weld defects in GTAW include: porosity, undercutting, lack of penetration, cold lap, and tungsten inclusions. Two problems that may occur while welding are metal oxidation and an unstable arc. These two problems are not weld defects. Oxidation shows there is a problem in the equipment or in the technique. An unstable arc is not a defect, but can cause defects.

Porosity is caused when shielding gas does not protect the weld area. If you find porosity in the weld area, check the flow rate of the shielding gas. Also, check all shielding gas connections for leaks. Keep a correct arc length. Too long an arc will allow oxygen from the air to get into the weld area.

Undercutting is caused by using incorrect torch angles. Use the proper torch angles and add filler metal to fill any undercutting.

Cold lap is caused when a large weld pool flows onto the base metal and becomes solid. The molten weld pool does not melt the base metal, it just lies on top of it. To prevent cold lap, keep the weld pool the proper size, use the correct torch angles, and do not add too much filler metal.

One defect that occurs only in GTAW is getting tungsten in the weld. When tungsten gets into the weld, it is called a *tungsten inclusion.* Tungsten inclusions

are caused by the following:
• Dipping the electrode into the weld pool.
• Using too much welding current.
• Using an electrode that is too small.

Tungsten inclusions can occur when too much current is used for the electrode diameter. To correct this, change to a larger diameter electrode or reduce the amount of current you are using.

Neither the weld bead nor the base metal should be oxidized. *Oxidized* means that oxygen has combined with the base metal. This is caused by a lack of shielding gas coverage. Check the shielding gas flow rate. Also make sure you use the correct electrode extension and the correct arc length. Remember that you must hold the torch over the end of the weld for the post flow of shielding gas.

When welding, the arc should be straight and concentrated between the end of the electrode and the base metal. When the arc jumps around the base metal, or moves up and down on the electrode in an uncontrolled manner, it is an *unstable arc.* An unstable arc is caused by one of the following:
• Electrode is contaminated.
• Electrode is too large in diameter.
• Weld joint is too narrow.
• Base metal is dirty.

REVIEW QUESTIONS

Write your answers on a separate sheet of paper.
1. List two ways to strike or start an arc.
2. What is the correct arc length for GTAW?
3. What are the two ways to increase the current while welding?
4. Why should the welding rod be held close to the arc while welding?
5. To what part of the weld pool is the filler metal added? Where should the torch be while the filler metal is added?
6. What should be done to prevent porosity in the weld?

Use the tables provided in this chapter and in Chapter 22 to answer questions 7 through 10.
7. Answer the following for welding a fillet on 1/8″ (3.2 mm) mild steel:
   Type of current_____
   Amount of current _____
   Electrode extension _____
      (beyond the nozzle)
   Type of electrode _____
   Time of post flow_____
8. Answer the following for welding a butt joint in 1/16″ (1.6 mm) 6061 aluminum.
   Type of current_____
   Amount of current _____
   When high frequency is used _____
   Type of filler metal _____

237

9. Answer the following for welding on 1/4" (6.4 mm) stainless steel:
   Type of welding current _____
   Size of electrode _____
   Type of shielding gas _____
   Shielding gas flow_____
   Shape of electrode tip _____

10. Answer the following for welding on nickel alloys:
    Type of current_____
    Shape of electrode tip _____
    Ways (2) to start arc _____

GTAW being used to weld an aluminum fishing boat tower. These welds are being made on square and round tubing. Since the weldment is too large to be moved, these welds are made in the vertical, flat, and overhead welding positions. (Miller Electric Mfg. Co.)

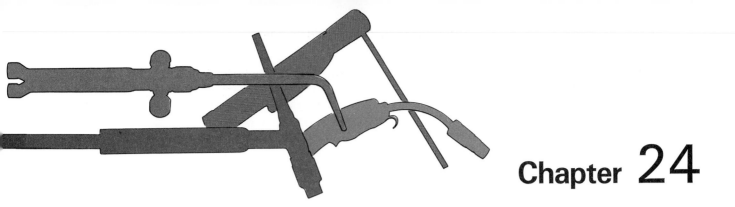

Chapter 24

GTAW: Horizontal, Vertical, and Overhead Welding Positions

After studying this chapter, you will be able to:
☐ Explain why out-of-position welding is often an important part of welder qualification tests.
☐ Identify the correct torch and filler rod angles for out-of-position welding.
☐ Weld in the horizontal welding position with GTAW.
☐ Weld in the vertical welding position with GTAW.
☐ Weld in the overhead welding position with GTAW.

TECHNICAL TERMS

Downhill, out-of-position welding, uphill.

OUT-OF-POSITION GTAW

Welding in a position other than flat is called *out-of-position welding*. Small assemblies usually can be placed so the welding is done in the flat position. Larger assemblies, which often cannot be moved, require the welder to weld out of position. See Fig. 24-1.

Many welder qualification tests are performed on out-of-position joints. If a welder can make good welds out of position, he or she can also weld properly in the flat welding position. Chapter 3 describes and illustrates the different welding positions.

WELDING IN THE HORIZONTAL WELDING POSITION

The weld axis and weld face are horizontal or close to horizontal when making a weld in the horizontal position. The weld axis can vary from $0° - 15°$. The weld face can vary from $80° - 150°$ or from $210° - 280°$.

SPECIAL PAPER SEALS

Fig. 24-1. A welder using GTAW to fabricate a stainless steel pipe assembly. On a fixed pipe, the welding position changes continually as the weld progresses. Note that the pipe is sealed. This permits inert gas to be used to shield the root side of the weld. (Hobart Brothers Co.)

Welding in the horizontal welding position can be more difficult than welding in the flat welding position. Gravity pushes the molten weld pool down. This causes the weld bead to sag and not properly fill the joint. It is important to use proper torch angles and techniques to obtain high-quality welds.

PREPARING TO WELD

Before beginning any welding, check that the equipment is assembled and set up correctly. Refer to Chapter 22. Select and set the type and amount of current, the electrode type and diameter, and the type and amount of shielding gas you will use. Select the correct filler rod. Refer to Chapters 22 and 23 for more information.

Pulsed GTAW may be used to make high quality welds in out-of-position joints. The pulsed arc allows the weld pool to cool between pulses. Good penetration is still obtained. Pulsed GTAW is discussed in Chapter 21.

Protective clothing worn for out-of-position welding is the same for flat welding. A long-sleeved shirt with the collar buttoned should be worn. Wearing a welding cap is a good practice. A leather coat or cape is recommended when welding overhead. Bare skin must be well-protected from falling molten metal.

Holding the torch and starting an arc is the same as in flat welding. Weld pools in out-of-position welding are kept the same size or slightly smaller than those used in flat welding. The current setting on the machine can be reduced. This keeps the weld pool (also referred to as the pool) smaller. Penetration is smaller, however, when current is reduced.

Fillet weld

Horizontal fillet welds are very common. They are made on lap joints, inside corner joints, and T-joints. Fig. 24-2 shows a welder making a fillet weld in the horizontal welding position.

When making horizontal fillet welds, the torch is held at 45° to the base metal. It is held at a 60°–75° angle to the weld axis. Filler metal is held at 15°–20° from the weld axis and in line with the axis of the weld. The torch and filler metal angles used to make a horizontal fillet weld are shown in Fig. 24-3. The torch is centered over the root of the weld, but pointed more toward the vertical piece. This helps prevent undercutting. Filler metal is added to the upper leading edge of the weld pool. When making a fillet weld in any position, a C-shaped weld pool is used. This shows that both pieces of metal are melting. This is similar to welding in the flat welding position.

The face of a completed fillet weld should be flat or slightly convex. See Fig. 24-4. The weld should not sag down onto the lower piece. Sagging of weld metal can cause undercutting, underfill, or overlap. If any of these happens, change your torch angles. Add the filler metal to the top edge of the weld pool. There should be no porosity. If there is, check that the shielding gas is set for the proper flow rate and that there are no leaks in hoses.

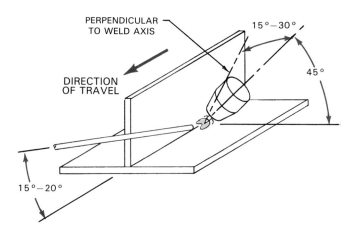

Fig. 24-3. The suggested electrode and filler rod angles for GTA welding a T-joint in the horizontal position. The angles are the same for forehand or backhand welding.

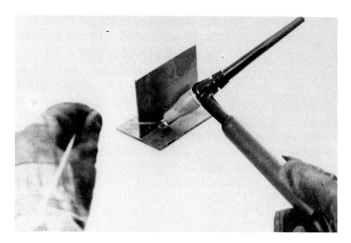

Fig. 24-2. A student welder uses GTAW to make a fillet weld on a horizontal T-joint.

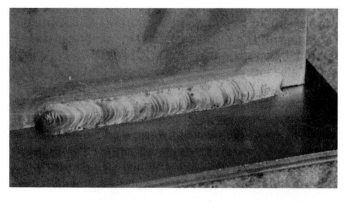

Fig. 24-4. An acceptable fillet weld on a T-joint. This weld was made on mild steel. It has a flat bead, even ripples, and an even bead width. There is no apparent undercutting, sag, or overlap.

EXERCISE 24-1 — MAKING A HORIZONTAL FILLET WELD ON A LAP JOINT

1. Obtain two pieces of mild steel measuring 1/8" x 1 1/2" x 6" (3.2 mm x 38 mm x 152 mm).
2. Clean one edge and one face of each piece.
3. Set up the welding machine for welding mild steel. Use DCEN (DCSP) and 100 A. Place the current switch in the remote position. Set the high frequency switch on start.
4. Set the shielding gas flow. Refer to Fig. 22-8 for shielding gas flow rates.
5. Obtain a 2% thoria tungsten electrode 1/16" (1.6 mm) in diameter. Grind the tip into a blunted point. Make sure the grind marks are lengthwise as shown in Fig. 22-15.
6. Select the correct size collet and collet body. Install these and the electrode in the torch. The electrode should extend 1/8"—3/16" (3.2 mm—4.8 mm) beyond the end of the nozzle.
7. Obtain a 3/32" diameter (2.4 mm) mild steel welding rod.
8. Clamp the pieces to form a lap weld. Tack weld the pieces together at three places on each side of the joint.
9. Hold the torch at the correct angle about 1/8" (3.2 mm) above the work.
10. Press the foot pedal or thumb switch. The arc will start.
11. Move the electrode closer to the work, so it is about 3/32" (2.4 mm) above the base metal. Watch the weld pool as it forms into a "C" shape. When it reaches the correct size and shape, add the filler rod. Begin moving the torch forward.
12. Maintain the same size weld pool, adding welding rod as needed, as you steadily move along the plate. Keep the electrode a constant distance above the work.
13. Stop welding at the end of the joint.
14. Weld the other side of the joint.

Inspection:
The weld bead should be straight with even ripples. There should not be any overlap or porosity.

EXERCISE 24-2 — MAKING A HORIZONTAL FILLET WELD ON A T-JOINT

1. Obtain two pieces of mild steel measuring 1/8" x 1 1/2" x 6" (3.2 mm x 38 mm x 152 mm).
2. Clean one edge of one piece and one face of the other piece.
3. Set up the welding machine for welding mild steel. Set the shielding gas flow.
4. Select, prepare, and install the electrode. Select the correct filler rod.
5. Tack weld the two pieces together to form a T-joint.
6. Weld both sides of the T-joint. Watch the weld pool to make sure both surfaces melt. Add welding rod

to form a 1/8" (3.2 mm) fillet weld. The bead should be slightly convex.

Inspection:
The completed weld should have even ripples. The bead should be even in width along the entire joint. There should not be any defects in the weld. Compare your completed weld to the one shown in Fig. 24-4.

Butt weld

Molten metal in a horizontal butt weld will sag downward due to the force of gravity. If the weld sags, it will have a poor appearance. It also can be weakened by undercutting or underfilling of the weld joint. Overlap may occur, as well.

The angles of the torch and filler rod are shown in Fig. 24-5. Notice that the welding rod is held 15°—20° above the axis of the weld. The welding rod is added to the front upper part of the weld pool. Adding the filler metal to the top of the pool reduces undercut and underfill. The back of the torch is tipped downward about 15°. Tilting the back of the torch down 15° allows the electrode to be pointed upward at a 15° angle. Pointing the electrode up helps keep the weld pool from sagging.

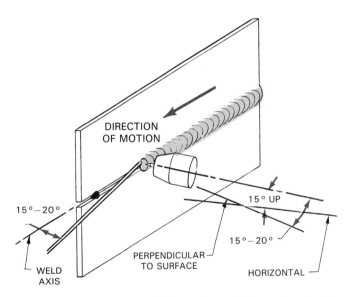

DIRECTION OF MOTION

15°—20°

15° UP

15°—20°

PERPENDICULAR TO SURFACE

HORIZONTAL

WELD AXIS

Fig. 24-5. The suggested electrode and filler rod angles for making a butt weld in the horizontal position. The electrode should point upward 15°—20° to keep the bead from sagging.

Welding with a small weld pool and a short arc length will help prevent these defects. Some common defects are shown in Fig. 24-6.

Butt joints require full penetration welds. Point the arc at the root of the weld. Heat the weld joint with the arc, so both pieces begin to melt. They should melt all the way to the root of the joint. You can then add the welding rod. Move the torch forward at a uniform speed, keeping both pieces molten.

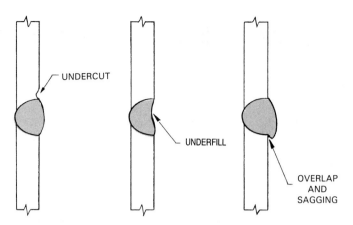

Fig. 24-6. Common weld defects that may occur on a horizontal butt weld. Undercut is caused by an incorrect torch angle. When not enough filler metal is added, underfill occurs. Overlap or sagging occurs when too much filler rod is used or if the bead is overheated.

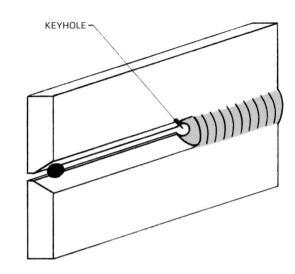

Fig. 24-7. The keyhole may be seen at the bottom of the groove weld when there is proper penetration.

The root of the weld will look like a keyhole, as shown in Fig. 24-7. The back and the sides of the weld are molten. Keep the keyhole size the same as the weld progresses.

A completed root pass will have even ripples on both the root side and the face side of the weld. To fill thick joints, use stringer bead or weaving bead passes. See Fig. 24-8.

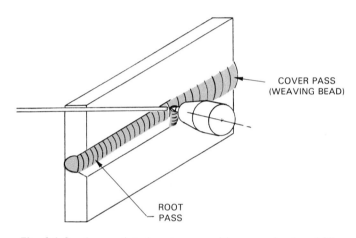

Fig. 24-8. A completed root pass, with a weaving bead filler pass being added to complete a V-groove weld.

EXERCISE 24-3 — MAKING A HORIZONTAL BUTT WELD

1. Obtain two pieces of mild steel measuring 1/8″ x 1 1/2″ x 6″ (3.2 mm x 38 mm x 152 mm).
2. Clean one edge of each piece.
3. Set up the welding machine. Set the shielding gas flow.
4. Select and prepare the electrode. Select the correct filler rod.
5. Tack weld the two pieces together to form a butt joint that has a root opening of 1/16″ (1.6 mm).
6. Weld the butt joint. Add filler metal to obtain a slightly crowned face on the weld. Watch the weld pool to make sure both pieces are melted to the root of the joint.

Inspection:
The completed weld should have even ripples on both the face side and root side of the joint. The bead should be even in width along the entire joint. There should not be any defects.

WELDING IN THE VERTICAL POSITION

When vertical welding with the gas tungsten arc process, center the electrode over the root of the weld joint. Point the torch upward so that it forms an angle of 60°−75° with the base metal. Fig. 24-9

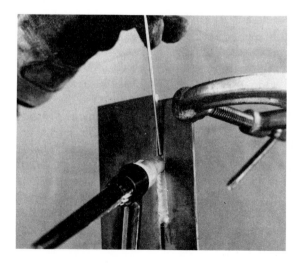

Fig. 24-9. A fillet weld on a lap joint being made uphill. Note that the filler rod is held at about 15°−20° to the weld axis. The electrode is pointed upward about 60°−75° from the base metal. It is pointing more at the surface than at the edge.

shows a welder welding a joint in the vertical position. Notice the angles of the torch and welding rod.

Thicker metals are welded with the weld pool moving from the bottom of the joint toward the top. This is called *uphill* ("vertically up") welding. Thin metals are welded *downhill* ("vertically down"). When welding downhill, the filler metal is added to the leading edge of the weld pool. Fig. 24-10 shows the angles used to weld downhill.

Fillet welds use a weld pool shaped like a "C." This is the same as in flat welding. Butt welds use the keyhole method to obtain full penetration. See Fig. 24-11.

Fig. 24-11. An acceptable butt weld that was made uphill. The bead is convex and straight, with evenly spaced ripples. There is no apparent undercut, underfill, or sag.

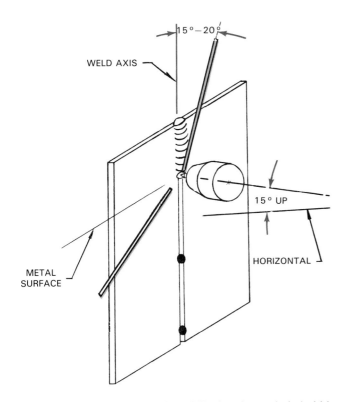

Fig. 24-10. When welding downhill, the electrode is held in line with the weld axis with a 15°−20° upward angle. The filler rod is held at 15°−20° from the base metal. It may be added from the top or side.

EXERCISE 24-4 — MAKING A VERTICAL FILLET WELD ON A T-JOINT

1. Obtain two pieces of mild steel measuring 3/16″ x 3″ x 6″ (4.8 mm x 52 mm x 152 mm).
2. Clean one edge of one piece and one face of the other piece.
3. Set up the welding station for welding mild steel.
4. Tack weld the two pieces together to form a T-joint. Use a welding positioner to hold them in the vertical position.
5. Weld both sides of the T-joint uphill. Each completed bead should be a 3/16″ fillet weld. This will require three stringer (straight) weld passes or one stringer bead pass and one weaving bead pass. Do not let the weld pool get too large, or it will sag.

Inspection:
The completed weld should be of high quality. The ripples should be evenly spaced and the completed weld bead even in width. There should not be any defects.

EXERCISE 24-5 — MAKING A VERTICAL BUTT WELD

1. Obtain two pieces of mild steel measuring 1/8″ x 1 1/2″ x 6″ (3.2 mm x 38 mm x 152 mm).
2. Clean one edge of each piece.
3. Set up the welding equipment for welding mild steel.
4. Tack weld the two pieces together to form a butt joint with a root opening of 1/16″ (1.6 mm).
5. Weld the butt joint uphill with the filler metal. Use the recommended torch and welding rod angles and the keyhole method. The weld will have complete penetration.

Inspection:
Examine the weld for complete penetration. There should be no porosity, underfill, or any other defects present in the weld. Ripples on the face and root of the weld should be even and straight. Compare your completed weld to the one shown in Fig. 24-11.

WELDING IN THE OVERHEAD WELDING POSITION

Overhead welding is the most difficult. It can be uncomfortable for the welder. The weld pool must be kept small when welding overhead. Use only stringer beads or very small weaving beads.

The amount of current set on the welding machine to weld thin metals overhead is usually 10 A less than when flat welding. Thicker metals use 20 A less current for overhead welding than for flat welding.

Pulsed welding gives very good results. It provides good penetration and gives the weld pool a chance to cool. Reducing current or using a pulsed arc keeps the weld pool smaller than the pool used for flat welding.

Shielding gas flow rates must be higher when overhead welding. Remember that argon gas is heavier than air. It will lie on a plate being welded in the flat position. Refer to Fig. 21-13. However, when welding overhead, the gas will fall away from the plate and not give good protection. Flow rates are increased by 3−5 cfh (1.4−2.4 Lpm) for thin metals. On metals 3/16" (4.8 mm) or thicker, an additional 5 cfh (2.4 Lpm) or more is used. If you get porosity when welding overhead, check to make sure you are keeping the correct arc length. If your arc length is correct and you are still getting porosity, increase the shielding gas flow.

The angles used to weld fillet or butt welds overhead are the same as those used in other positions. Filler metal is added to the leading edge on a fillet weld. Fig. 24-12 shows a welder making a fillet weld in the overhead welding position. Fig. 24-13 shows a butt weld being made in the overhead welding position.

The torch is often held like a hammer when overhead welding. See Fig. 23-4. This provides flexibility and control.

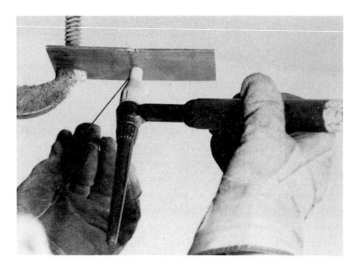

Fig. 24-13. A bead being laid in a butt joint in the overhead welding position. Note how the torch is being held.

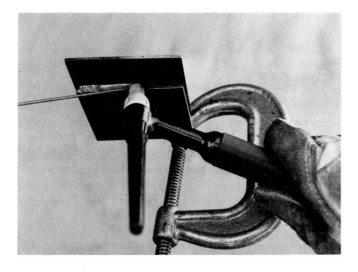

Fig. 24-12. A fillet weld being made on a lap joint in the overhead welding position. The angles of the electrode and welding rod are the same as those used in the flat welding position.

EXERCISE 24-6 — MAKING AN OVERHEAD FILLET WELD ON A T-JOINT

1. Obtain two pieces of mild steel measuring 3/16" x 3" x 6" (4.8 mm x 52 mm x 152 mm).
2. Clean one edge of one piece and one face of the other piece.
3. Set up the welding station for welding mild steel. Use 85 A—90 A current. Use 18 cfh—20 cfh (8.5 Lpm—9.4 Lpm) argon shielding.
4. Tack weld the two pieces together to form a T-joint. Clamp the tack welded pieces in the overhead position.
5. Weld both sides of the T-joint. Keep the molten weld pool small, so the metal does not fall or sag.

Inspection:

The completed weld should be of the same high quality as a weld made in the flat position. Examine the completed weld for evidence of porosity. Also look for any metal sagging that might cause underfill or overlap.

EXERCISE 24-7 — MAKING AN OVERHEAD BUTT WELD

1. Obtain two pieces of mild steel measuring 1/8" x 1 1/2" x 6" (3.2 mm x 38 mm x 152 mm).
2. Clean one edge of each piece.
3. Set up the welding equipment for welding mild steel. Adjust the current and shielding gas flow for overhead welding.
4. Tack weld the two pieces together to form a butt joint with a root opening of 1/16" (1.6 mm). Clamp the tack welded pieces in the overhead position.
5. Weld the butt joint. Use the correct torch and welding rod angles. Watch the weld pool. It must melt both pieces and form the keyhole shape. Obtain complete penetration.

Inspection:

The completed weld should have even ripples. The edges of the weld should be the same width along the entire length. The root side of the weld should show consistent penetration. There should not be any porosity along the weld.

WELDING OTHER METALS

Stainless steel and aluminum are metals very commonly welded with gas tungsten arc welding. You should develop proper skills for welding both. Chapter 23 discusses how welding these metals differs from welding mild steel.

After learning to weld mild steel out of position, you should practice welding these metals in the horizontal, vertical, and overhead positions. Figs. 22-9—22-12 list current and shielding gas settings for metals other than mild steel. Each of the exercises in this chapter can be done with aluminum, stainless steel, or any other base metal weldable with GTAW.

Backings are often used to control penetration when welding out of position. See Chapter 23 for more information.

REVIEW QUESTIONS

Write your answers on a separate sheet of paper.
1. What safety precautions should a welder take to prevent being burned by falling molten metal?
2. How does the size of the weld pool used for out-of-position welding compare to the size of the weld pool for flat welding?
3. Where is filler metal added to a horizontal fillet weld?
4. What are three common defects that occur when welding in the horizontal welding position?
5. What method is used to obtain complete penetration on a butt weld?
6. Are thin metals welded uphill or downhill?
7. How much is current increased or decreased when welding thick metals in the overhead welding position?
8. Why is the flow of argon shielding gas increased when welding overhead?
9. When welding overhead, is the torch most often held like a pencil or like a hammer?
10. In out-of-position welding, which of the following can be done to keep the weld pool small?
    a. Change to AC from DC.
    b. Reduce the amount of current.
    c. Change to DC from AC.
    d. Use a pulsed arc.
    e. Just a and b.
    f. Either b or d.

GTAW being used to fabricate a stainless steel vessel.   (Miller Electric Mfg. Co.)

SECTION 6

RESISTANCE WELDING

Lors Machinery, Inc.

A resistance spot welder designed to make nine welds at one time on a computer room floor panel.   (LORS Machinery, Inc.)

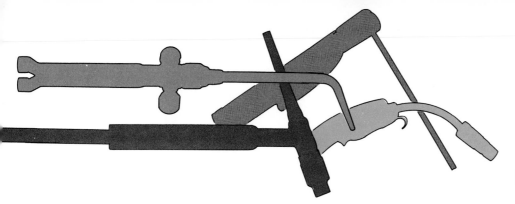

Chapter 25

Resistance Welding Equipment and Supplies

After studying this chapter, you will be able to:
- ☐ Explain the principle of electrical resistance and how it is used in resistance welding.
- ☐ Describe the differences between the four most common resistance welding machine designs.
- ☐ Explain how a step-down transformer affects voltage and current.
- ☐ List the properties of a material suitable for use as an electrode in resistance welding.
- ☐ Describe the regular checks to be made for safe operation of a resistance spot welding machine.

TECHNICAL TERMS

Expulsion weld, hold time, horn spacing, off time, percent heat control, resistance, spot weld, squeeze time, step-down transformer, tap settings, throat depth, weld nugget, weld time, wheel electrode.

PRINCIPLES OF RESISTANCE SPOT WELDING

When electricity flows through metal, the metal heats up. This occurs because there is *resistance* (opposition) to the flow of electricity through the metal. For example, when electricity flows through the metal coil in an oven, the coil gets red-hot. The heat from the glowing coil is used to cook foods.

If two metals are touching and an electrical current is passed through them, heat will be generated. The largest amount of heat is created at the surfaces where the two metals are touching.

In resistance spot welding, two or more metal pieces are stacked on top of one another. Pressure is applied to the pieces. Then electrical current flows through the metal.

As the current flows, heat develops at the surfaces where the metal pieces touch. When enough heat is present, the metal will melt in the area where the pieces are touching. After the current stops, the molten metal becomes solid again. The parts are welded together. The metal that is welded together is called a *spot weld* or a *weld nugget*. Fig. 25-1 illustrates the steps that take place when a spot weld is made.

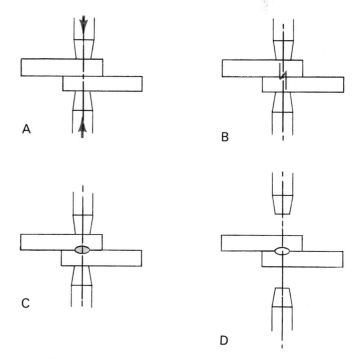

Fig. 25-1. Steps in making a spot weld. A—The metal is clamped between the electrodes under pressure. B—Current flows. C—Spot weld nugget is formed. D—Electrodes are removed after the weld becomes solid.

WELDING MACHINES

A resistance welding machine must accurately apply the desired force to the metal to be welded. Different methods are used to apply force to the metal. They are *air pressure, hydraulic pressure,* and *mechanical leverage.*

These force systems are used by various types of resistance welding machines. The most common machine designs are:
- Mechanical leverage.
- Rocker arm.
- Press.
- Gun.

A welding machine of the mechanical leverage type is shown in Fig. 25-2. Rocker arm, press, and gun welding machines may use either air or hydraulic pressure. Fig. 25-3 shows these types of welding machines.

Two important dimensions must be determined when a welding machine is selected. They are the throat depth and the horn spacing. **Throat depth** is the distance from the electrodes to the frame of the machine. It limits how far the electrodes can reach over a piece to be weld-

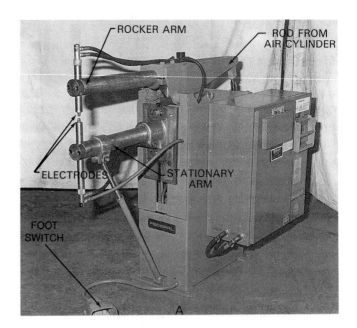

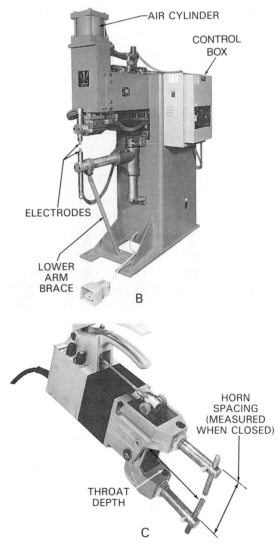

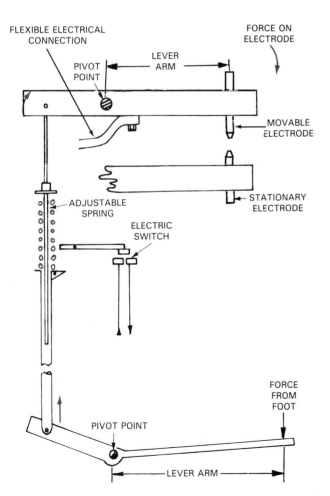

Fig. 25-2. A mechanical leverage spot welding machine. The electrode force is controlled by adjusting the spring and by the force applied by the operator.

Fig. 25-3. A—A rocker arm-type spot welding machine. Force is applied to the upper arm by an air cylinder. Notice the stationary arm brace. (Berkeley-Davis, Inc.) B—A press-type spot welding machine. The force from the air cylinder acts to push the upper electrode down. (LORS Machinery, Inc.) C—A gun-type welding machine. The lower arm is movable. This machine is generally suspended from overhead, so that the operator does not have to support its weight. Note how the horn spacing and throat depth are measured. (LORS Machinery, Inc.)

ed. *Horn spacing* is the distance between the arms of the welding machine, measured when the electrodes are touching. This dimension determines how tall a part can be welded. Throat depth and horn spacing are illustrated in Figs. 25-3C and 25-4.

Fig. 25-4. A gun-type welding machine with a long throat depth. A long throat measurement is required to place spot welds at the back of this cabinet. (LORS Machinery, Inc.)

EXERCISE 25-1 — LEARNING THE PARTS OF A SPOT WELDING MACHINE

1. Your instructor will assign you to study a specific piece of spot welding equipment.
2. Make sure the equipment is off. You will not need to operate the equipment for this exercise.
3. Sketch the equipment. Show each of the items listed, if present on your machine:
   a. Electrical connections.
   b. Water connection and drain.
   c. Air or hydraulic connections.
   d. Regulators for controlling the air or hydraulic pressure.
   e. Air or hydraulic cylinder, or force spring.
   f. Electrodes.
   g. Tap switch or switches.
   h. Control panel.

TRANSFORMERS

Most resistance welding uses alternating current (AC). Power supplied to a welding machine can vary from 115 V to 460 V, with current varying from a few amps up to 200 A to 400 A.

A transformer is used to reduce the supplied high voltage to low voltage. The same transformer also increases the supplied low current to a very high current.

Transformers have a primary winding and a secondary winding, as shown in Fig. 25-5. The primary winding is connected to the supplied power, usually 230 V or 460 V. The secondary winding is connected to the electrodes.

Transformers that reduce the supplied voltage are called *step-down transformers.* In a spot welding machine, a step-down transformer may reduce the voltage supplied to less than 1 V or as high as 15 V. The current will increase to thousands or tens of thousands of amps.

Changing the *voltage* of the transformer secondary results in large changes in welding *current*. The voltage is changed by using *tap settings.* Thus, when you change the tap setting on a machine, you are changing the secondary voltage and current. Some machines have a high and a low position, others have multiple settings; some have both. Fig. 25-6 shows a tap switch on a resistance welding machine.

PROCESS CONTROLS

There are five variables that must be taken into account when spot welding. These process controls or variables are:
• Time.

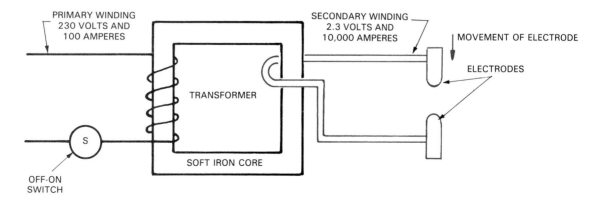

Fig. 25-5. A schematic drawing of the primary and secondary circuits of a spot welding machine transformer. This is a step-down transformer. When the voltage is reduced from 230 V to 2.3 V, the amperage increases from 100 A to 10,000 A.

Fig. 25-6. A tap switch is sometimes used to set the voltage and current in the secondary circuit of the transformer. Not all resistance welding machines use tap switches.

- Current.
- Electrode force.
- Electrode type.
- Type of welding machine.

In resistance spot welding, four *times* are used. These are:

- Squeeze time.
- Weld time.
- Hold time.
- Off time.

Squeeze time is the period when the welding electrodes close to apply the correct force to the metal to be welded. *Weld time* is the time when the electrical current flows. *Hold time* occurs after the weld is made. It allows the molten metal to solidify (become solid). At the end of the hold time, the electrodes lift off the metal. The spot weld is completed.

Off time is the time between one weld sequence and the next. During this time, the welded parts are removed and new parts are placed between the electrodes for welding. In some systems, either the electrodes or the metal parts are moved to a new weld location during this time period. Off time is used when spot welds are made automatically over and over. In manual operations, when a welder is required to align parts between the electrodes, off time is not used.

Force is required to press together the parts to be welded. This force creates a path for the electricity to flow. It also keeps the molten metal within the weld area. If there is not enough force, or if the electrode tip diameter is too small or the current is too high, molten metal will squirt out of the weld. This is called an *expulsion weld.* This is not a desirable weld.

Controllers

Spot welding is a fast welding process. A spot weld is often started and completed in less than one second. As discussed in the preceding section, there are different times (squeeze, weld, hold, and off) that must be controlled.

In spot welding, a controller is used to correctly set the length of each of these. Times used in spot welding are set in cycles (one cycle = 1/60 sec.). Squeeze time must be long enough for the electrodes to close on the work and apply the correct force. Squeeze time is usually 10 – 30 cycles. Weld time can vary from 1 – 99 cycles; 5 – 30 cycles is typical. Hold time allows the molten metal to cool. It is often 10 – 30 cycles. Off time is the time needed by automatic systems to unload parts that have been welded and load new parts. Off time may also be the time required to relocate a part under the electrodes. Off time must be set to allow enough time to accomplish these tasks safely. An example of a complete weld schedule is shown in Fig. 25-7.

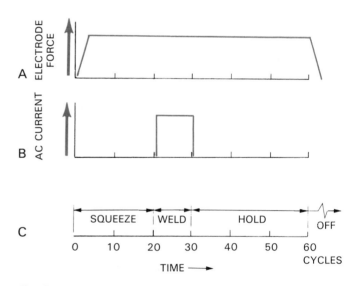

Fig. 25-7. A complete welding schedule. A—The electrode force plotted against time. B—Welding current versus time. C—The desired times are set in cycles (1 cycle = 1/60 sec.).

In addition to controlling times, a controller regulates the current within the tap setting. Large changes in current are made by changing the tap setting. Within a given tap setting, however, the controller changes the current by using the *percent heat control.* Percent heat is the amount of the available current that will be used. For example, a 100% heat setting will use all the available current. A 50% heat setting will use only half the available current. NOTE: Do not use a percent heat setting of less than 20%. Instead, lower the tap setting and use a higher percent heat.

Many controllers used in industry have additional features that improve the quality of a weld. These features include *upslope* and *downslope.* Upslope gradually increases the current at the beginning of the weld. Downslope gradually decreases the current at the end of the weld. Another feature is *multiple pulses.* This uses more than one pulse of weld current to make a single spot weld. The short intervals between pulses allow the weld to cool.

Manual control

Mechanical leverage-type spot welding machines do not use an electric controller. The welding operator controls all times, using a foot pedal. The pedal has two positions and two switches. When the pedal is pressed to the first position, the first switch closes. The electrodes close and force is applied to the work. This is the squeeze time. When the foot pedal is pressed to the second position, current flows. This is the weld time. Most manual equipment has a weld current timer that controls the length of the weld time. Some equipment requires the operator to control the length of the weld time. On these machines, current continues to flow until the operator raises the foot pedal to the first position. The current stops, but force is still applied. This is the hold time. When the operator releases the foot pedal, the electrodes open and the weld is complete.

ELECTRODES

Electrodes used in resistance welding do not become part of the weld. Instead, they are used to squeeze the metal together and carry the current into the metal. A material used as an electrode should have the following properties:

- Be a good conductor of electricity (electrical conductivity).
- Be a good conductor of heat (thermal conductivity).
- Have good mechanical strength and hardness.

A material that conducts electricity and heat well often will not have good strength and hardness. An example is pure copper. A material that is very strong and hard often will not conduct electricity and heat well. Examples are tungsten and molybdenum.

The most common electrode materials are copper alloys. They are used to weld mild steel, stainless steel, nickel, and other metals. There are different types of copper electrodes. As the hardness and strength of a copper alloy increases, its ability to carry electricity and heat decreases. Other electrodes are made from molybdenum or tungsten. These electrodes have better strength than copper alloy electrodes, but do not carry current as well.

Electrode classification

Electrodes have been grouped into classes by the Resistance Welder Manufacturers' Association (RWMA). The most common electrode materials are listed below.

Class 1. Copper-cadmium alloy. This class has the highest thermal and electrical conductivities. Its electrical conductivity is 90% (annealed copper has a conductivity of 100%), and has the highest thermal conductivity. It is also the softest material, with a Rockwell hardness of 70B.

Class 2. Copper-chromium alloy. This class has good electrical conductivity (85%) and good thermal conductivity. It has a Rockwell hardness of 83B. Electrodes in this class are the most widely used. They are used to weld most steels, stainless steels, nickel, and other metals with low electrical and thermal conductivities.

Class 3. Copper-cobalt-beryllium alloy. These electrodes have an electrical conductivity of 48%, with poorer thermal conductivity than Class 2. They have a Rockwell hardness of 100B.

Classes 4 and 5. These electrodes are made from copper alloys that are very hard, but have poor electrical and thermal conductivity.

Class 11. Copper-tungsten alloy. Conductivity is 46%; hardness is Rockwell 90B. These electrodes are used to replace Class 1, 2, or 3. They usually do not stick to the work.

Class 13. Tungsten electrode. This electrode has a conductivity of 32% and is very hard. Tungsten electrodes are used to weld copper, brass, and other very conductive metals.

Class 14. Molybdenum electrode. This electrode has a conductivity of 31% and has a hardness of Rockwell 90B. Like the Class 13 electrodes, these electrodes are used on nonferrous metals (metals that do not contain much iron) with high electrical and thermal conductivities.

Electrode shapes

Electrodes can be machined, cast, or forged into complex shapes. Fig. 25-8 shows many different shapes of electrodes. Electrodes also are available as caps and adaptors. Caps are much less expensive than large electrodes; when they wear, they can be replaced separately. Fig. 25-9 shows caps and adaptors.

Because of the heat developed in a resistance weld, most electrodes are water-cooled. There is a metal tube in the electrode that carries water very close to the tip. Water flows all the time that the machine is on, removing the heat from the electrode. See Fig. 25-10.

SEAM WELDERS

In addition to making individual spot welds, resistance welding is used to make seam welds. Seam welds can be continuous, or a series of individual spots. Seam welding machines use a round electrode that operates like a wheel. It rolls along the weld joint. The electrode is often called a **wheel electrode.** See Fig. 25-11.

The current used to make a seam weld can flow constantly as a seam weld is made, or it can turn on and off numerous times. When the current turns on and off, it actually makes individual spot welds. These spot welds can overlap to form a continuous seam, or can be a

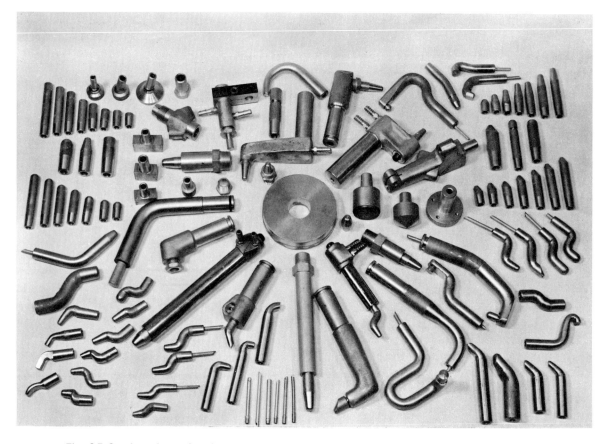

Fig. 25-8. A variety of resistance welding electrode shapes. (Hercules Welding Products)

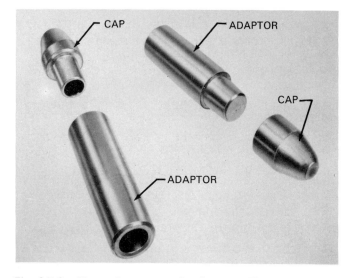

Fig. 25-9. Electrode caps and adaptors. These caps are generally held in place by a taper fit.   (Tuffaloy Products, Inc.)

series of individual spots that do not overlap. Spot welds that do not overlap form an *intermittent* seam weld. See Fig. 25-12.

SAFETY

Resistance welding machines can be much more complex than arc welding machines. When using a

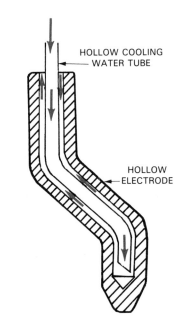

Fig. 25-10.   The cooling water tube carries water to the tip of the electrode.

resistance welding machine, you should regularly:
- Check all electrical connections.
- Check the electrodes and other parts that carry electricity to be sure they are installed correctly.
- Check all air and hydraulic connections and hoses. Look and listen for any leaks.

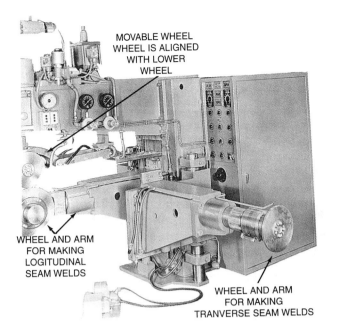

Fig. 25-11. *A large universal seam welding machine. The electrode wheels are set up for longitudinal welding. The wheel and arm at the right may be swung into position for transverse welding. The upper wheel may be rotated as needed.*
(Sciaky Bros., Inc.)

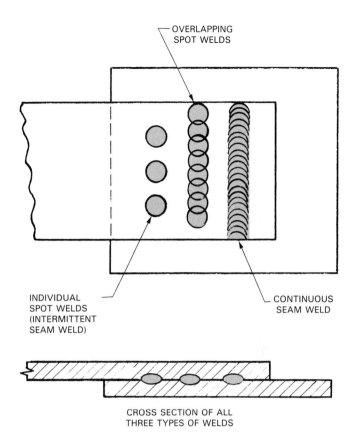

Fig. 25-12. *Welds that are possible with a seam welding machine: individual spots (intermittent seam weld), overlapping spots, or a continuous seam.*

- Check all water connections and hoses. Make sure water is flowing to cool the welding machine.

The voltages and currents used in resistance welding can be high enough to shock you and possibly kill you. The pressures used in resistance welding can crush any part of your body that may get in between the electrodes. If a problem develops with a machine you are working on, *turn it off.* Call your instructor or supervisor.

When resistance welding, molten metal can squirt out of a spot weld and be thrown into the air. Always wear safety glasses when spot welding or when working near spot welding equipment.

You should also wear gloves when spot welding. This will prevent cuts from sharp metal edges and will protect your hands from molten metal that may squirt out during a weld. You should wear long pants without cuffs, a long-sleeved shirt, and steel-toed work shoes.

REVIEW QUESTIONS

Write your answers on a separate sheet of paper.
1. What causes the heat used to make a resistance spot weld?
2. Where is most of the heat developed when electricity flows through two or more pieces of metal?
3. List the five variables, or process controls, involved in spot welding.
4. Explain the difference between hold time and squeeze time.
5. What happens to the current in a step-down transformer? Does it increase or decrease? What happens to the voltage? Does it increase or decrease?
6. List two ways to change the current used for a spot weld.
7. Which class of electrode material is most commonly used? What metals are they used to weld?
8. What method is used to remove the heat developed in electrodes during spot welding?
9. What is the electrode used in seam welding often called?
10. If a problem develops with a resistance welding machine, what is the first thing you should do?

A resistance spot weld being made. Safety goggles or a safety shield must be worn as protection from flying sparks.   (The Nippert Co.)

Chapter 26

Resistance Welding Procedures

After studying this chapter, you will be able to:
☐ Set up and adjust a spot welding machine.
☐ Describe the methods used to determine the correct force for spot welding.
☐ Calculate the weld time needed for resistance welding mild steel.
☐ Describe the means used to test for a good spot weld, and the signs that indicate a weld is of the desired quality.

TECHNICAL TERMS

Electrode tip diameter, force gauge, newton, projection welding.

SELECTING THE WELDING MACHINE

There are several types of resistance spot welding machines, as discussed in Chapter 25. Some are very large and can handle large pieces of material. Some are small and can weld only small parts. Some machines have wheel electrodes and are used to make seams. Refer to Chapter 25 for help in selecting a machine that is the right size for the welding you will be doing.

The type and the thickness of the metal you will weld dictates how much force and current the machine must apply. Metal 1/16″ (1.6 mm) or less in thickness can be welded with a mechanical leverage-type machine, but a machine with an air cylinder is often used for these jobs. Thick metals require a machine with an air or hydraulic cylinder. More current is needed to weld thick metals than thin metals. Larger welding machines usually will supply more force and more current than smaller ones. Select the size machine that will supply enough force and current for the job you are doing.

SELECTING AND PREPARING THE ELECTRODES

Once the welding machine is selected, there are four additional variables that must be set. These are:
• Time.
• Current.
• Electrode force.
• Electrode type.

Two important areas must be considered when selecting the electrode type. First, what type of electrode material you will use. Chapter 25 discusses electrode materials. Select the correct material for the metal you are welding.

Second, the electrode tip diameter. This helps determine the size of the spot weld. A spot weld cannot be larger than the diameter of the electrode. Select a tip that is slightly larger in diameter than the spot weld you want to make.

When welding mild steel, use this formula to determine the correct tip diameter.
Tip diameter in inches = .18″ + (total thickness, in inches, of the two pieces being welded)
Tip diameter in mm = .454 mm + (total thickness, in mm, of the two pieces being welded)
Example: Two pieces of .031″ (0.8 mm) mild steel will use an electrode tip diameter of about .18″ + .062″ = .242″ or 4.54 mm + 1.6 mm = 6.14 mm.

After calculating the correct diameter, select a tip that is approximately the size you need. Not every size tip is available, so use one that is close to the size you need. In the example above, you could use a 9/64″ (.234″) or a 1/4″ (.250″) tip. In metric sizes, you could use a 6.0 mm or a 6.2 mm tip.

If the electrodes are not parallel and in line, poor welds will result. If the electrodes are not parallel, or are not in line, then only *part* of the tip will contact the metal. See Fig. 26-1. After installing the electrodes into the welding machine, adjust them so they meet correctly.

Electrode tips come in different designs. Different designs are needed to reach into some areas. Some of the more common shapes are shown in Fig. 25-10.

After making a number of welds, the tips of the electrodes will have to be cleaned. Cleaning may be needed after as few as 10 welds or as many as 500. Cleaning removes dirt, oxides, and metal particles from the end of the electrodes. Electrodes that are used with too much current and/or pressure, or for too long a period, will flare out (mushroom) at the end. See Fig. 26-2.

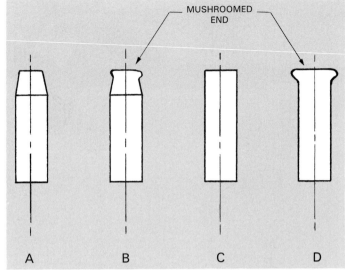

Fig. 26-2. Mushroomed electrodes. A and C—These electrodes are in good condition. B and D—These electrodes are flared or mushroomed on the end from too much heat and pressure.

Because the tips become larger in diameter, the current used to make the spot weld is spread over a larger area. This causes a smaller and weaker weld. If electrodes mushroom, replace them or return them to the correct shape by filing or machining.

Remember, when choosing an electrode, you must select the correct tip diameter with the tip design you need. You must also select the correct electrode material for the job you are welding.

SELECTING AND SETTING THE FORCE

After selecting the welding machine and the electrodes, the variables remaining are force, time, and current.

Force is applied by the electrodes to push together the metals being welded. As noted in Chapter 25, the applied force creates a path for the electrical current to flow through the metal. The force cannot be too great, or it will damage the metal. The force will indent the metal slightly, but you do not want to crush the area under the electrodes. Fig. 26-3 shows two welds: one made with the correct force and one made with excessive force.

On machines that use an air cylinder, force is increased or decreased by adjusting the air pressure. The higher the pressure going into the air cylinder, the higher the force at the electrodes. Machines that use *hydraulic* pressure are adjusted similarly.

Force is measured by a *force gauge.* Place the gauge between the electrodes, then close them. The gauge will measure the force applied by the electrodes. See Fig. 26-4. Force is measured in pounds or *newtons.*

On machines that use mechanical leverage, a spring is usually adjusted to obtain the desired force. By compressing the spring more, the force increases. Adjust

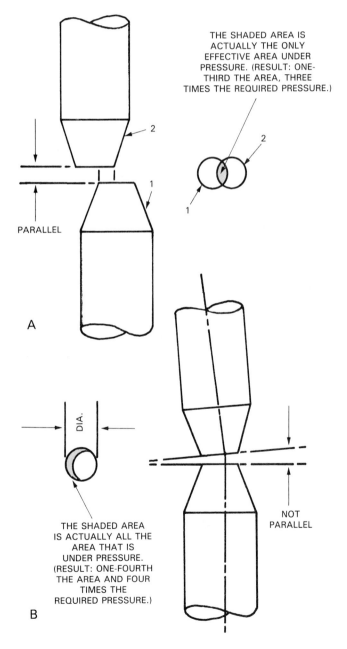

Fig. 26-1. Effects of misaligned electrodes. A—These electrodes are off-center. B—These electrodes are not parallel. In either case, pressure on the reduced contact area is too great.

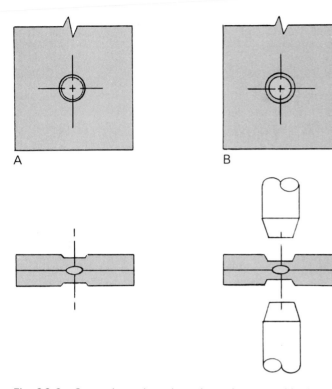

Fig. 26-3. Correctly made and poorly made spot welds shown in cross section. A—This weld was made with the correct heat and force. B—This weld is deeply indented because too much force was used.

Fig. 26-4. A force gauge being used to measure the force applied by the electrodes. The amount of force is changed by varying the pressure to the air or hydraulic force cylinders.

the compression of the spring to obtain the desired force.

When welding mild steel up to 1/8″ (3.2 mm) thick, use the following formula to approximate the correct force.

Force in pounds = 8500 × (total thickness, in inches, of the two pieces being welded)

Force in newtons = 1500 × (total thickness, in mm, of the two pieces being welded)

Example: Two pieces of .031″ (0.8 mm) mild steel requires about 8500 × .062″ = 527 pounds or 1500 × 1.6 mm = 1800 newtons of force.

Another way to approximate the correct force on mild steel is to use the welding machine. Set the weld time to zero so that no welding current will flow. If the machine has a switch that has a "No Weld" position, placing the switch in this position will have the same effect.

Place two pieces of metal between the electrodes. When you press the foot pedal, the electrodes will close on the metal but no welding will occur. Release the pedal, so the electrodes lift off. Look at the metal. If there is no indentation from the electrodes, increase the force until a slight dent is made. Then, reduce the force about 10 percent so no dent is made. This provides a starting point. When welding, you may need to readjust the force to obtain a good quality weld.

SELECTING AND SETTING THE TIME AND CURRENT

Time and current are the two variables that have the greatest effect on the quality of a weld. A high quality spot weld required that all the variables be set correctly, but time and current are the most important. If they are not set correctly, the weld may be too small and too weak, or it may be too large and squirt out liquid metal. When a weld squirts out liquid metal it is called an expulsion weld.

Making time settings

As noted in Chapter 25, there are four *times* that make up a complete weld schedule. They are the squeeze, weld, hold, and off times. Squeeze time must be long enough to allow the electrodes to close and apply the correct pressure on the metal being welded. Typical squeeze time settings are from 20 to 40 cycles.

Weld time, when the current flows, must be long enough to allow the metal to melt and the weld to grow to the correct size. The weld time varies, depending on the kind and thickness of the metal being welded. For welding mild steel, the following formula can be used to approximate the correct weld time:

Time in cycles = 120 × (total thickness in inches, of the two pieces being welded)

Time in cycles = 4.73 × (total thickness, in mm, of the two pieces being welded)

Example: Two pieces of .050″ (1.27 mm) mild steel

require 120 × .100 = 12 cycles or 4.73 × 2.54 mm = 12 cycles of weld time.

Hold time must be long enough for the molten metal to become solid. This time can vary from 20 to 60 cycles. Off time is used only when spot welds are made automatically, one after the other. The off time must be long enough to relocate the parts being welded, or to remove one part and put the next one between the electrodes. Set the off time appropriately. If it is not used, set it to zero.

When operating a mechanical leverage system, *you* control the squeeze and hold times. Press the foot pedal down so the electrodes close on the work. Then, press down farther so the weld current will flow. Keep the electrodes closed on the work for one second or longer. Release the pressure on the foot pedal, and the electrodes will open.

Making current settings

Current is changed by changing the tap setting or the percent heat setting. Tap changes are for large increases in current; percent heat is for fine adjustments. Several methods are used to properly set the current. If you have operated the equipment before, you know the approximate settings required. You can also use a trial-and-error method to determine the current. A third way is to calculate the required current, and then use special equipment to set and monitor it.

If you use a resistance welding machine for some time, you will become familiar with it. You will know what tap and percent heat settings are needed to weld a particular combination of metals.

When you are first learning about spot welding, the trial-and-error method of setting the current is useful. First, set the tap switch. A higher setting is used for thicker metals; a lower setting for thin metals. Set the percent heat to 70%. Make two spot welds, as shown in Fig. 26-5. If metal squirts out, turn down the tap setting. If the metal pieces do not weld together, turn up the tap setting. If the pieces *do* weld together, tear them apart and look at the spot welds. The weld size should be slightly smaller than the diameter of electrode being used. A good weld will tear some metal away from one of the pieces as they are pulled apart. Figs. 26-6 and 26-7 show pieces of metal pulled apart

to check welds.

Adjust the tap setting and percent heat up or down. This will increase or decrease the current and affect the size of the weld.

Make two more spot welds. Tear them apart and look at the weld quality. Adjust the current up or down. Continue this process until a good quality weld is made.

An approximate current can be calculated for welding mild steel up to 1/8″ (3.2 mm) thick. Use the following formula:
Current in Amps = 125,000 × (total thickness, in inches, of the two pieces being welded)
Current in Amps = 4935 × (total thickness, in mm, of the two pieces being welded)
Example: Two pieces of .062″ (1.6 mm) mild steel require 125,000 × .124 = 15,500 amps or 4935 × 3.2 mm = 15,792.

WELDING MILD STEEL

Properly setting up the equipment is the most important part of resistance welding. This involves choos-

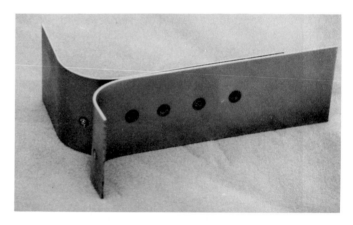

Fig. 26-6. A spot welding sample that was bent prior to being pulled apart.

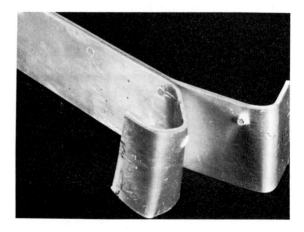

Fig. 26-7. A spot weld nugget torn out of one piece during a destructive peel test. Test welds are made to determine the correct machine settings to get the proper spot weld size and strength.

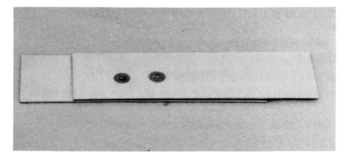

Fig. 26-5. Two pieces of mild steel overlapped about seven inches (178 mm) and spot welded.

ing the right electrodes and setting the correct current, times, and pressure. Once the equipment is set, it will produce quality spot welds if these two conditions are met.
- The surface of the metal to be welded is thoroughly cleaned and free of oil, and rust.
- The electrodes are kept in good condition by cleaning and resurfacing as needed.

Each weld schedule is set for one combination of metals. If a spot welding machine is set up to weld two pieces of 1/16″ (1.6 mm) mild steel, the same setting cannot be used to weld 1/8″ (3.2 mm) mild steel. That setting cannot be used to weld aluminum, either. Each type and thickness of metal requires a different setting or weld schedule.

The formulas that were discussed earlier in the chapter are shown in Fig. 26-8. These are starting points. You will need to adjust the machine as needed to obtain good quality spot welds.

EXERCISE 26-1 — SPOT WELDING MILD STEEL

1. Obtain ten or more pieces of mild steel measuring approximately 1 1/2″ x 8″ x 1/16″ (40 mm x 200 mm x 1.6 mm).
2. Clean both surfaces of the pieces. This can be done with steel wool, a wire brush, or a degreasing solution supplied by your instructor.
3. Calculate the following, using the formulas given in this chapter. Write your answers on a separate sheet of paper.
   Electrode diameter_____
   Electrode force _____
   Weld time_____
   Weld current _____
4. Select the proper electrodes and install them in the machine.
5. Set the force, by adjusting the air or hydraulic pressure, or by adjusting the force spring.
6. If you are using a machine with a controller, set the

squeeze time to 30 cycles. Set the weld time to the time calculated in Step 3. Set the hold time to 30 cycles. Set the off time to zero.
7. Set the starting weld current as described in this chapter.
8. If the machine is water-cooled, turn on the water.
9. Overlap two pieces of mild steel about seven inches, as shown in Fig. 26-5. Make a spot weld. Adjust the current as required. Make additional welds, adjusting the force, time, and current as needed, until you obtain a spot weld that is about 1/8″ (3.2 mm) in diameter. You may make more than one weld on each sample while setting the machine.
10. Once you have set the machine correctly, weld two pieces together. Make welds, one every inch (25 mm), along the overlapped portion. Do not weld closer than 1″ (25 mm) to the end of each piece.
11. On a separate sheet of paper, record the settings you used to make these welds.
    Electrode tip diameter _____
    Air or hydraulic pressure (if used)_____
    Weld time_____
    Tap setting _____
    Percent heat setting _____
12. Tear or peel apart at least five spot welds. See Figs. 26-6 and 26-7. Measure and record, on a separate sheet of paper, the size of five different spot welds.
    1. _____ 2. _____ 3. _____
    4. _____ 5. _____ .

WELDING ALUMINUM

Aluminum is softer than steel. It is also more electrically and thermally conductive than steel. This means that electricity and heat will flow through aluminum more easily than they do through steel. For these reasons, the welding setup for aluminum is different from the setup for steel. Less force is used. Much more current is required, but the weld time is the same or even less than is needed for steel.

CALCULATING RESISTANCE SPOT WELDING VARIABLES				
Variable	Equation (Conventional)	Units of Measure	Equation (Metric)	Units of Measure
Contact Tip Dia.	.18 + (total sheet thickness)	inches	4.54 + (total sheet thickness)	mm
Weld Time	120 × (total sheet thickness)	cycles	4.73 × (total sheet thickness)	cycles
Current	125,000 × (total sheet thickness)	amps	4935 × (total sheet thickness)	amps
Electrode Force	8500 × (total sheet thickness)	lbs.	1500 × (total sheet thickness)	newtons
Weld Size	The weld size is the same size or slightly smaller than the contact tip diameter.			

Fig. 26-8. This table of equations approximates the values of the variables used when resistance spot welding low carbon steel. The total sheet thickness is the combination of the two thicknesses being welded.

Class 11, 13, or 14 electrodes are used to weld aluminum. Refer to Chapter 25 for more information about these electrodes.

Aluminum must be cleaned to remove oxides from the surfaces before welding. A stainless steel brush or a cleaning solution can be used.

EXERCISE 26-2 — SPOT WELDING ALUMINUM

1. Obtain ten or more pieces of aluminum measuring approximately 1 1/2'' x 8'' x 1/16'' (38 mm x 200 mm x 1.6 mm).
2. Clean both surfaces of the pieces. This can be done with steel wool, a wire brush, or a degreasing solution supplied by your instructor.
3. Select proper electrodes and install them in the machine.
4. Set the electrode force. Adjust the air or hydraulic pressure, or the force spring, depending on the machine. The force should be less than is used for welding steel of the same thickness.
5. If you are using a machine with a controller, set the squeeze time to 30 cycles. Set a weld time. The time should be only a few cycles. Set the hold time to 30 cycles. Set the off time to zero.
6. Set the tap switch to high. Set the percent heat on the controller.
7. If the machine is water-cooled, turn on the water.
8. Overlap two pieces of aluminum about seven inches, as shown in Fig. 26-5. Make one or two spot welds. Follow the trial-and-error approach described earlier in this chapter, to determine the weld time and weld current. Make additional welds, adjusting the force, time, and current, until you have spot welds that are about 1/8'' (3.2 mm) in diameter.
9. Once you have set the machine correctly, weld two pieces together. Make welds, one every inch (25 mm), along the overlapped portion. Do not weld closer than 1'' (25 mm) to the end of each piece.
10. On a separate sheet of paper, record the settings you used to make welds.
    Electrode tip diameter _____
    Air or hydraulic pressure (if used) _____
    Weld time_____
    Tap setting _____
    Percent heat setting _____
11. Tear apart five spot welds. See Figs. 26-6 and 26-7. Measure the size of the spot welds. Write them down on a separate sheet of paper.
    1. _____ 2. _____ 3. _____
    4. _____ 5. _____.

PROJECTION WELDING

In *projection welding,* welds similar to spot welds are made at locations where projections or bumps have been formed in one of the pieces. See Figs. 26-9 and

26-10. The pieces that are to be welded are forced together by the electrodes. When the current flows, a weld is formed at each point where there is a projection or bump. The projections accurately locate the weld points.

A welding schedule for projection welding is similar to spot welding the same material and thickness without a projection. More than one projection weld can be made at a time. This requires more force and more current than is needed to make a single weld.

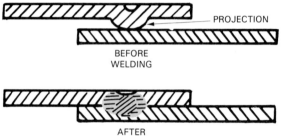

Fig. 26-9. A drawing of metal before and after projection welding. The metal is stamped to form a projection. Part of the depression from stamping remains after the weld is completed.

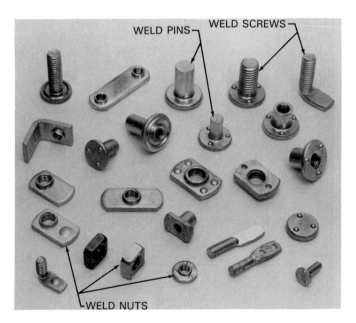

Fig. 26-10. Nuts, screws, and pins with projections formed on them for projection welding. Notice the different projection types and sizes. (The Ohio Nut and Bolt Co.)

RESISTANCE SEAM WELDING

Seam welding is done with two wheel electrodes. These wheels roll along opposite surfaces of the metal to be welded. A seam weld can be continuous, or can be a series of spot welds that are evenly spaced. Such a line of evenly spaced spot welds is called an intermittent seam weld.

The pressure needed to make a seam weld is the same or slightly higher than that used to make individual spot welds. The current required to make a continuous seam weld is much higher than is necessary to make individual spot welds. An intermittent seam weld will require more current than individual spots, but less than a continuous seam weld.

REVIEW QUESTIONS

Write your answers on a separate sheet of paper.
1. What are the five variables in resistance welding?
2. Which two variables are the most important?
3. To produce a good weld, what must be done to the electrodes after they are installed into the welding machine?
4. Why must the electrode tips be cleaned?
5. If the air or hydraulic pressure is increased, does the electrode force increase or decrease?
6. When welding two pieces of .032″ (0.8 mm) mild steel, calculate the following, using the formulas given in this chapter.
   Electrode tip diameter _____
   Electrode force _____
7. If one piece of .045″ (1.14 mm) mild steel is welded to a piece of .062″ (1.6 mm) mild steel, calculate the following:
   Weld time _____
   Weld current _____
8. After a welding machine is set up properly, what two conditions must be met to continue to make high-quality spot welds?
9. Does aluminum conduct heat and electricity more or less easily than steel?
10. What are the two types of seam welds that can be made?

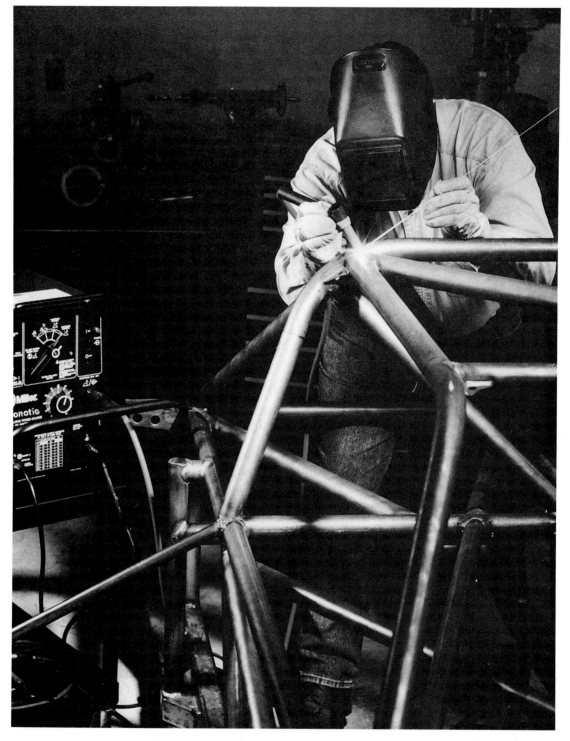

A complex tubing assembly being welded, using the GTAW process.   (Miller Electric Mfg. Co.)

SPECIAL PROCESSES

ABB Flexible Automation, Welding Systems Division

Two plasma arc torches cutting duplicate patterns on a water-filled cutting table. Cutting under water reduces the sound level.   (Thermadyne Industries, Inc.)

Chapter 27

Special Welding and Cutting Processes

After studying this chapter, you will be able to:
☐ List the advantages of some special welding and cutting processes that are used in industry.
☐ Identify several special welding processes used in industry for unusual metals or unusual positions.
☐ Identify several special cutting processes used in industry.

TECHNICAL TERMS

Arc stud welding, chemical flux cutting, electrogas welding, electron beam welding, electroslag welding, explosion welding, friction welding, kerf, kinetic energy, laser, laser beam cutting, laser beam welding, metal powder cutting, nontransferred arc, orifice, orifice gas, oxygenarc cutting underwater, oxyfuel gas cutting underwater, photons, plasma, plasma arc cutting, plasma arc welding, restricted nozzle, solid state welding, sonotrode, submerged arc welding, transferred arc, ultrasonic transducer, ultrasonic welding, upset, vacuum.

FREQUENTLY USED SPECIAL PROCESSES

The welding, cutting, brazing, soldering, and surfacing processes discussed in earlier chapters of this book are not the only processes available. The American Welding Society lists 94 different welding and cutting processes. Refer to Chapter 1 (Fig. 1-6) for a chart listing all these processes. Among them are special processes used on unusual metals, or in locations that are unusual. Extremely thin or thick metals

require special welding and cutting processes, as well.

Many of the special processes discussed in the following sections are not found in small production shops, body shops, home or farm shops, or school shops. These special welding and cutting processes are, however, being found in ever-increasing numbers in large manufacturing and repair businesses. The terms and abbreviations used in this chapter to describe these processes are those approved by the American Welding Society in ANSI/AWS A3.0-94, "Standard Welding Terms and Definitions."

SPECIAL ARC WELDING PROCESSES

There are many welding situations that require development of a special form of welding. Welding on metal over 1 ft. (305 mm) thick is very difficult to do using SMAW, GMAW, or GTAW. Electrogas arc welding and electroslag welding were developed to meet this need.

When scientists created plasmas experimentally, they discovered that extremely high temperatures and heat outputs were also created. Welding engineers developed the plasma arc welding and cutting processes to take advantage of this new heat source. Arc stud welding makes it possible to attach fastening devices to steel structures very easily. Specially designed equipment makes it possible to arc weld under water.

Electrogas welding (EGW)
The *electrogas welding* process was developed as a means of welding extremely thick metal. Thick metal sections are hard to weld, since the weld area must be

kept near the melting point at all times. This is difficult with regular welding equipment.

Metal of almost any thickness can be welded with the electrogas process. See Fig. 27-1. The weld is always made in a vertical position. One or more consumable electrodes are fed into the joint from above. To make the weld, the welding arc is struck, and the consumable electrode melts in the joint to form the weld. Solid electrode wire or flux-cored wire may be used. Two water-cooled copper shoes act as dams. They contain the molten metal in the weld area and cool the completed weld. As the copper shoes and electrode move up, the welding continues to the top of the joint.

The weld area is protected from contamination by a shielding gas. The gas may be carbon dioxide, an inert gas, or a combination of two gases. It is supplied from above and covers the molten metal.

Electroslag welding (ESW)

The *electroslag welding* process is similar to electrogas welding. A thick layer of powdered flux is placed in the joint before welding begins. Once welding starts, the powdered flux melts. It forms a protective, floating slag above the weld area as shown in Fig. 27-2. Impurities in the molten metal will float to the top and mix with the flux. The molten flux protects the weld metal from impurities in the air above the weld.

Submerged arc welding (SAW)

Submerged arc welding has several advantages over other welding processes:

- It is faster than manual arc welding.
- There is no visible arc.
- There is no spatter.
- The weld is of high quality.
- The welding groove angle may be smaller.

A consumable electrode is fed into the weld joint, as in GMAW. Ahead of the electrode, as shown in Fig.

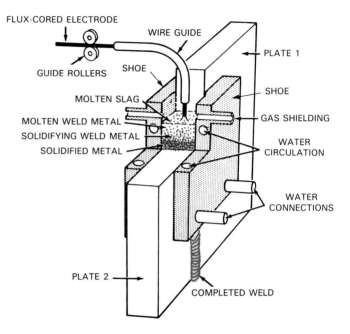

Fig. 27-1. A schematic drawing of the electrogas welding process. The shoes are water-cooled and move up along with the weld. A shielding gas protects the weld and molten metal.

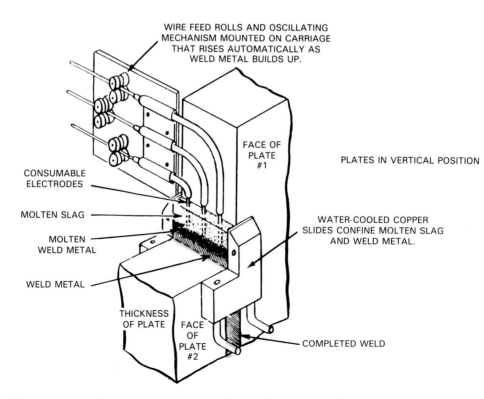

Fig. 27-2. A schematic drawing of an electroslag weld in progress. Three consumable electrodes are used in this application. The molten slag floating above the weld prevents oxidation.

27-3, a thick layer of a powdered flux is deposited. The arc between the electrode and base metal occurs beneath this thick flux layer. The larger electrodes, higher currents, and narrower groove angles used in SAW permit faster welding. The flux covers the arc and prevents any splatter. This is normally a fully automated machine welding process, Fig. 27-4. It may be done semiautomatically, as well. In that case, the welder guides the wire and flux-feeding mechanism.

Plasma arc welding (PAW)

Scientists say that there are four states (conditions) of matter: solid, liquid, gas, and plasma. *Plasma* is an ionized gas (a gas made up of atoms that have lost or gained electrons). Ionized gas or plasma is extremely hot—temperatures as high as 43,000°F (24,000°C) have been reached. The plasma arc is an excellent heat source for welding or cutting. Cross sections of the two types of arc plasma torches, transferred and nontransferred, are shown in Fig. 27-5.

An inert gas, such as helium, argon, or nitrogen, is used to create the plasma. The gas, which flows through the *restricted nozzle,* is called the *orifice gas.* The inert *shielding* gas is passed around the restricted nozzle to shield the weld area. A nonconsumable tungsten electrode is used in the torch.

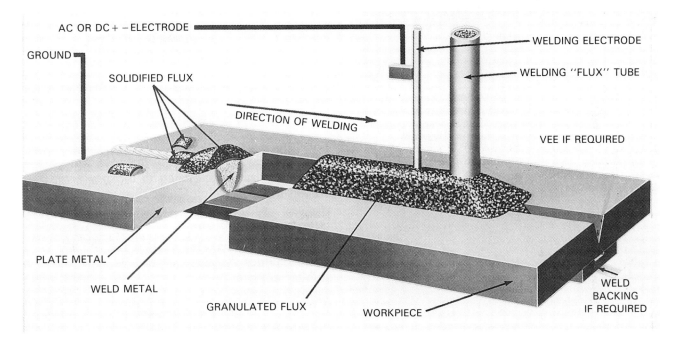

Fig. 27-3. A diagrammatic view of a submerged arc weld in progress. Note that the flux granules are deposited ahead of the consumable electrode.   (ESAB Welding and Cutting Products)

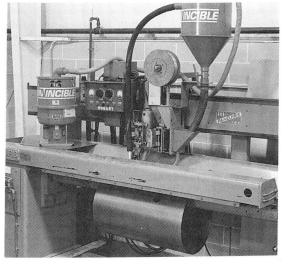

Fig. 27-4. A submerged arc welding outfit. Automated flux recovery equipment is in use. A butt weld is being made on a cylindrical shape that is mounted below the welding fixture. (Invincible Airflow Systems)

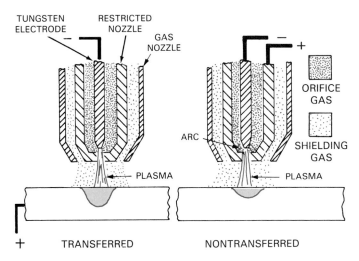

Fig. 27-5. Cross sectional drawing of the transferred and non-transferred plasma arc welding torch.

In the *transferred arc* process, an arc is struck between the electrode and the work. The inert gas is turned into a superheated plasma as it passes through the arc at the *orifice* (hole) in the restricted nozzle. It is the plasma that melts the base metal.

In a *nontransferred arc* plasma torch, the arc occurs between the tungsten electrode and the restricted nozzle. See Fig. 27-5. Edge and flanged joints are welded without filler metal. When automatic plasma arc welding is done, the filler wire is fed into the weld pool automatically. See Fig. 27-6.

Arc stud welding

Arc stud welding was developed to weld threaded studs, location pins, or nails to metal plates. The process is fast and simple. It requires little skill.

To perform the weld, the stud is loaded into a stud welding gun. See Fig. 27-7. The gun and stud are placed against the base metal, as shown in Fig. 27-8. When the gun's trigger is pulled, electricity flows through the stud to the base metal. The stud is then automatically pulled away from the plate, and an arc is struck. Electricity flows for only a fraction of a second. The gun

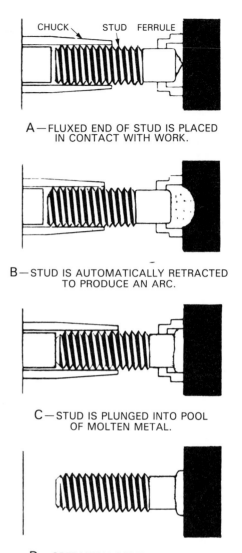

A—FLUXED END OF STUD IS PLACED IN CONTACT WITH WORK.

B—STUD IS AUTOMATICALLY RETRACTED TO PRODUCE AN ARC.

C—STUD IS PLUNGED INTO POOL OF MOLTEN METAL.

D—OPERATION COMPLETED—STUD IS WELDED TO WORK.

Fig. 27-8. The steps that take place during an arc stud weld on a steel plate. (TRW, Nelson Stud Welding Div.)

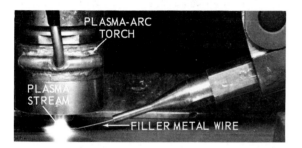

Fig. 27-6. Welding wire being fed into the weld pool as a plasma arc weld progresses. (ESAB Welding and Cutting Products)

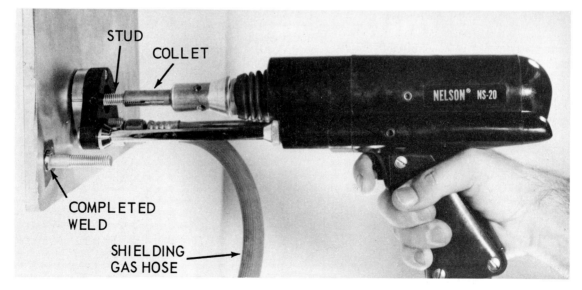

Fig. 27-7. An arc stud welding gun. This gun is being used to weld aluminum. Shielding gas is fed into a chamber surrounding the stud. (TRW, Nelson Stud Welding Div.)

then forces the molten tip of the stud into the molten base metal. It is held for a second or two and the weld is completed. On non-ferrous metals, a shielding gas is used to produce good arc stud welds. See Fig. 27-7.

SMAW underwater (SMAW)

Shielded metal arc welding is done underwater to repair bridges, ships, and oil drilling rigs. The process is similar to regular SMAW. The electrode holder for underwater welding, however, has especially good electrical insulation. The coating on the electrode is waterproof. See fig. 27-9. Special training, equipment, and safety procedures are required to weld underwater.

Fig. 27-9. Underwater welding electrode holder. Note that the holder is well-insulated, with no bare metal areas exposed. (Broco, Inc.)

SOLID STATE WELDING PROCESSES

Solid-state welding (SSW) processes do not use a flame, an arc, a beam, or resistance to heat the base metal. The American Welding Society lists nine forms of solid state welding. Refer to Fig. 1-6. Depending on the process. SSW may be done cold, warm, or hot. However, the temperature of the base metal is never higher than its melting temperature.

Explosion welding (EXW)

To prepare an explosion weld, the metals are placed at an angle. A protective layer is applied to the outer metal surface, then a layer of explosive material is added. An example of an *explosion welding* setup is shown in Fig. 27-10.

Detonation of the explosive causes small ripples on the surfaces of the base metals. The pieces are welded as these ripples lock together. This process is normally done in specially designed chambers. Explosion welding has been used on many metals. It has been used to weld a layer of one metal to another. Special permits are required from federal, state, and local authorities before this process can be used. The explosion welding process is very dangerous. It should be done only by experts in handling explosive material.

Friction welding (FRW)

When two objects are rubbed together, they generate heat. This principle is used to do *friction welding.* Fig. 27-11 shows the steps in a friction weld. One part is held stationarey, while the other part is held in a chuck and rotated rapidly. The parts are pressed tightly together. Friction will heat the two parts to their welding temperature. When the welding temperature

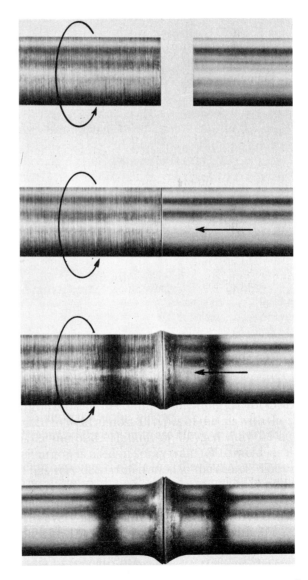

Fig. 27-11. The steps in making a friction weld on two pieces of 1'' (25.4 mm) diameter carbon steel. The rod at the left is spun at a high speed. The rod at the right is then forced against it. Friction creates enough heat to produce a strong weld.

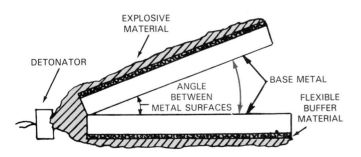

Fig. 27-10. A schematic drawing of a setup for explosion welding.

271

is reached, the rotation is stopped. The parts are then suddenly forced together under heavy pressure. After the heavy pressure is applied, the parts are held firmly until they cool. The finished weld is strong, with complete fusion over the entire joining surface. As shown in Fig. 27-11, the weld is *upset* (enlarged) where the parts meet.

Friction welds are often produced in less than 15 seconds. Friction welding has been used successfully to join some dissimilar (unlike) metals that normally cannot be welded.

Ultrasonic welding (USW)

Sound waves can cause vibrations in objects. For example, a loud stereo can make the walls vibrate. In *ultrasonic welding,* a very high-pitched sound is used to vibrate the surfaces of the metals to be welded.

Ultrasonic welding has many advantages. Since there is literally no heat, there is no metal distortion. Fluxes and filler metal are not needed. Very thin metals can be joined easily. Ultrasonic welds are normally small welds like resistance spot welds. Seam welds can also be made.

Fig. 27-12 shows a schematic of an ultrasonic weld in process. The parts to be joined are placed between two tips, as in resistance spot welding. An electronic device called an *ultrasonic transducer* causes either a wedge-reed or a lateral drive to vibrate extremely fast. This, in turn, causes a *sonotrode* (sound electrode) to vibrate. A slight force is applied to the parts through the sonotrode. The vibrations break up surface films, causing the parts to bond together without heat. An ultrasonic weld is completed faster than a resistance spot weld. Ultrasonic welds can be made on similar or dissimilar metals.

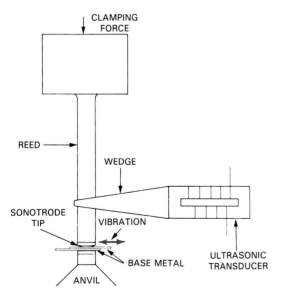

Fig. 27-12. A schematic drawing of the wedge-reed ultrasonic welding principle. A small spot weld is made by vibrating the electrode while they are in contact with the base metal. (Sonobond Ultrasonics, Inc.)

OTHER WELDING PROCESSES

There are several welding processes that do not use an arc, oxyfuel gas, or electrical resistance to weld parts together. The AWS identifies these as "Other Welding Processes."

Laser beam welding (LBW)

The *laser* is a high-energy-density beam of light. This light energy is emitted from the *laser beam welding* machine as energy particles called *photons.* The photons produce heat at the weld joint by thermal radiation. The most popular lasers are the NI:yag (neodymium-doped yttrium aluminum garnet) crystal and the gas CO_2. A schematic view of a CO_2 laser is shown in Fig. 27-13. Electrical energy is generally used to excite the photons in the atoms of these layers. Photon energy is allowed to build up in the crystal or gas until the desired level is reached. The energy is then released to fuse the weld joint.

Fusion occurs through thermal (heat) radiation. Laser beams may be continuous or pulsed. They can be focused with lenses to a very small and accurate beam. Their direction can be changed through the use of mirrors.

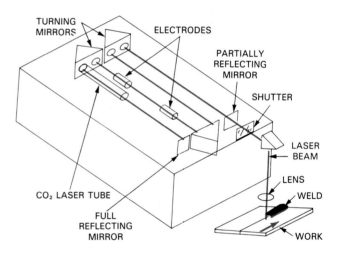

Fig. 27-13. A carbon dioxide (CO_2) laser weldling machine. Mirrors reflect and aim the laser beam. A lens is used tofocue it. The beam is released to make the weld when the shutter opens.

Electron beam welding (EBW)

Fig. 27-14 shows the basic parts of an *electron beam welding* machine. Electrons are emitted from the electron gun. They are then focused and directed at the weld joint. The kinetic energy of the electrons creates the heat for welding. *Kinetic energy* is the energy of an object in motion. In this case, the objects in motion are the electrons.

Electron beam welding may be done in a high vacuum, a partial vacuum, or under normal atmosphere pressure. A *vacuum* is a condition in which

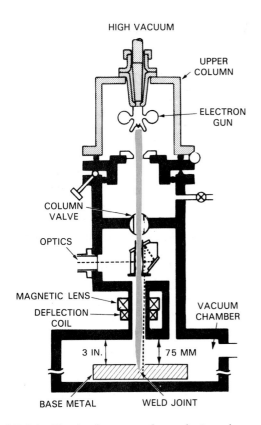

Fig. 27-14. The basic parts of an electron beam welding machine shown in schematic form. Note that, in this machine, the weldment is mounted in a vacuum chamber. (PTR - Precision Technologies, Inc.)

atmospheric pressure in a closed vessel has been decreased. This is done by pumping out air. Fig. 27-15 shows an electron beam welder in use in industry. The beam energy is easier to direct under high vacuum conditions. Also, less energy is lost in a vacuum.

Advantages of laser and electron beam welding

Laser and electron beam welding machines are made in a variety of energy levels. The greater the electrical energy input, the greater the energy output. ND:yag lasers are produced in sizes from 100 W – 600 W (watts). CO_2 lasers are produced in sizes from 500 W – 25,000 W. A 600 W ND:yag laser will penetrate steel .01″ (2.5 mm) in thickness.

Electron beam welding machines are produced in sizes from 1 kW – 100 kW (kilowatts). A 100 kW EB welding machine will produce 100 percent penetration on steel 10″ (254 mm) thick. See Fig. 27-15. Laser beam and electron beam welding offer these advantages:
- Low heat input.
- Controlled welding atmospheres.
- The ability to weld dissimilar metals.
- The ability to weld parts as thin as .001″ – .002″ (.025 mm – .050 mm).
- No need for filler metal.
- Very precise aiming of beams to produce accurate and repeatable welds.

SPECIAL CUTTING PROCESSES

Special cutting processes using oxyfuel gas and oxygen arc cutting equipment have been developed for use in underwater salvaging and repairs. Other special thermal cutting processes were developed to permit the cutting of nonferrous metals and even concrete. These processes make it possible to cut metals that cannot be cut using regular oxyfuel gas cutting.

Fig. 27-15. An operator aiming the electron beam prior to making a weld. (PTR - Precision Technologies, Inc.)

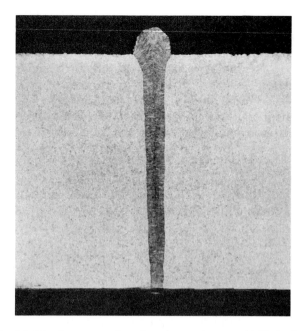

Fig. 27-16. A cross section of an electron beam weld. Notice how thin weld is in relationship to the metal thickness. (PTR - Precision Technologies, Inc.)

Oxyfuel gas cutting underwater (OFC)

Underwater cutting is often done to repair underwater structures. It is also used to cut and salvage sunken ships. *Oxyfuel gas cutting* can be done underwater, using a special torch. The torch is similar to the regular OFC torch, with two major changes. First, an air jacket is installed around the cutting tip. Second, a tube is added to carry compressed air to the air jacket. See Fig. 27-17.

Oxygen and fuel gas are mixed in the torch and burned at the tip, as in a regular cutting torch. A cutting oxygen valve and lever directs pure oxygen through the center orifice for cutting. The compressed air in the air jacket keeps water away from the cutting tip.

Oxygen arc cutting underwater (AOC)

Oxygen arc cutting uses an electrode and electrode holder. The electrode holder is equipped with an oxygen passage. The oxygen flow is controlled by a lever. In Fig. 27-18, note that the electrode is hollow. The oxygen flows through this hollow electrode to the *kerf* (cut).

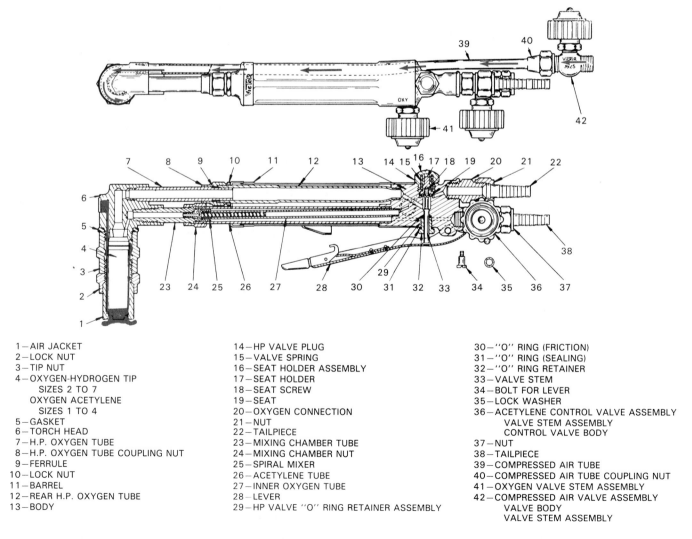

1—AIR JACKET	14—HP VALVE PLUG	30—"O" RING (FRICTION)
2—LOCK NUT	15—VALVE SPRING	31—"O" RING (SEALING)
3—TIP NUT	16—SEAT HOLDER ASSEMBLY	32—"O" RING RETAINER
4—OXYGEN-HYDROGEN TIP	17—SEAT HOLDER	33—VALVE STEM
SIZES 2 TO 7	18—SEAT SCREW	34—BOLT FOR LEVER
OXYGEN ACETYLENE	19—SEAT	35—LOCK WASHER
SIZES 1 TO 4	20—OXYGEN CONNECTION	36—ACETYLENE CONTROL VALVE ASSEMBLY
5—GASKET	21—NUT	VALVE STEM ASSEMBLY
6—TORCH HEAD	22—TAILPIECE	CONTROL VALVE BODY
7—H.P. OXYGEN TUBE	23—MIXING CHAMBER TUBE	37—NUT
8—H.P. OXYGEN TUBE COUPLING NUT	24—MIXING CHAMBER NUT	38—TAILPIECE
9—FERRULE	25—SPIRAL MIXER	39—COMPRESSED AIR TUBE
10—LOCK NUT	26—ACETYLENE TUBE	40—COMPRESSED AIR TUBE COUPLING NUT
11—BARREL	27—INNER OXYGEN TUBE	41—OXYGEN VALVE STEM ASSEMBLY
12—REAR H.P. OXYGEN TUBE	28—LEVER	42—COMPRESSED AIR VALVE ASSEMBLY
13—BODY	29—HP VALVE "O" RING RETAINER ASSEMBLY	VALVE BODY
		VALVE STEM ASSEMBLY

Fig. 27-17. A cross sectional drawing of an underwater oxyfuel gas cutting torch. The air valve, part #42, controls the air flow to the air jacket, part #1. The air flow is shown in red. This high-pressure air keeps water away from the cutting tip. (Victor Equipment Co.)

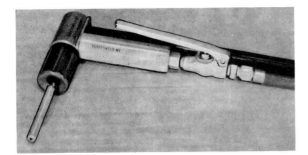

Fig. 27-18. An arc-oxygen underwater cutting torch. (Craftsweld Equipment Corp.)

Preheating is done by striking an arc between the hollow electrode and the metal to be cut. Once the arc is struck, the cutting oxygen lever is depressed and the cutting begins. The arc and oxygen flow must be maintained as the torch is moved along the cutting line.

Oxygen arc cutting may be done above or below water. Oxygen arc and plasma arc are two of the most effective ways of cutting underwater.

Plasma arc cutting (PAC)

Plasma arc cutting was developed to cut nonferrous metals cleanly and efficiently. An inert gas is often used when cutting nonferrous metals. However, dry, filtered compressed air may be used as the orifice gas to cut most ferrous and nonferrous metals. To protect the cut metal from contamination, water or a shielding gas is used. Shielding gas or water flows around the restricted nozzle, as shown in Fig. 27-19. Clean, narrow kerfs are possible with plasma arc cutting. Cuts can be made at the rate of several hundred inches per minute.

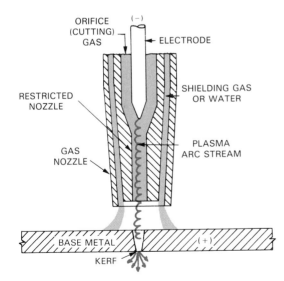

Fig. 27-19. A dual-flow plasma arc cutting torch nozzle. Note the restricted nozzle. The orifice gas becomes a plasma as it passes through the arc stream at the restricted nozzle. The shielding gas or water flows out through several holes around the plasma stream.

Plasma arc cutting is noisy. It can be quieted by cutting on a water table. The metal to be cut is placed under 2″ − 3″ (50mm − 75 mm) of water. The plasma arc torch is submerged while making the cut. See Fig. 27-20. When a water table is not used, industrial quality ear protection must be worn.

Fig. 27-20. A plasma arc torch cutting steel under a few inches of water. Cutting under water makes the process much quieter and almost eliminates fumes and ultraviolet radiation. (ESB Cutting and Welding Products)

Metal powder cutting (POC)

Metals such as aluminum, bronze, nickel, and alloy steel are difficult to cut. They do not oxidize (burn) readily when oxyfuel gas cutting is attempted.

Metal powder cutting (POC) uses an iron oxide or aluminum oxide mixture together with an oxyfuel gas cutting torch. These metal oxides burn easily in the cutting flame.

During the cutting operation, the metal oxide mixture is fed into the oxyfuel gas flame. The burning metal oxide creates the high heat needed to melt the material being cut. Metal powder cutting can even be used to cut through reinforced concrete. There are several methods used to introduce the metal oxide powder into the flame. One method, shown in Fig. 27-21, uses a separate tube to carry the powder under pressure to a point near the cutting flame.

Chemical flux cutting (FOC)

Chemical flux cutting is similar to metal powder cutting. In chemical flux cutting, chemicals are added to the flame (rather than iron or aluminum oxides). These chemicals improve the cutting action. In chemical flux cutting, the chemicals are fed into the cutting oxygen stream. Refer to Fig. 27-22.

Laser beam cutting

Lasers (particularly the CO_2 laser) can cut mild steel, stainless steel, and even titanium cheaply and quickly. *Laser beam cutting* produces clean-cut edges. Numerically controlled pulsed laser beams can also be used to drill or pierce clean holes of extremely small diameter in production parts. An oxygen jet is sometimes used with laser beam equipment to drill or

pierce. Holes larger than .014″ (.36 mm) are generally made with conventional drill bits.

REVIEW QUESTIONS

Write your answers on a separate sheet of paper.
1. How many welding and cutting processes are listed by the AWS?
2. How does EGW differ from ESW?
3. List three gases that may be used to form the plasma for PAW.
4. What process is used to rapidly weld nails, bolts, and location pins to metal plates?
5. List at least four advantages of SAW.
6. List three or more processes used to weld or cut underwater.
7. What can be done to quiet the PAC process?
8. What cutting process may be used to cut through reinforced concrete?
9. Which special process requires government permission to use, because of its dangers?
10. In LBW, the metal is heated by _____ radiation. In EBW, the metal is heated by _____ energy.

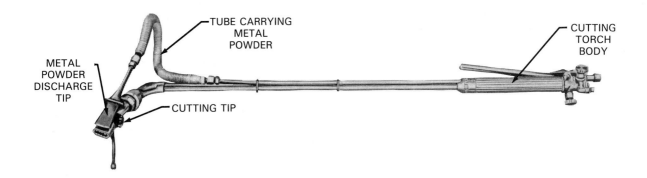

Fig. 27-21. A metal powder cutting torch. The metal powder is carried to the cutting area by a large tube. (ESAB Cutting and Welding Products)

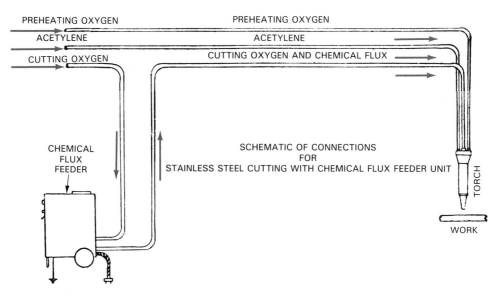

Fig. 27-22. A schematic drawing of a chemical flux cutting outfit.

Chapter 28

Robotics in Welding

After studying this chapter, you will be able to:
□ Cite advantages of using robotic welding equipment in manufacturing.
□ Identify the main parts of a robot and the components of a robotic welding station.
□ Describe the use of a teach pendant in programming a robot to perform its designated tasks.
□ Discuss the safety precautions to be taken when working around robots.

TECHNICAL TERMS

Actuator, axis, elbow (joint), operator controls, parts positioner, repeatability, robot, robot cell, robot controller, robotic welding system, robotic welding work station, servomotor, shoulder (joint), teach pendant, waist (joint), working volume, wrist (joint).

THE REASON FOR ROBOTS

Industrial *robots* are essentially mechanical devices that can be programmed to perform a task or several tasks under the control of a computerized program. They are used because they save the manufacturer money. Robots save money for these reasons:
• They can weld at a far faster speed than a human welder. A speed of 800″ (20.3 m) per minute is not uncommon.
• They can weld continuously.
• They can maintain the same arc length and travel speed over the entire weld.
• They will produce fewer defects than a human welder.
• They can be programmed to do many welds in a variety of locations. Robotic welding equipment can repeat programmed welds with an accuracy of less than ±.004″ (.10 mm).

Robots are useful and cost-effective only when large numbers of welds must be made. They are not practical for all welding applications. Low-volume production and most repair welding are not cost-effective for robotic welding.

TYPES OF INDUSTRIAL ROBOTS

A robot can move in a number of directions. These movements are made possible by the use of mechanical joints and actuators. An *actuator* is a device that will cause something to move. The mechanical joints on robots are moved by either *electric* actuators or by *hydraulic* actuators. Robots are thus classified as either hydraulically actuated robots or electrically actuated robots.

Hydraulic actuators may be hydraulic motors or hydraulic cylinders. A hydraulic motor causes rotation. A hydraulic cylinder causes a shaft to move in or out. Each joint requires its own hydraulic actuator. See Fig. 28-1. Hydraulic robots can carry and move heavier loads than electric robots.

Electric actuators are electric motors called servomotors. *Servomotors* are motors that rotate very precise distances. These motors are used to rotate the joints of a robot. Each joint requires its own servomotor. See Fig. 28-2. Some robots use a rotating screw shaft to achieve rotation about the shoulder or #2 axis. Screw shafts may be used to move the arm in and out between the elbow and the wrist. These screw shafts are rotated by servomotors. See Fig. 28-3. Electric robots are more accurate in their movements than hydraulic robots.

Repeatability refers to a robot's ability to accurately perform the same movement again and again. An electric robot can repeat a movement with an accuracy of

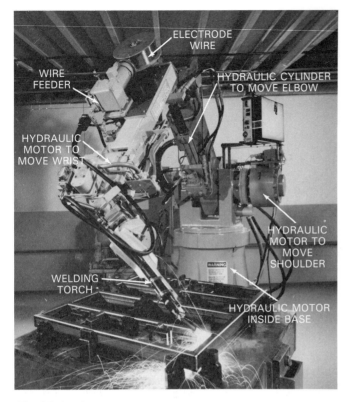

Fig. 28-1. A robot operated by hydraulic motors and cylinders. Hydraulic robots can carry more weight than electric robots. Notice the electrode wire supply, wire feeder, and torch mounted on the robot arm.

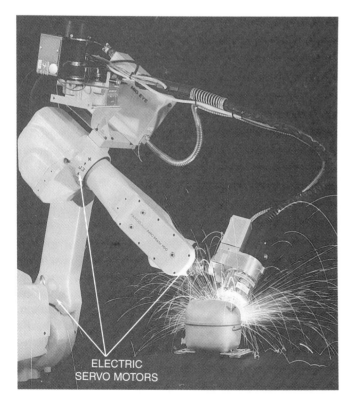

Fig. 28-2. An electrically actuated robot. The servomotors that move the robot may be seen at some joints. Other servomotors are built into the robot arms and joints.
(ABB Robotics, Inc., Welding Systems Div.)

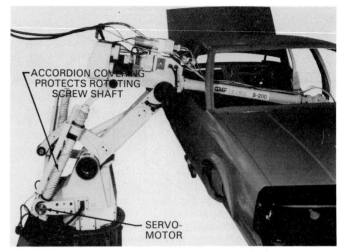

Fig. 28-3. Rotating screw jacks are used on this robot to move it around the waist axis. Accordion-type covers protect the screws. The screws are turned by servomotors.
(FANUC Robotics North America, Inc.)

± .004″ (.10 mm). A hydraulic robot's repeatability is usually accurate to .015″ − .030″ (.38 mm − .51 mm).

ROBOTIC WELDING SYSTEMS

A robot cannot work by itself. It must function together with other parts to form a *robotic welding system.* The parts of a robotic welding system are:
- The robot itself.
- Automatic welding or cutting equipment.
- The robot controller.
- A parts positioner.
- Operator controls.

The robotic welding system also makes up the *robotic welding work station.* See Fig. 28-4.

THE ROBOT

When choosing a robot for any welding application, a manufacturer must consider a number of factors, such as:
- The weight of the welding equipment to be carried by any part of the robot. Hydraulic robots will carry heavier loads.
- The location of the welds to be made. This will determine the number of axes required.
- The accuracy of repeatability required.
- The amount of money that can be spent.
- The future and alternate uses for the robot.

Robot movements

Each mechanical joint on a robot will permit a movement about one *axis* (straight line about which rotation takes place). The location of the welds to be made will determine the number of axes required. Most industrial robots move about four, five, or six axes. Each axis is given a name or number. These axes are naed

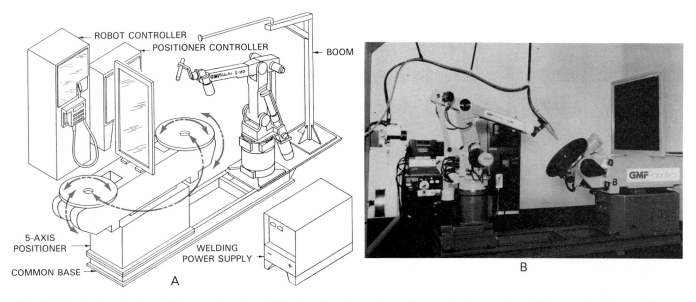

Fig. 28-4. A—A robotic welding work station. This drawing shows the robot controller, parts positioner controller, positioner, robot, welding machine, and GMAW torch. The positioner has two work stations separated by an arc filtering plastic partition. B—The actual work station shown in the drawing. Safety note: This photo does not show the fence normally installed to keep people out of the robot's working volume. (FANUC Robotics North America, Inc.)

after parts of the human body: the *waist, shoulder, elbow,* and *wrist.* Refer to Fig. 28-5.

Wrist movement often takes place around three axes, as shown in Fig. 28-6. Robot manufacturers often refer to these wrist axes by number. They are called simply axes 4, 5, and 6.

The robot must often reach out toward the welding area. Most of this reaching motion is done by rotating

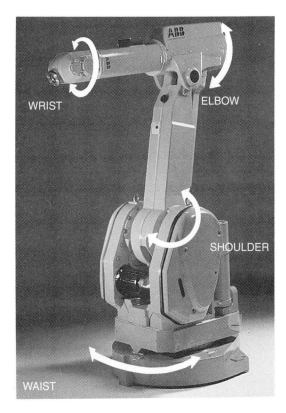

Fig. 28-5. Four of the robot axes are given the names of human parts. They are also referred to by numbers. They are the waist (#1), shoulder (#3), and wrist (#4). (ABB Robotics, Inc., Welding Systems Division)

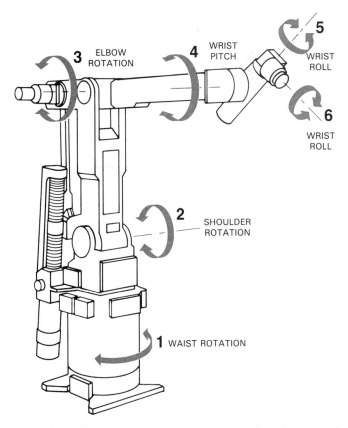

Fig. 28-6. The numbers and names assigned to the axes of a six-axis robot. (FANUC Robotics North America, Inc.)

the ffl2 and ffl3 axes. Hydraulic cylinders or servo motors with screw jacks may also be used to extent the robot arm. The extension is a straight line (linear) motion between the elbow and the wrist.

WELDING OR CUTTING EQUIPMENT

Any form of welding or cutting equipment can be mounted on a robot. However, cutting is usually done on cutting tables using electronic pattern followers.

The most common welding and cutting processes used with robotic systems are:
- Resistance spot welding.
- Gas metal arc welding.
- Gas tungsten arc welding.
- Plasma arc cutting.
- Laser beam welding and cutting.
- Surfacing.

ROBOT CONTROLLER

The *robot controller* is the "brains" of the robot welding system. The robot controller includes a computer that controls the movement of the robot in all axes. It controls all the variables in the welding equipment, and may also control the parts positioner. Sometimes, however, a separate controller is used for the parts positioner.

The robot controller stores the program of movements and actions that the robot must follow to weld a part. The robot controller also checks or monitors all aspects of the welding operation. The robot controller also monitors the operating temperature of all motors and hydraulic pumps.

If a problem is detected, the robot controller will alert the operator, or—if programmed to do so—may shut down the entire robotic welding system. To program the robot controller either a teach pendant or a numerical control (NC) program is used.

Teach pendant

The robot controller must be programmed before a robotic welding system can be used to make accurate, repeatable welds. The computer in the robot controller is programmed to move the robot to different locations by means of a *teach pendant*. See Fig. 28-7. The pendant is connected to the controller by an electrical cable. See Fig. 28-2.

The teach pendant is used to record in the robot controller's memory the step-by-step procedures for completing one or more tasks. First, the parts to be welded are placed in jigs or fixtures on the positioner and clamped in place. The operator then positions the robot and welding torch at a beginning point on the part. This point is placed into the computer memory when the operator presses a button on the teach pendant. The operator then sets the welding conditions (parameters) into the computer memory, using the

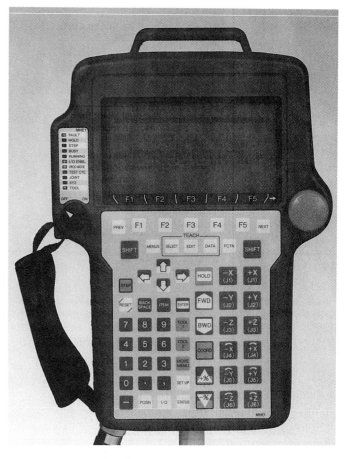

Fig. 28-7. A robot teach pendant. The teach pendant is used to move the robot through all the movements required to make all the welds. Variables such as amperage, voltage, wire speed, and welding speed are set and shown on the visual display. They are then placed into the controller's memory. (FANUC Robotics North America, Inc.)

teach pendants or robot controller. These parameters include:
- Welding current or wire feed speed.
- Welding voltage.
- Nozzle-to-work distance.
- Welding speed.

The welding torch is moved along the first weld line by the operator, using the controls on the teach pendant. When the end of the weld is reached, that point is placed into the computer memory. The robot and torch are then moved through any additional welds. Each movement or point on every weld is recorded into the robot controller's memory. When completely stored in memory, the program will repeat itself each time the operator presses the start button on the operator's control panel.

Parts positioner

The *parts positioner* is used to bring the part to be welded within reach of the robot and welding torch. The positioner also rotates the part, where possible. (If the part is small enough to rotate, a robot with fewer axes may be used.) The positioner also may be used to move the completed part away from the robot. Once clear of the robot, the completed part is unloaded and a new part is loaded for welding. See Figs. 28-4, 28-8, and 28-9.

Parts positioner controller

A separate controller with its own computer may be used for the parts positioner. See Fig. 28-4. This computer contains the program of motions for the parts positioner. The same teach pendant is used to program the parts positioner controller as is used for the robot controller.

Fig. 28-9. Railings or fences are used to keep people out of a robot's work area, or robot cell. This robot, outfitted with GMAW equipment, is being used to weld office furniture. The fixture used to hold the parts in position for welidng is on a turntable for ease of loading and unloading by the operator. (FANUC Robotics North America, Inc.)

Fig. 28-8. A robot welding a lawnmower housing. The positioner moves the housing to align the welds under the torch. (FANUC Robotics North America, Inc.)

OPERATOR CONTROLS

The *operator controls* are on a panel that generally contains the welding sequence start button, emergency stop button, and system monitoring warning lights. The operator controls are located at a safe distance from the working robot and parts positioner. Once the program is completed, welding stops. The positioner controller then directs the positioner to move the part to a safe area away from the robot. The completed part is unloaded and a new part is mounted on the positioner.

To start the welding program again, the operator pushes the sequence start button on the operator control panel. The parts positioner controller then moves the positioner back to the starting point near the robot, and the sequence repeats. Operator controls can also be used to shut down the system in an emergency.

SAFETY AROUND ROBOTS

A moving robot is extremely dangerous. For this reason, the *robot cell,* which is the floor area containing all the parts of the robotic welding system, is often fenced off. See Fig. 28-9. The robot cell may be fenced

with metal hand rails or meshed wire. This is done to keep unauthorized persons from accidentally entering the space.

Robot working volume

The *robot working volume* is the three-dimensional space occupied by the robot at the extremes of its movements. Fig. 28-10 shows the working volume of a typical robot.

The robot's movements can be erratic and unpredictable. It can move with great power and speed, and could cause serious injury if it should strike a person.

The robot working volume is also protected from unauthorized entry. This may be done with a fence or electronic devices. One system uses infrared light beams reflected by mirrors around the outside of the working volume. If anything or anyone passes through these beams, the robot automatically shuts down. See Fig. 28-11.

Caution: Only trained technicians should enter the robot working volume, and then only under controlled conditions, usually with the robot de-energized.

REVIEW QUESTIONS

Write your answers on a separate sheet of paper.
1. Name the two types of actuators used on robots.
2. List three reasons why robots save money for a manufacturer.
3. An _____ actuated robot will repeat welds with an accuracy of ± .004″ (.10 mm).

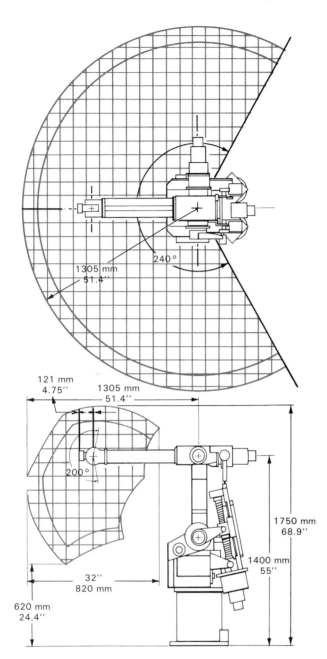

Fig. 28-10. The robot working volume is the three-dimensional space through which the robot may move during its operation. (Advanced Robotics Corp.)

4. Name the robot axes shown in the sketch below:
5. Number the axes shown in Question 4.

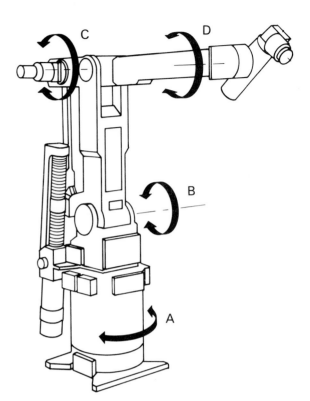

6. The operator starts the robot controller welding program by pressing the start button on the _____ _____ _____.

7. What device can an operator use to program robot movements into the robot controller's computer?

8. Name two devices used to keep unauthorized persons out of the robot working volume.

9. Completed welds are unloaded and new parts loaded by moving the _____ _____ out of the robot working volume.

10. What is the most important safety precaution to take when working around a robot?

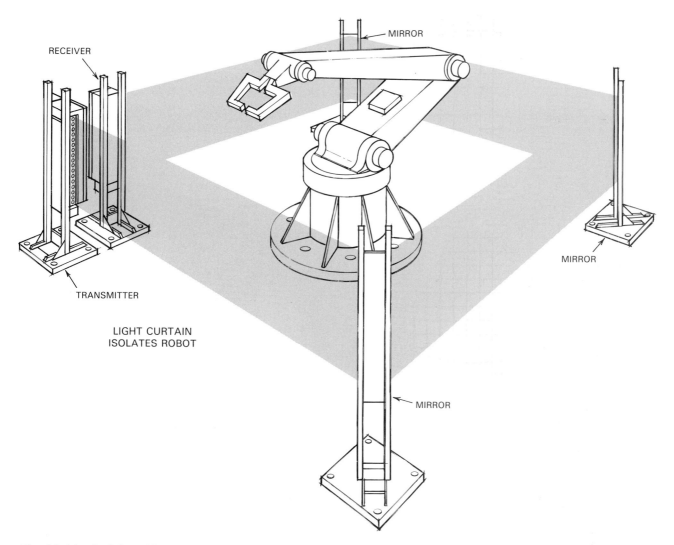

RECEIVER

MIRROR

TRANSMITTER

LIGHT CURTAIN
ISOLATES ROBOT

MIRROR

MIRROR

Fig. 28-11. An infrared light curtain used to isolate the robot's working volume. Anything passing through this light curtain will immediately shut down the robot. (Welding Design and Fabrication)

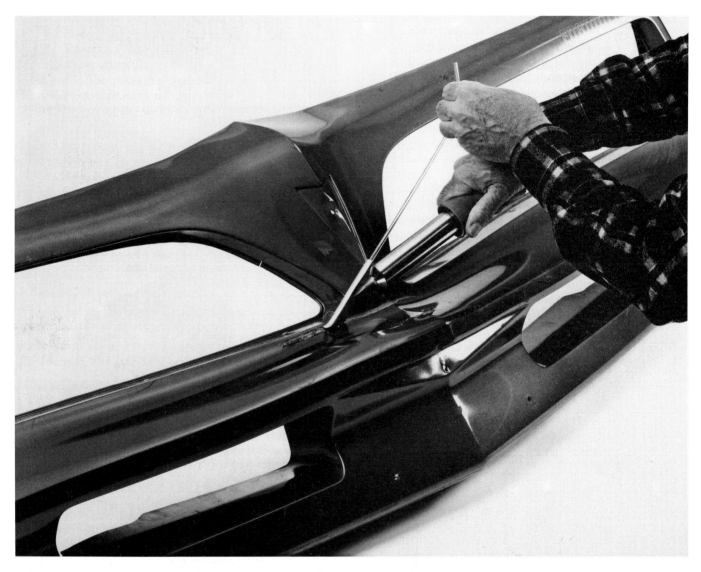

A damaged plastic grille and bumper from a Pontiac automobile being repaired by plastic welding. (Seelye, Inc.)

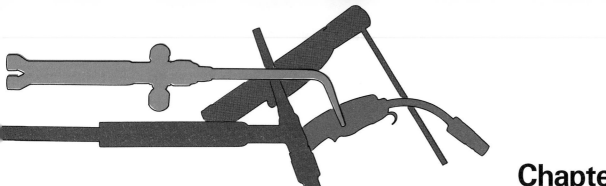

Chapter 29

Welding Plastics

After studying this chapter, you will be able to:
- ☐ Distinguish between the two basic groups of plastics and name representative examples of each.
- ☐ Identify various thermoplastics by their flame and odor characteristics.
- ☐ Describe the procedure used to join plastics through use of heated gas and a filler rod.

TECHNICAL TERMS

ABS, plastics, polyethylene, polypropylene, polyurethane, polyvinyl chloride, speed welding tip, tack welding tip, thermoplastic, thermosetting.

Plastics are synthetic organic materials that can be molded and shaped into a desired form. Often, this molding or shaping is done under heat and pressure.

PLASTICS IN MANUFACTURING AND CONSTRUCTION

The use of plastics in manufacturing and construction has increased dramatically in the past twenty years. The use of plastic materials has been attempted in the construction of practically every imaginable object. See Fig. 29-1.

Automobile bodies and bumpers, engines, firearms, aircraft, pipes and ducts, furniture, teeth, houses, lug-

Fig. 29-1. The body panels of this Pontiac Fiero are made of plastic.   (Pontiac Motor Div., General Motors Corp.)

gage, floors, moldings, sinks, skis, and boots are all being made of weldable plastics. The list of articles made of plastics is almost endless. See Figs. 29-2 and 29-3.

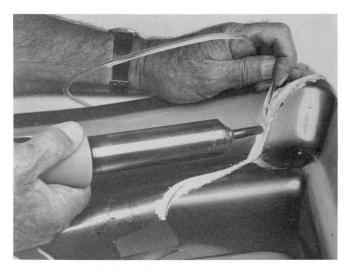

Fig. 29-2. A plastic automobile bumper being welded with a regular welding tip and a clear, round welding rod. (Seelye, Inc.)

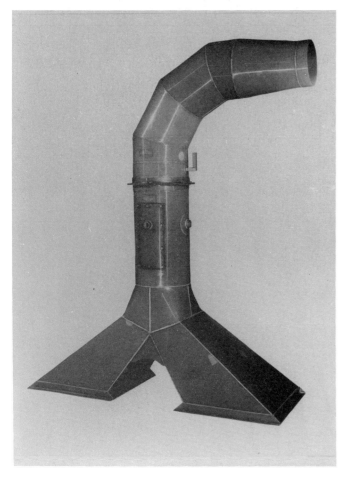

Fig. 29-3. A complicated duct fabricated from plastic parts and joined by plastic welding. (Kamweld Products Co., Inc.)

TYPES OF PLASTICS

There are many types of plastics that have been created to meet specific needs. However, all plastics fall into one of two basic groups. They are either thermosetting or thermoplastic materials.

Thermosetting plastics cannot be welded by heat and fusion. Plastics in this group are initially heated and formed into a desired shape. After they have once cooled, however, they cannot be reheated and reformed. Some plastics in the thermosetting group are: alkyd, allylic, aminoplastic, bakelite, casein, epoxy, phenolic, and silicone.

Thermoplastic materials may be joined by welding in the same manner as metal. Plastics in this group become soft and may be formed when heated. Thermoplastics may be reheated and reformed repeatedly. Important thermoplastics include: ABS (acrylonitrile butadiene styrene), polyethylene (PE), polypropylene (PP), polyurethane (TPUR), and polyvinyl chloride (PVC).

How to identify thermoplastics

To make certain that a plastic is a weldable thermoplastic, you must test it. Using the following test to identify thermoplastics:

1. From the back side of the part, remove a small sliver of the plastic to be tested.
2. Hold the plastic sliver with pliers. Ignite the sliver with a flame.
3. Observe the plastic as it burns.
4. Compare the appearance and odor of the burning sample with those of known thermoplastics.

The flame characteristics and odors of the most common thermoplastics are listed below.

ABS: Burns with a thick, black, sooty smoke and continues to burn after the flame is removed. It has a sweet odor that cannot be described, but is peculiar to ABS plastic.

Polyethylene: It melts and drips; drips may continue to burn. It will continue to burn after the ignition flame is removed. Polyethylene smells like burning wax. Drips will float on water.

Polypropylene: It burns with no visible flame, and has its own peculiar acid-like odor.

Polyurethane: It burns with a yellow-orange flame and produces black smoke. It will continue to burn with a sputtering flame after the ignition flame is removed.

Polyvinyl chloride: PVC will char and give off a gray smoke, and will self-extinguish when the ignition flame is removed. It has its own unusual odor.

WELDING EQUIPMENT AND SUPPLIES

All the joint designs used when welding metals may be used successfully with thermoplastics. See Chapter 3 for information on joint designs.

THE WELDING TORCH

The fusion welding of thermoplastics is achieved by a pressurized gas heated in the welding torch. The gas used may be air, or an inert gas (nitrogen is usually used). When an inert gas is used, it is supplied under pressure from a cylinder. If air is used as the gas for heating, it is pressurized by a motor- or engine-driven compressor. The pressurized gas is carried to the torch by a hose. Fig. 29-4 lists the recommended heating gas for welding various thermoplastics.

TYPE OF PLASTIC	TYPE OF WELDING GAS
ABS (Acrylonitrile Butadiene Styrene)	Compressed Air
Polyethylene (PE)	Nitrogen
Polypropylene (PP)	Nitrogen
Polyurethane (TPUR)	Compressed Air
Polyvinyl Chloride (PVC)	Compressed Air

Fig. 29-4. Welding gases recommended for use with various thermoplastics.

The gas is heated by electric heating coils enclosed in the special welding torch. See Fig. 29-5. The heated gas is then directed at the surfaces of the plastic joint. A welding outfit used for plastics is shown in Fig. 29-6. The welding gas temperature is adjusted by regulating the flow of the gas. A flow regulator is shown in Fig. 29-6.

WELDING ROD

Most joints require a welding rod. The rod must be made of the same plastic as the part being welded. The cross-sectional shape of a plastic welding rod is usually round. See Fig. 29-7. Round plastic welding rods

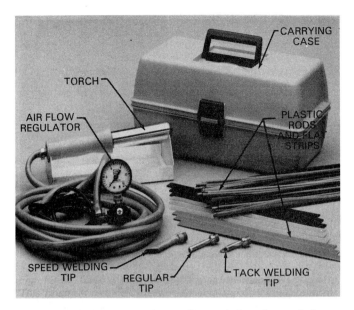

Fig. 29-6. A plastic welding outfit consisting of a torch, hose, regulator, tips, rods, and carrying case. (Seelye, Inc.)

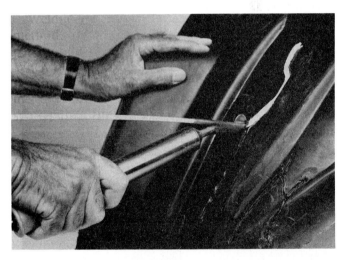

Fig. 29-7. A round plastic welding rod being used with a speed welding tip to repair cracks in a plastic inner body panel. A white rod is used here for better photographic contrast. (Seelye, Inc.)

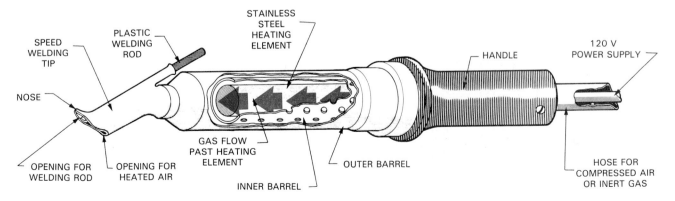

Fig. 29-5. A diagrammatic cross section of a welding torch for plastics. The gas is heated by an electric coil as it passes through the torch body. Notice that one passage in the tip heats the base material and the other heats the welding rod. (Kamweld Products Co, Inc.)

287

are generally produced in the following diameters: 1/8″ (3.2 mm), 5/32″ (4.0 mm), and 3/16″ (4.8 mm). Fig. 29-8 is a table of rod diameters recommended for use with various base material thicknesses.

Plastic rods may be oval or triangular, or may be wide rectangular strips. See Figs. 29-9 and 29-10. The welder's choice of welding rod is determined by the shape of the joint, the equipment available, and the thickness of the plastic to be welded. See Fig. 29-11. A triangular or round welding rod may be used for a fillet weld. See Fig. 29-12.

PLASTICS-WELDING PROCEDURE

Before welding, a joint is often tack welded. Tack welding tips are shown in Figs. 29-6 and 29-13. The

pointed *tack welding tip* is pressed into the joint to fuse (weld) the parts together. A tack weld on plastic is generally thin, shallow, and relatively weak. It is, however, strong enough to hold the parts in alignment

Fig. 29-10. Flat plastic welding rod was used to join the parts of these pipe joints.   (Kamweld Products Co., Inc.)

BASE MATERIAL THICKNESS	WELDING ROD DIAMETER
1/16″ (1.6 mm)	1/8″ (3.2 mm)
1/8″ (3.2 mm)	1/8″ (3.2 mm)
3/16″ (4.8 mm)	3/16″ (4.8 mm)
1/4″ (6.4 mm)	3 rods of 3/16″ (4.8 mm) diameter
For thicknesses larger than 1/4″ (6.4 mm), multiple beads are used.	As many 5/32″ (4.0 mm) or 3/16″ (4.8 mm) diameter rods as required to fill the joint.

Note: Try to select a welding rod diameter close to the thickness size of the base material. For thicknesses larger than 3/16″ (4.8 mm), use several layers of welding rod.

Fig. 29-8. A table of recommended welding rod sizes for various base material thicknesses.

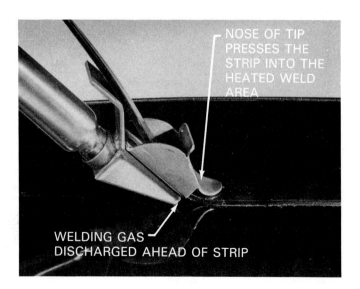

Fig. 29-9. A plastic butt joint being welded with a wide flat welding rod. Note the shape of the speed welding tip. (Laramy Products Co., Inc.)

Fig. 29-11. This double-V-groove joint was welded with round welding rods. Several passes were necessary. (Laramy Products Co., Inc.)

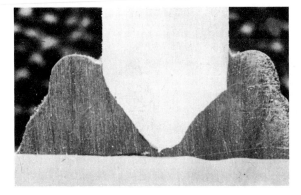

Fig. 29-12. The first pass on each side of this double-bevel-groove corner joint was apparently welded with a triangular welding rod. Additional passes used round welding rods. (Laramy Products Co. Inc.)

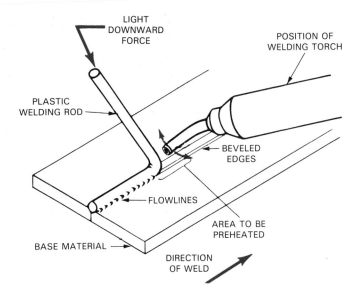

Fig. 29-14. The correct procedure for welding a plastic butt joint, using a regular welding tip. Note the positions of the rod and welding tip. (Kamweld Products Co., Inc.)

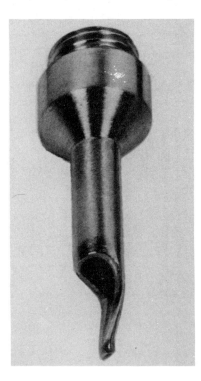

Fig. 29-13. A regular tack welding tip. (Kamweld Products Co., Inc.)

the joint while welding. As the tip is moved along the joint, the rod is pulled out of the feed tube. A speed welding tip permits welding 4-6 times faster than a regular tip. Fig. 29-7 shows a speed welding tip being used to weld a butt joint. The speed welding tip in Fig. 29-15 has a roller to press the heated rod into the joint.

The heat of the welding gas is controlled by the gas flow regulator. The less gas per minute that flows over the heating element, the higher the temperature to which it will be heated. If more gas is fed through the torch per minute, it will pick up less heat. Larger heating elements can be used to provide more heat, as well.

while they are welded. The tack welded parts may be easily disassembled for realignment.

The welding may be done with a regular tip or a speed welding tip. When a regular welding tip is used, the torch is held in one hand. The plastic welding rod is fed into the weld joint with the other hand.

Figs. 29-2 and 29-14 show a regular welding tip in use. The hand holding the welding rod is used to lightly press the heated rod into the softened plastic joint.

Fig. 29-7 illustrates a speed welding tip in use. In this tip, there are two openings. The heated gas from one opening heats the plastic to the welding temperature. The welding rod is fed into the joint through the second opening. The *speed welding tip* allows one-handed welding. It is held tightly against

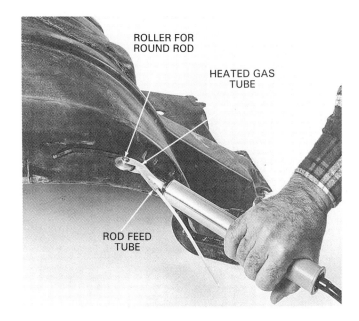

Fig. 29-15. A roller on this speed welding tip presses the rod into the joint. (Seelye, Inc.)

The plastic joint and welding rod are heated to a temperature between 450°F—800°F (232°C—427°C) for welding. Fig. 29-16 shows welding temperatures for various plastics. At the welding temperature, the plastic surface and welding rod become soft. While in this condition, the rod is forced into the surface with a light pressure. When a speed welding tip is used, this pressure is applied by the nose at the end of the tip. See Fig. 29-7. When a regular welding tip is used, the welding rod is lightly pressed into the plastic joint by the hand holding the rod. As the parts of the joint and rod are heated and pressed together, a fusion weld occurs.

TYPE OF PLASTIC	WELDING TEMPERATURE
ABS (Acrylonitrile Butadiene Styrene)	500°F (260°C)
Polyethylene (PE)	550°F (288°C)
Polypropylene (PP)	575°F (302°C)
Polyurethane (TPUR)	575°F (302°C)
Polyvinyl Chloride (PVC)	525°F (274°C)

Fig. 29-16. A welding temperature chart for various plastics.

REVIEW QUESTIONS

Write your answers on a separate sheet of paper.
1. List five uses for plastics that are not mentioned in this chapter.
2. List four plastics that cannot be welded by heat and fusion.
3. The heated gas used to weld plastics may be air or an inert gas. True or False?
4. To increase the temperature of the welding gas, you would (raise/lower) _____ the flow rate of the gas.
5. _____ _____ in the welding torch are used to heat the gas as it flows through the torch.
6. Welding rods may be round, triangular, or rectangular in shape. True or False?
7. The welding gas used for polypropylene is _____.
8. The welding temperature for ABS thermoplastic is _____ °F.
9. What is used to press the welding rod into the heated plastic joint when a speed welding tip is used?
10. What is used to push the welding rod into the heated plastic joint when a speed welding tip is no' used?

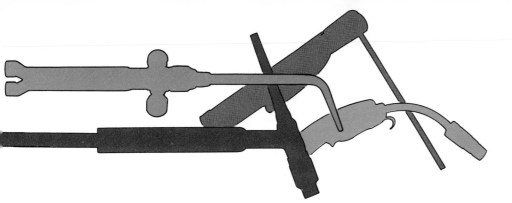

Chapter 30

Welding Pipe and Tube

After studying this chapter, you will be able to:
☐ Distinguish the difference between pipes and tubes.
☐ List the names of the passes used in welding pipe with walls more than 3/16'' (4.8 mm) in thickness.
☐ Describe the procedures used to weld pipes or tubes using SMAW.
☐ Discuss the differences in technique involved in uphill and downhill welding.

TECHNICAL TERMS

Backing ring, convex bead, cover pass, downhill welding, filler pass, hot pass, ID, keyhole, mandrel, OD, pipe schedule number, plastic state, root pass, socket connection, uphill welding.

PIPES AND TUBES

Pipes and tubes are hollow cylinders, normally used to carry liquids and gases. They also are sometimes used as structural elements; for example, as railings or scaffolding. Pipes and tubes are also used in the fuselages (bodies) of small airplanes, and in frames for race cars and bicycles.

Pipes are ordered by the inside diameter *(ID)* measurement. The wall thickness on a pipe is ordered by a *pipe schedule number.* These numbers stand for different sizes (wall thicknesses). Generally, six to eight schedule sizes are available for each diameter of pipe. Some pipe diameters, however, have as many as fourteen pipe schedule sizes. See Fig. 30-1.

Tubes are ordered by the outside diameter *(OD)* size. Tubes can be ordered in virtually any wall thickness. The wall thickness of a tube is usually less, though,

than the wall thickness of a pipe with the same outside diameter. See Fig. 30-2.

Pipes may be joined by means of threaded fittings. The thicker walls of pipe allow threads to be used. Tubes are often joined by fittings that are not threaded. These fittings are held together by soft- or hard-soldered joints. Both tubes and pipes can be joined by welding. Tubes and pipes may be made from virtually any metal. Low-carbon steel, high-carbon steel, stainless steel, copper, brass, and aluminum are common metals used for pipes and tubes.

Both tubes and pipes can be ordered in either seamless or welded form. Tubes also may be ordered in a variety of cross sections other than round. Three such shapes are square, rectangular, and oval.

SEAMLESS PIPES AND TUBES

The wall of a seamless tube or pipe is made from a single piece of metal without a seam or joint. A seamless pipe or tube begins as a solid rod or bar of metal. The metal is heated until it is in the *plastic state* (almost molten). A steel *mandrel,* or forming tool, is then forced through this solid bar. The mandrel forms the inside diameter of the pipe or tube. See Fig. 30-3. The pipe or tube is then run through forming rolls to make the outside round.

WELDED PIPES AND TUBES

Welded pipes and tubes are formed from flat stock. The flat metal is formed into a round pipe or tube by a series of rolls. As the completed tube shape exits the last roll, the seam is welded. Resistance welding is most common, but GMAW or some other process may be used.

DIMENSIONS OF SEAMLESS AND WELDED STEEL PIPE

PIPE SCHEDULES AND WALL THICKNESSES															
Pipe Size	OD in inches	5	10	20	30	40	STD.	60	80	E.H.	100	120	140	160	DBLE. E.H.
1/8	.405	.035	.049			.068	.068		.095	.095					
1/4	.540	.049	.065			.088	.088		.119	.119					
3/8	.675	.049	.065			.091	.091		.126	.126					
1/2	.840	.065	.083			.109	.109		.147	.147				.187	.294
3/4	1.050	.065	.083			.113	.113		.154	.154				.218	.308
1	1.315	.065	.109			.133	.133		.179	.179				.250	.358
1 1/4	1.660	.065	.109			.140	.140		.191	.191				.250	.382
1 1/2	1.900	.065	.109			.145	.145		.200	.200				.281	.400
2	2.375	.065	.109			.154	.154		.218	.218				.343	.436
2 1/2	2.875	.083	.120			.203	.203		.276	.276				.375	.552
3	3.500	.083	.120			.216	.216		.300	.300				.437	.600
3 1/2	4.000	.083	.120			.226	.226		.318	.318					.636
4	4.500	.083	.120			.237	.237	.281	.337	.337		.437		.531	.674
5	5.563	.109	.134			.258	.258		.375	.375		.500		.625	.750
6	6.625	.109	.134			.280	.280		.432	.432		.562		.718	.864
8	8.625	.109	.148	.250	.277	.322	.322	.406	.500	.500	.593	.718	.812	.906	.875
10	10.750	.134	.165	.250	.307	.365	.365	.500	.593	.500	.718	.843	1.000	1.125	
12	12.750	.165	.180	.250	.330	.406	.375	.562	.687	.500	.843	1.000	1.125	1.312	
14	14.000			.250	.312	.375	.437	.375	.593	.750	.500	.937	1.093	1.250	1.406	
16	16.000			.250	.312	.375	.500	.375	.656	.843	.500	1.031	1.218	1.437	1.593	
18	18.000			.250	.375	.437	.562	.375	.750	.937	.500	1.156	1.375	1.562	1.781	
20	20.000			.250	.375	.500	.593	.375	.812	1.031	.500	1.280	1.500	1.750	1.968	
24	24.000			.250	.375	.562	.687	.375	.968	1.218	.500	1.531	1.812	2.062	2.343	

Fig. 30-1. A table of standard pipe sizes. The inside diameter (ID), outside diameter (OD), and wall thickness is given for each schedule size. (The Gage Co., Taylor Engineering Div.)

Outside Diameter, Inches		Wall Thickness							
		Inch	.028	.035	.049	.065	.083	.095	.109	.120
		Gauge	22	20	18	16	14	13	12	11
1/4	.250									
5/16	.312									
3/8	.375									
1/2	.500									
5/8	.625									
3/4	.750									
7/8	.875									
1	1.000									
1 1/8	1.125									
1 1/4	1.250									
1 3/8	1.375									
1 1/2	1.500									
1 3/4	1.750									
2	2.000									

Fig. 30-2. A table of common tubing sizes. The outside diameter (OD) is given in fractions and decimals. The wall thickness is given in fractions of an inch and gauge size. (The Gage Co., Taylor Engineering Div.)

WELDING PROCESSES

Tubes and pipes may be joined by virtually any welding process. The processes generally used are shielded metal arc welding (SMAW), gas metal arc welding (GMAW), gas tungsten arc welding (GTAW), and oxyfuel gas welding (OFW). Electroslag welding has been used to weld pipes with thick walls, as shown in Fig. 30-4.

Both manual and automatic welding equipment can be used to join pipe and tubing. Fig. 30-5 shows an automatic GTAW machine used to weld a butt joint

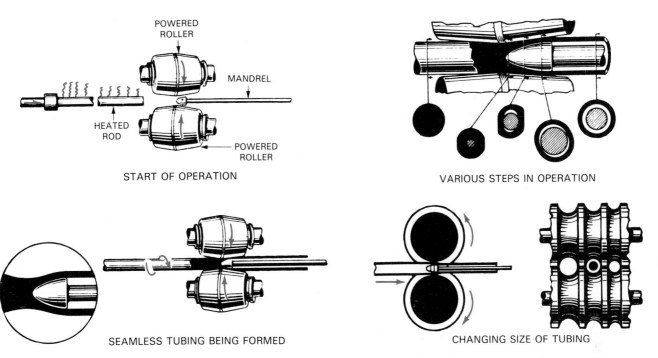

Fig. 30-3. Forming seamless tubing from a solid rod. The length of the tubing is determined by the length of the mandrel. (Michigan Seamless Tube Co.)

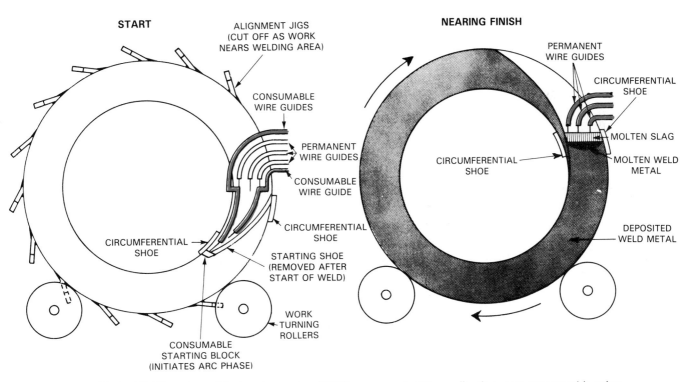

Fig. 30-4. Welding pipe with the electroslag (ESW) process. In this application, two consumable wires are used at the beginning of the weld. Three wires are used to complete the weld.

on pipe. The welding torch and wire feed mechanism moves around the pipe on a track. Together, the electric drive motor and wire feed have a very thin profile. See Fig. 30-6. This allows welds to be made where clearances around the pipe are small. Automatic welders can also be used to make welds other than butt joints. See Fig. 30-7.

TYPE OF JOINTS

To weld pipes end to end, a V-groove butt joint is most often used. Connections are often necessary at a number of points along the length of a pipe. These connections are referred to as T, K, and Y connections because of their shapes. See Fig. 30-8.

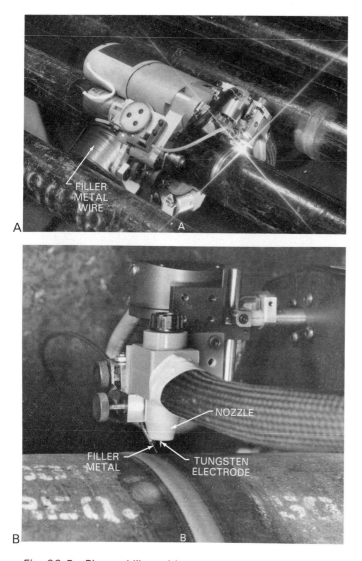

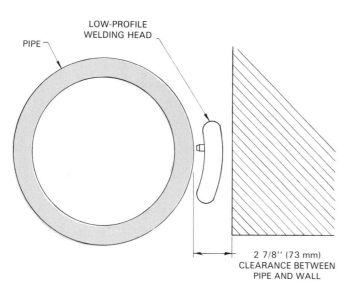

Fig. 30-5. Pipe welding with an automatic, traveling GTAW machine. A—Notice the filler metal wire feed and the machine's thin profile. B—Note the perfect bead, the tungsten electrode, nozzle, and filler metal. (Magnatech Limited Partnership)

Fig. 30-6. A simplified profile of the machine shown in Fig. 30-5. It can be used in spaces as narrow as 2 7/8" (73 mm).

Fig. 30-7. An automatic machine used to weld T-type pipe joints. (ESAB Welding and Cutting Products)

To weld T, K, or Y connections, all pieces must be accurately cut. Automatic cutting equipment may be used to cut these shapes. These automatic cutting machines can be programmed to cut special shapes. Metal templates may be used to draw the required shape on a pipe before the metal is cut. See Fig. 30-9. Special devices are also available to mark a pipe before cutting a desired shape, as shown in Fig. 30-10.

Fittings are also available for many of the standard pipe connections. See Fig. 30-11. The edges of these preshaped fittings are often beveled and ready for welding.

Thin-wall tubing connections are often made using preformed fittings. These fittings are usually socket connections. A socket connection has the OD of the pipe fitted into the ID of the fitting.

On thin-wall tubing, solder is used with socket-type joints. To permit the solder to flow into the joint, the fittings provide a small amount of clearance space. See Fig. 30-12. Socket connections may also be used when joining pipe. In this case, a lap weld is used to join the pipe to the socket. See Fig. 30-13. Tubing with thicker walls may be welded in the same manner as pipe.

PIPE OR TUBE WELDING POSITIONS

Pipes and tubes may be welded in any position. Whenever possible, the material is rotated so that all welding is done in the flat welding position. Most welds, however, must be made with the pipe or tube in a fixed

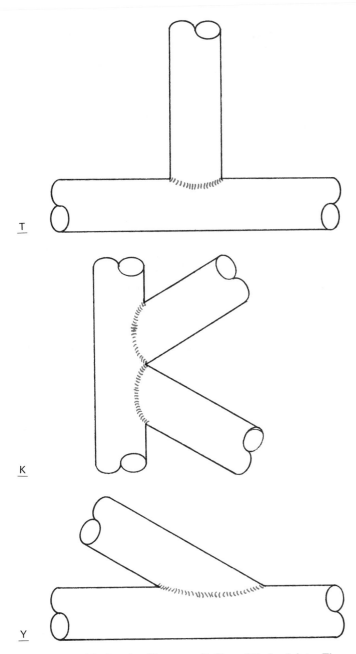

Fig. 30-8. This drawing illustrates T, K, and Y pipe joints. The pipes may be of different diameters.

ing, a square-groove butt joint is usually used. Generally, a V-groove butt joint is used on pipe and thick-walled tubing. The angle of the groove is given

Fig. 30-9. A metal template is sometimes used as a guide to mark a ipe prior to cutting. A soapstone may be used to mark the pipe. (ESAB Welding and Cutting Products)

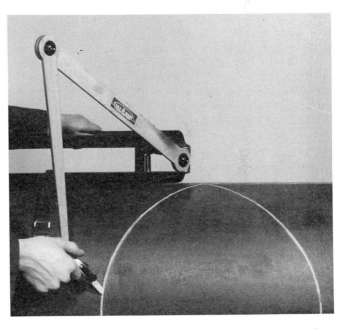

Fig. 30-10. A special device used to mark a pipe before cutting.

position. In a fixed position, the weld may change from vertical, to horizontal, to flat, to overhead as it progresses around the pipe or tube.

Welding test positions for pipes and tubes are described by numbers and letters assigned by the American Welding Society (AWS). These same number and letter combinations are often used to describe welding positions on the job. The AWS test positions for butt welding are shown in Fig. 30-14. The AWS test positions for lap welds on pipe are shown in Fig. 30-15.

PREPARATIONS FOR WELDING

The edges of pipe and tube must be prepared prior to welding. When a butt weld is made on thin-wall tub-

on the welding symbol. The pipe or tube must be cut to the required angle. The ends can be cut manually, semi-automatically, or automatically.

Fig. 30-16A shows a motorized pipe cutter traveling around a pipe on a track. This special tool cuts the pipe to length and cuts the angle at the same time. Fig. 30-16B shows this type of pipe cutter in use on a pipe repair job. Two other pipe-beveling cutters are shown in Fig. 30-17. Pipe ends may also be cut using oxyfuel gas cutting torches that travel around the pipe or tube. See Fig. 30-18. The pipe or tube must be aligned and held in position prior to welding. This may be done

90° ELBOWS	90° ELBOWS	90° REDUCING ELBOWS	3 R ELBOWS 45° AND 90°	90° ELBOWS	45° ELBOWS	180° RETURNS
180° RETURNS	TEES	CROSSES	CONCENTRIC REDUCERS	ECCENTRIC REDUCERS	CAPS	LAP JOINT STUB ENDS
LATERALS	SHAPED NIPPLES	SLEEVES	SADDLES	WELDING RINGS	EXPANDER FLANGES	VENTURI EXPANDER FLANGES

Fig. 30-11. Pipe joint fittings. Such fittings are commercially made in a variety of sizes and thicknesses. The edges are often beveled and ready for welding.

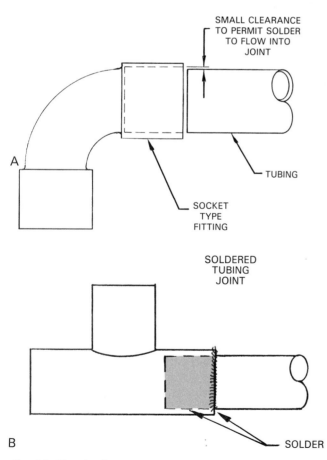

Fig. 30-12. A—A schematic drawing of a typical solder-type socket connection used on tubing. B—A completed soldered joint showing how the solder penetrates into the socket connection.

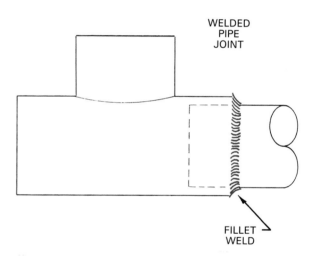

Fig. 30-13. Socket connections in tube and pipe joints. Solder is often used in and around the socket when joining tubing. On pipe, a fillet weld is made completely around the lap joint on the socket.

using special clamps and alignment fixtures. Two alignment fixtures are shown in Fig. 30-19.

BACKING RINGS

Backing rings may be used to align the pipe and control penetration in the butt joint. They may also be used to provide an even root opening while the joint is tack welded. The pins on the backing ring are removed after tack welding is completed. Fig. 30-20 shows a typical backing ring. Fig. 30-21 shows the backing ring in position inside a pipe.

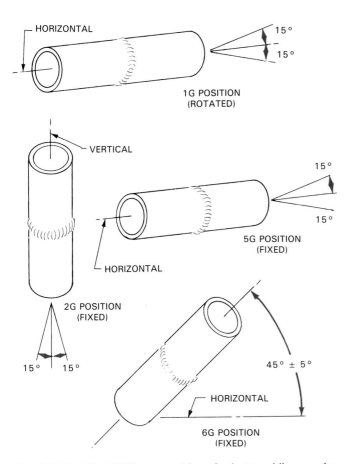

A

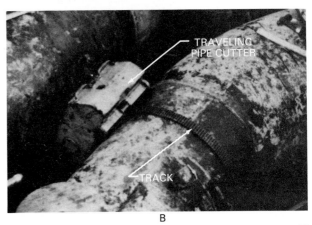

B

Fig. 30-14. The AWS test positions for butt welding on pipe or tubing. In the 1G and 5G positions, the centerline of the pipe/tube must be within 15° of horizontal.

Fig. 30-16. A—A motorized traveling pipe cutter. This machine can cut and bevel the pipe in one operation. B—A similar cutter in use on a pipe repair. Notice the close spaces in which it can operate. (E.H. Wachs Company)

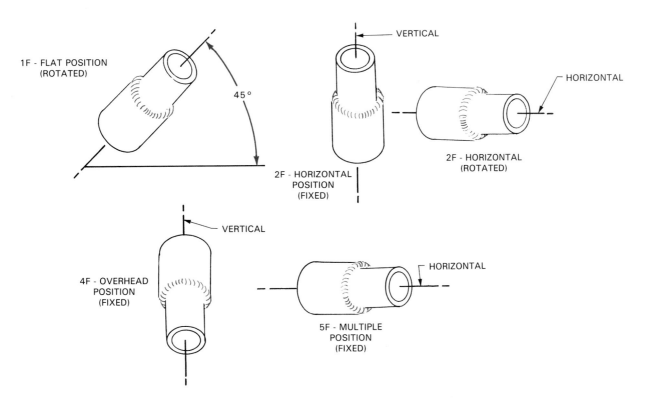

Fig. 30-15. The AWS test positions for lap welding pipe or tubing. The 2F position may be requested with the pipe or tube fixed or rotated.

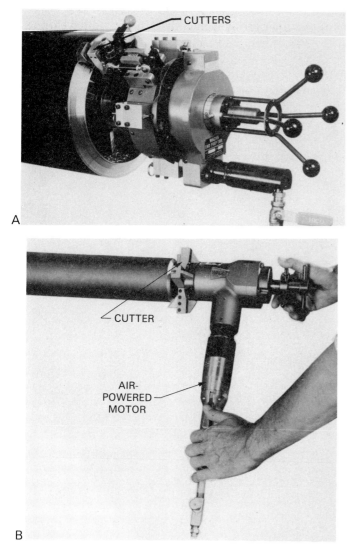

Fig. 30-17. Two models of air-powered pipe cutters. A—Machine cutting a 12'' (305 mm) pipe. This machine can cut the pipe to length and make compound bevel cuts in one operation. B—A smaller air-powered pipe cutter and beveler. This machine can be used on pipe with up to a 4'' (102 mm) ID. (E.H. Wachs Company)

Fig. 30-18. A hand-operated flame-cutting and pipe beveling machine. Note the gearing and angle adjustment for the oxyacetylene cutting torch. A hand crank is used to move the torch around this 22'' (559 mm) pipe.

Fig. 30-19. Two pipe alignment devices. A—A rigidly made pipe alignment device. Chain clamps are used to position these pipes for a T-joint weld. (Jewel Mfg. Co.) B—A pipe or tube may be clamped to this protractor at any desired angle prior to welding. (Strippit, Inc., Unit of IDEX Corp.)

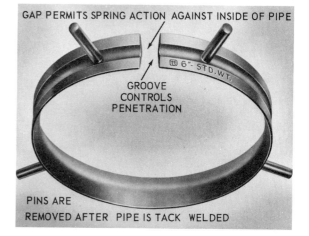

Fig. 30-20. A backing ring of the type used in pipe butt welds. This device aligns the pipe, provides a uniform root opening, and controls the amount of penetration. (Tube Turns, Div. Chemetron Corp.)

WELDING PROCEDURES

Pipe and tube welding using SMAW may be done in any position. If a pipe or tube is capable of being rotated, the welding may be done in the flat welding position.

Fig. 30-21. A pipe and an elbow being assembled with a backing ring in place.

When a pipe is fixed in a vertical position, the weld must be made in the horizontal position around the entire cylinder. A pipe or tube that is fixed in the horizontal position will be welded in the flat, vertical, and overhead welding positioins. These positions change as the weld moves around the outside of the pipe or tube. The vertical welding may be done uphill or downhill.

EXERCISE 30-1 — BUTT WELDING PIPE WITH SMAW

1. Obtain two lengths of pipe that are 3'' (76.2 mm) or larger in diameter. The instructor will determine the diameter and schedule size. Also obtain six 1/8'' (3.2 mm) diameter E6011 electrodes.
2. Set the arc welding machine for the proper amperage and polarity. See Chapter 13.
3. Bevel one end of each piece to a 45° angle. Leave a 1/16''–1/8'' (1.6 mm–3.2 mm) root face (thickness).
4. Tack weld the two sections in four places 90° apart. The root opening should be between 1/16''–1/8'' (1.6 mm–3.2 mm), depending on the pipe size and wall thickness.
5. Weld the root pass. The pipe may be rotated.
6. Thoroughly clean the root pass.
7. Have the root pass inspected and approved by your instructor.
8. Weld the hot pass and the filler passes.
9. Use a weaving bead to complete the cover pass.
Note: Thoroughly clean each tack weld and weld bead before starting the next pass.
Inspection:
   The root pass must penetrate uniformly through the pipe wall. Each filler pass must be free of pits and inclusions. The cover pass must be a weaving bead. It must have uniform ripples and be free of gas pockets.

UPHILL WELDING

Uphill welding will generally produce the soundest welds. This welding direction is recommended for pipe and tube with wall thicknesses greatere than 1/2'' (12.7 mm). A larger groove angle than in downhill welding is recommended, with a larger root opening. The electrode diameter should be 1/8'' (3.2 mm) or 5/32'' (4.0 mm). The amperage should be on the lower end of the recommended range. Travel speed should be fairly slow. If the weld pool becomes overhead, a whip motion may be used. See Fig. 30-22. In the whip motion, the electrode and arc are rapidly moved ahead of the weld pool about 1'' (25.4 mm), then returned to the pool. During this up-and-back motion, the pool has

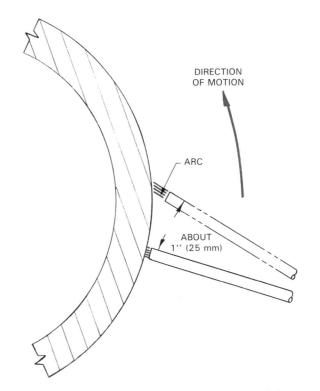

Fig. 30-22. The whip motion. To control the heat of the crater, move the electrode rapidly forward and then back to the crater. The arc should never be broken (stopped) in this motion.

a chance to cool. The weld is made from the 5 o'clock to the 11 o'clock position on each side of the pipe or tube. See Fig. 30-23. A 10° forward electrode angle is used for welding both uphill and downhill. The angle is measured from lines through the center of the cylinder and the center of the electrode. See Fig. 30-24.

VERTICAL DOWN WELDING

On tube or pipe with a wall thickness up to 1/2” (12.7 mm), *vertical down welding* is often used. As in all vertical down welding, the arc crater must be kept ahead of the slag. The weld will become contaminated if slag runs into the crater.

The groove angle used may be less than for a vertical up weld. A 3/16” (4.8 mm) or 1/4” (6.4 mm) elec-

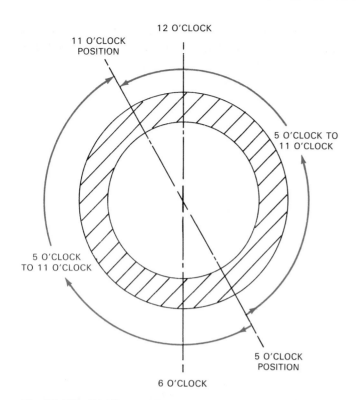

Fig. 30-23. *Welding uphill. The weld is started at the 5 o'clock position and is continued to the 11 o'clock position on both halves of the pipe or tube.*

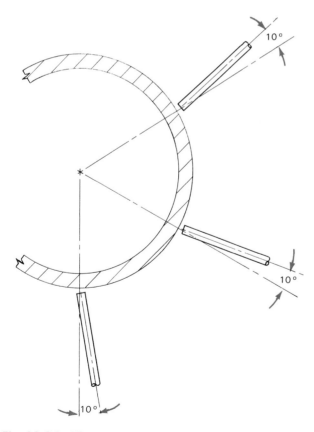

Fig. 30-24. *The suggested electrode angle for welding large diameter pipe is about 10° forward. The angle is measured between the electrode and a line drawn to the center of the pipe. This 10° angle is continually changing as the weld is made.*

trode should be used. The amperage should be on the high side of the suggested range for the electrode being used. Travel speed must be rapid, so that the weld pool stays ahead of the slag. The downhill weld is generally made from the 11 o'clock to 5 o'clock position on each half of the pipe or tube. See Fig. 30-25.

A weaving motion is suggested for the cover pass. The forward motion of the electrode weave should be between 1 and 1 1/2 electrode diameters. See Fig. 30-26.

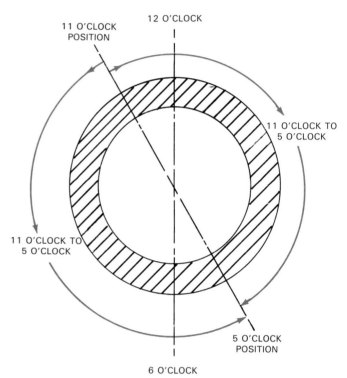

Fig. 30-25. *Welding downhill from the 11 o'clock to the 5 o'clock position. The joint should be tack welded in at least three places. It is advisable to weld past the 12 o'clock and 6 o'clock positions. Test samples are taken from these positions.*

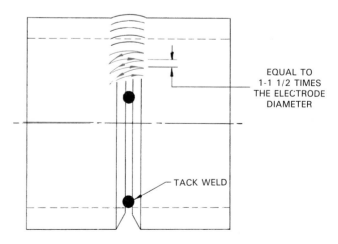

Fig. 30-26. *A weaving bead is recommended for the cover pass on large groove joints. The electrode should be moved forward a distance equal to 1 to 1 1/2 electrode diameters during each swing of the weave pattern.*

KEYHOLE

The size of the keyhole is an important indicator of the depth of the penetration. It will also indicate the quality of the penetration.

A keyhole that is too large indicates too much penetration or burn-through. It also indicates possible undercutting of the penetration bead on the inside of the pipe. A small keyhole is a sign of lack of penetration or uneven penetration.

MULTIPLE PASSES

Multiple passes are generally required to complete a weld except on wall thicknesses under 3/16″ (4.8 mm). Fig. 30-27 shows the names and relationships of the various passes. The names of the passes are:

1st pass ---------------------------------Root pass
2nd pass --------------------------------Hot pass
3rd and succeeding passes ------------Filler passes
Final pass -----------------------Cover (cap) pass
Passes should not be thicker (deeper) than 3/16″ (4.8 mm). Such thin passes allow gases and impurities to reach the surface before the weld cools.

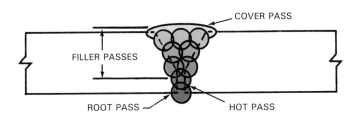

Fig. 30-27. The names given to the various passes in a weld joint. Note that there may be more than one hot pass and several filler passes.

Weld passes on alloy steel pipe or tube should be made smaller in width and depth, preferably less than 1/8″ (3.2 mm). Smaller passes put less heat into the pipe or tube and reduce the amount of alloy loss due to welding.
Note: When SMA welding pipes or tubes, each bead must be thoroughly cleaned before the next bead is started. This will reduce the possibility of slag inclusions in the next bead.

The root pass should be made using an electrode 1/8″ (3.2 mm) in diameter. This is the most important of all the passes. The root pass must penetrate through the pipe or tube wall. The penetration bead may be seen from the inside of the pipe or tube. The heights of the root reinforcement should be about 1/16″ (1.6 mm). See Fig. 3-3B. A higher root reinforcement will interfere with flow through the pipe or tube.

The hot pass or second pass is made with the same size electrode. However, a higher amperage should be used. This pass eliminates any undercutting that may

have occurred in the root pass. Filler passes should be made with an electrode larger than 1/8″ (3.2 mm). The cover pass is also made with an electrode larger than 1/8″ (3.2 mm). It must be made with a weaving motion. This pass must tie in all the filler passes and form a single, wide **convex** (curved outward) **bead.**

RESTARTING AND ENDING A WELD

With SMAW, you must stop occasionally to change electrodes. The correct method of restarting is extremely important. If the restart is not done properly, a weakness may be created in the weld bead.

RESTARTING THE WELD

To restart, the arc is struck about 1/2″ − 1″ (12.7 mm − 25.4 mm) ahead of the previous crater. The electrode and new weld pool are then moved backward until the ripples of the new bead just touch the old bead at the rear of the old crater. The electrode and bead are then moved forward again at the correct speed.

To restart a root pass when a backup ring is not used, the procedure is slightly different. Strike the arc, as explained above. When the electrode reaches the keyhole, push the tip of the electrode through the keyhole at a 45° angle. See Fig. 30-28. Hold the electrode there momentarily, then withdraw it to the normal distance and angle. Inserting the electrode through the keyhole ensures that the penetration bead inside the pipe is continuous.

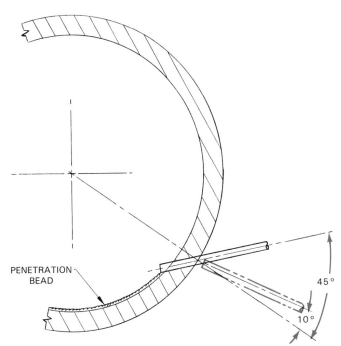

Fig. 30-28. Restarting the root pass without a backing ring. The electrode is momentarily inserted through the keyhole at a 45° to ensure that the penetration bead is continuous. It is withdrawn and the weld continued.

ENDING THE WELD

Run the electrode and weld pool to the end of the weld. Then, reverse the bead for a short distance, as you slowly increase the electrode gap until the arc stops. This reversal of the electrode causes the weld pool at the end of the weld to fill. The end of the bead will be relatively smooth, without a depression.

GTAW AND GMAW OF PIPES AND TUBES

Both GTAW and GMAW may be used to weld pipes and tubes. GTAW is often used to make the very important root pass on both tubes and pipes. GTAW will probably produce the strongest possible weld in such applications, Fig. 30-29. Bead cleaning between passes is unnecessary. GTAW is, however, a slow and expensive process.

GMAW is growing in popularity as a pipe and tube welding process. It is fast and clean. Bead cleanup between passes is usually not necessary. GMAW out of

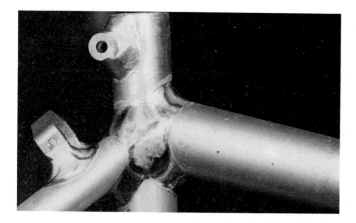

Fig. 30-29. Gas tungsten arc welding (GTAW) was used to produce these strong, well-made welds on an aluminum bicycle frame.   (Miller Electric Mfg. Co.)

position may be done using either the short circuiting or pulsed spray transfer methods. Completed welds that are properly made are usually free of inclusions and other flaws.

REVIEW QUESTIONS

Write your answers on a separate sheet of paper.
1. Pipe is ordered by ID size. The wall thicknesses must be chosen from a list of _____ _____ numbers.
2. A mandrel and rolls are used to form the hole in seamless pipes and tubes. True or False?
3. When butt welding pipe, the horizontal fixed position is called the _____ position.
4. List two reasons for using a backing ring.
5. Name three different types of weld passes used when pipe or tube welding.
6. Uphill welding is done from the _____ o'clock to the _____ o'clock position on both sides of the pipe or tube.
7. When using a _____ motion to complete a cover pass, the electrode should move forward in steps of 1 to 1 1/2 electrode diameters.
8. A _____ keyhole indicates a lack of penetration.
   a. small
   b. large
   c. elongated
   d. distorted
9. When restarting a bead, the arc should be struck 1/2″−1″ (12.7 mm−25.4 mm) behind the old weld _____.
10. Each bead must be cleaned before the next bead is started when _____.
    a. SMAW
    b. GTAW
    c. GMAW
    d. OFW

SECTION **8**

TECHNICAL INFORMATION

Nederman, Inc.

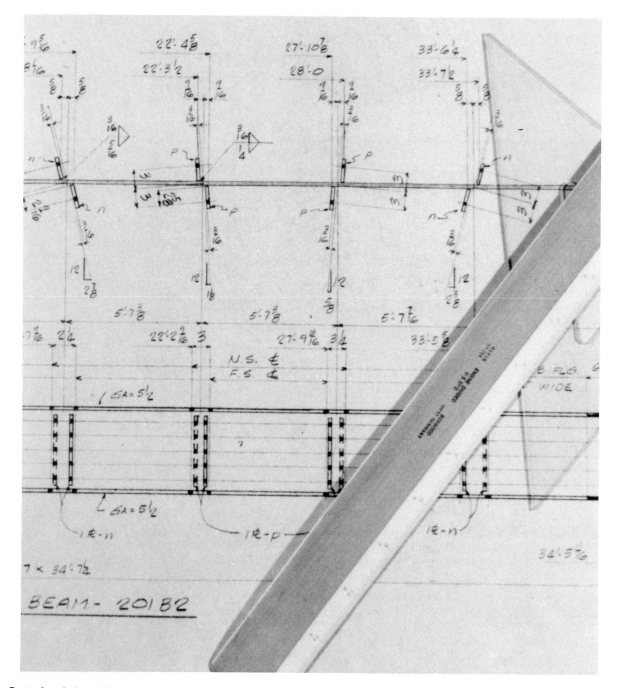

Part of an industrial mechanical drawing for some structural steel weldments. Notice the welding symbols used to indicate the types of welds required.

Chapter 31

Welding Symbols

After studying this chapter, you will be able to:
☐ Describe the method for making a mechanical drawing of a three-dimensional object, using the orthographic projection process.
☐ List the names of the views used in an orthographic projection.
☐ Identify the basic types of welds indicated on the AWS welding symbol.
☐ Locate information on the weld symbol to determine the size of the root opening, the groove angle, and the desired size, contour, and finish of the weld.

TECHNICAL TERMS

Arrow side, AWS welding symbol, backing weld symbol, basic weld symbol, chain intermittent welding, depth of preparation, dimensions, drafting machine, effective throat, field weld symbol, finish symbol, groove weld size, intermittent welding, legs, mechanical drawings, melt-through symbol, orthographic projection, other side, plug weld, reference line, root opening, size of weld, sketches, slot weld, spot weld symbol, staggered intermittent welds, tail, weld all-around symbol, weld contour symbol, weld size.

THE THREE-VIEW DRAWING

Parts and assemblies used in industry are produced from mechanical drawings or sketches. They are called *mechanical drawings* because they are made by mechanical means, using drafting tools or computers.

For high-production parts, drawings have traditionally been made with compasses, triangles, and a T-square, or with a *drafting machine* (a device attached to a drawing board that combines the functions of T-square and triangle). In recent years, computers running CADD (computer-aided drafting and design) programs have become widely used for making drawings. Low-production or one-of-a-kind parts are often made from *sketches* — simple drawings made without drafting tools.

These sketches or mechanical drawings must show the shape and size of the part or assembly. All the size information needed to make the part must be given on the drawing or sketch. These sizes are called *dimensions.* Dimensions may be given in US conventional or SI metric sizes.

ORTHOGRAPHIC PROJECTION

To completely describe the shape of a part, it may be necessary to look at it from a number of viewpoints. Almost all industrial drawings are done using a process called orthographic projection.

In *orthographic projection,* as shown in Fig. 31-1, the object to be drawn is treated as if it were inside a clear plastic cube. All points and lines are "projected" onto the surface of the cube by the viewer's eyes, as shown in Fig. 31-2. The object is drawn on the cube as it is seen from the front, back, top, bottom, left, or right sides. Thus, the object may be seen or viewed from these six different sides. Normally, no more than three views are required to fully describe the shape of an object. The imaginary cube is then unfolded until it lies flat like a sheet of paper. See Fig. 31-3.

In reality, of course, industrial drawings are not made on a cube, but rather on flat sheets of paper or plastic. However, the top, front, and side views are in the same positions that they would be on the cube.

To obtain information from a drawing, you must

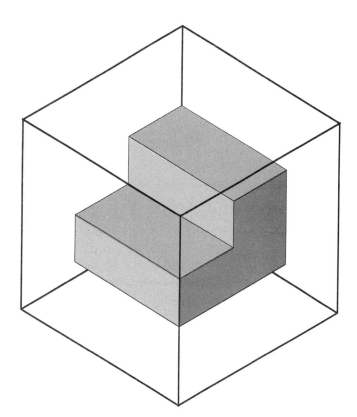

Fig. 31-1. *The theory of orthographic projection. A part, such as this step block, is viewed from all sides as if it were in a plastic box. Six different views are possible.*

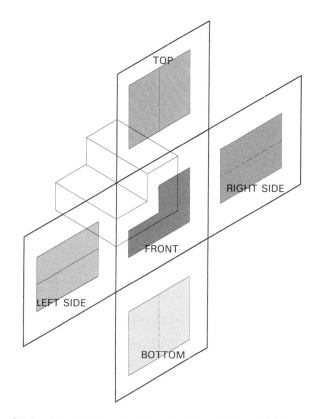

Fig. 31-3. *The plastic cube shown in Figs. 31-1 and 31-2 can be unfolded so that it lies flat. This is the way the various views of a part will appear on a sheet of paper. The rear view is seldom used.*

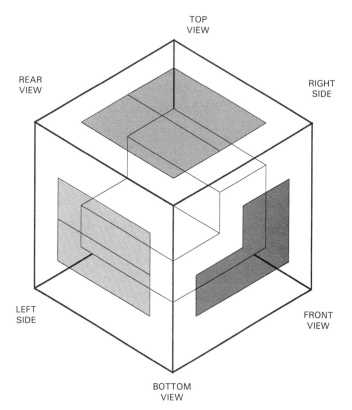

Fig. 31-2. *Each view of the part is "projected" onto the cube's surfaces by the person viewing it. The front view is the one that shows the most about the shape of the part. All other views are named in relation to the front view. The right side view is always to the right of the front view.*

often follow a line or a point from one view to another. Straight lines have two ends. The ends of these lines may be identified by letters or numbers. When viewed from the end, a line will look like a point or a corner. See Fig. 31-4.

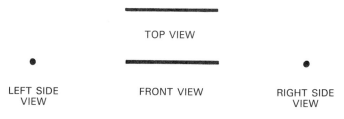

Fig. 31-4. *A three-view orthographic drawing of a wire. This drawing illustrates that a point in one view will appear as a line in the other two views.*

EXERCISE 31-1 — IDENTIFYING LINES AND POINTS ON A THREE-VIEW DRAWING

On the three-view drawing above, the lines in the top and front views have been identified with numbers or letters. All points on the upper surfaces have been identified with *numbers*. All points on the lower surface have been identified with *letters*.

NOTE: A point or corner in one view is usually a line in

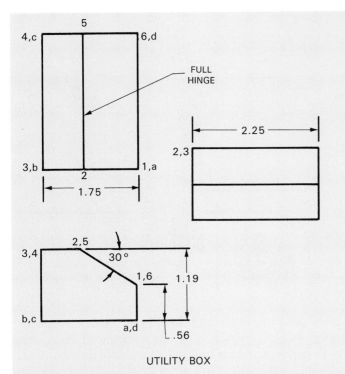

UTILITY BOX

the other two views. For example, *line* 1,6 in the top view appears as *point* 1,6 in the front view.

On a separate sheet of paper, sketch the right side view as shown above. Next, identify all points and lines in this view. The number or letter of the point closest to you must be given first (2,3 has been identified as an example). Show your completed sketch to your instructor for checking.

THE AWS WELDING SYMBOL

Each weld joint must be described in detail. This is done so that the welder will know precisely how to make each weld. The American Welding Society (AWS) has developed a welding symbol to convey all the information needed to properly make the weld on a joint. The AWS welding symbol is used on mechanical drawings or sketches of welded parts. The basic weld joint to be used is shown by the position of the metal parts on the mechanical drawing or sketch. There are five basic weld joints: butt, lap, edge, T-, and corner.

Information given on the welding symbol must always be shown in the designated location on the symbol, as shown in Fig. 31-5. Important rules for using the AWS welding symbol will be shown in italic type in this chapter. For example: *Each time the weld joint changes direction, a new welding symbol must be drawn. There are a few exceptions to this rule. They will be described later.*

THE REFERENCE LINE AND ARROW

The **reference line** *is always drawn horizontally.* See Fig. 31-6. On the drawing, it is placed as close as possible to the weld joint it describes. All other information is placed above or below this horizontal reference line, as shown in Fig. 31-5.

An **arrow** *may be drawn from either end of the reference line. The arrowhead must touch the line to be welded.* See Fig. 31-6. The arrowhead may touch

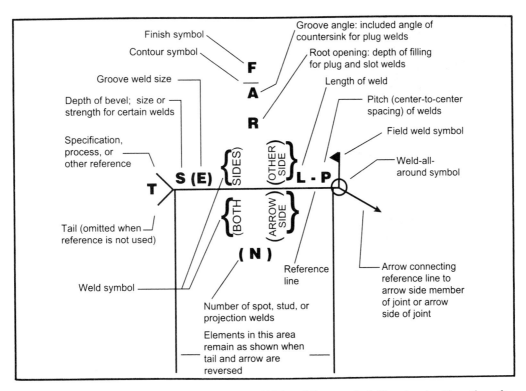

Fig. 31-5. *The American Welding Society (AWS) welding symbol. The standard locations for information on the welding symbol are marked and explained.* (From ANSI/AWS A2.4-93)

307

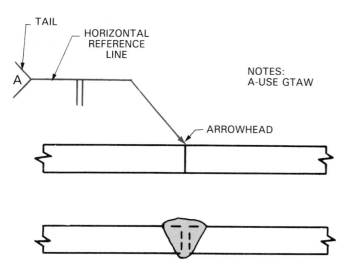

Fig. 31-6. The arrow, reference line, and tail. The arrowhead must touch the line to be welded, usually in the view where it appears as a point. The reference line is always drawn horizontally. The tail may be used to hold a symbol for a note elsewhere on the drawing.

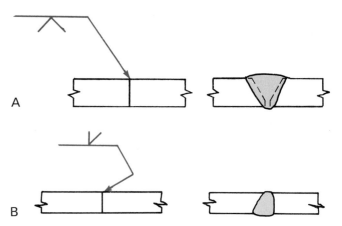

Fig. 31-7. Groove welds. A—The symbol is for a V-groove joint. B—The symbol is for a bevel-groove joint. The arrow is bent to point to the piece of metal that will be beveled. The arrow is bent to the left, so the left piece is beveled.

the weld line in any of the views. However, the arrowhead usually touches the weld line in a view where it appears as a point or corner.

Some joints—the bevel, J, and flare-bevel groove—have only one edge prepared for welding. For these welds, the arrow is bent to point at the edge that is cut, ground, or bent in preparation for welding. See Fig. 31-7. When the same preparation is used for both edges, the arrow is not bent. *The AWS welding symbol may appear in any view, but must appear* once *for each weld joint.*

The tail

The *tail* is added to the symbol only when special notes are required. See Fig. 31-6. A number or letter

code used inside the tail directs the welder to special notes located elsewhere on the drawing. These notes may specify the heat treatment, welding process used, or other information not given on the welding symbol.

BASIC WELD SYMBOLS

The basic weld symbol describes the type of weld to be made. This symbol is a miniature drawing of the metal's edge preparation prior to welding. See Fig. 31-8. The basic weld symbol is only part of the entire *AWS welding symbol.* Fig. 31-9 illustrates the basic weld symbols used by the AWS.

THE ARROW SIDE AND OTHER SIDE

On the drawing, the arrowhead of the welding symbol always touches the line or joint to be welded. The joint, however, has two sides. A weld may be needed

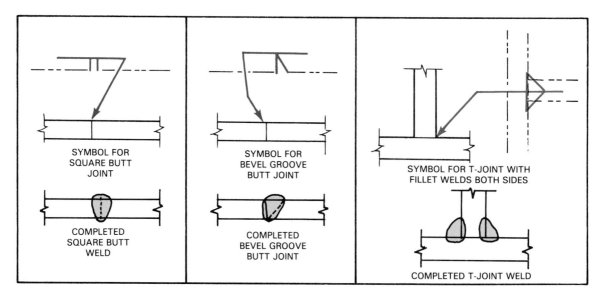

Fig. 31-8. Weld symbols. The weld symbol is a miniature drawing of the joint prior to welding. Phantom lines have been added to these drawings to show the joint better. They are not used on a real welding symbol. The vertical line in the bevel-groove and fillet weld symbols is always drawn at the left-hand side of the weld symbol.

on one side or the other. In some cases, both sides of a joint are welded.

The side of the joint that the arrow touches is always called the **arrow side.** The opposite side of the joint is called the **other side.**

On the welding symbol, weld information for the arrow side is shown below the horizontal reference line. Weld information for the other side is shown above the reference line. See Fig. 31-10 for examples of arrow side and other side information.

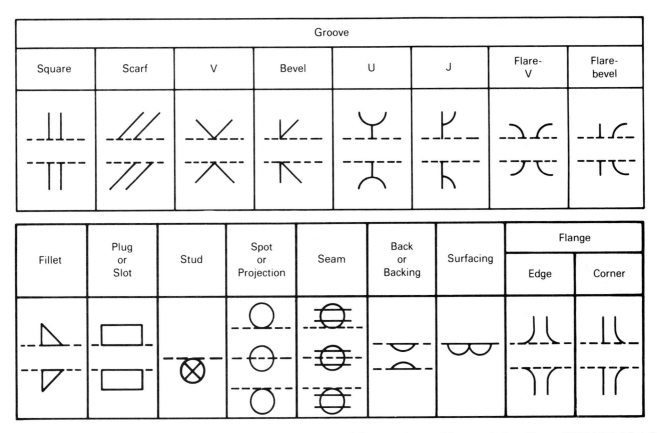

Groove							
Square	Scarf	V	Bevel	U	J	Flare-V	Flare-bevel

Fillet	Plug or Slot	Stud	Spot or Projection	Seam	Back or Backing	Surfacing	Flange	
							Edge	Corner

Fig. 31-9. Basic weld symbols. The weld symbol is only one part of the complete welding symbol.   (From ANSI/AWS A2.4-93)

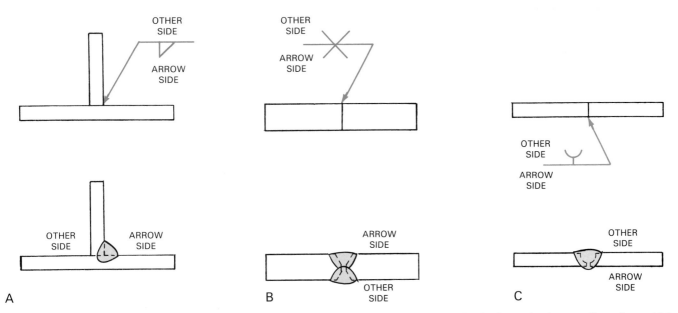

A   B   C

Fig. 31-10. Arrow side and other side. A—Since the weld symbol is drawn below the horizontal reference line, the weld is made on the arrow side. B—Welds are made on both sides of the joint. C—The weld symbol is above the reference line, so the weld is made on the side opposite the arrow.

309

ROOT OPENING AND GROOVE ANGLE

The *root opening* is the space between the base metal pieces at the root (bottom) of the joint. The root opening is held constant in a jig or fixture, or by tack welding. On the welding symbol, the size of the root opening may be given in fractions of an inch, metric units, or decimals. Root opening size is given inside the basic weld symbol. See Fig. 31-11.

The edge of the metal for a groove-type weld is flame cut or machined to a specific angle before welding. The angle at which the metal is cut is shown on the welding symbol. If a V-groove weld is being made, the angle cut on each piece is half that shown on the welding

symbol. The total angle of the groove-type welded joint is shown just beyond the outer edge of the basic weld symbol. See Fig. 31-11.

WELD CONTOUR SYMBOL

To indicate the desired contour for the completed weld, a curved or straight line is used on the welding symbol. The contour (shape) of the face of the completed weld may be convex (curved outward), flat, or concave (curved inward).

The contour line is drawn between the basic weld symbol and the finish symbol, as shown in Fig. 31-12. If no *contour symbol* is shown, the weld face should have the normal convex shape.

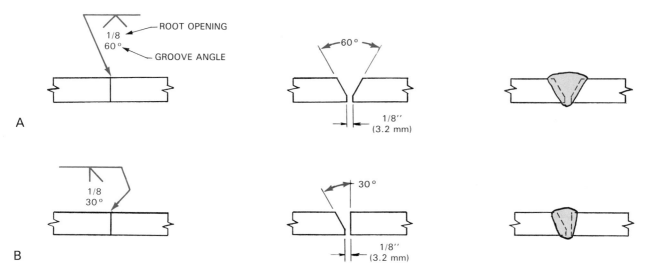

Fig. 31-11. Groove weld preparation. The root opening size is given inside the weld symbol. The groove angle is shown just outside the weld symbol. A—V-groove welding symbol, the joint preparation, and the completed V-groove weld, B—Bevel-groove joint symbol, joint preparation, and the finished weld.

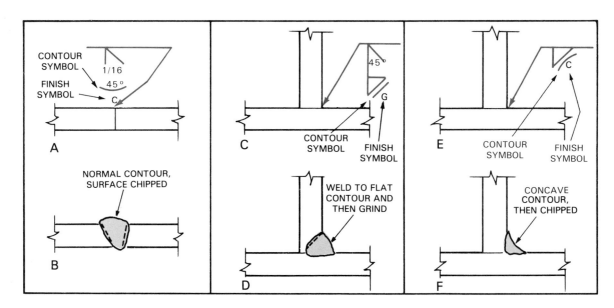

Fig. 31-12. Weld contour symbols and finish symbols. These symbols are shown just beyond the weld symbol. The upper part of each pair of illustrations shows the contour and finish symbols on a welding symbol. The shapes and finishes of the completed welds are shown at bottom.

FINISH SYMBOL

If the weld face is not to be left "as welded," a symbol designating the required finish is shown on the welding symbol. The American Welding Society lists the following methods of finishing a completed weld:
- C (chipping).
- G (grinding).
- M (machining).
- R (rolling).
- H (hammering).

The *finish symbol* designating the finish is placed beyond the contour symbol, as shown in Fig. 31-12. Users of finish symbols may create their own symbols.

Sometimes, all welds are to be finished in the same manner. In this case, a general note may be placed on the drawing. This avoids the need for a finish symbol on each welding symbol.

DEPTH OF PREPARATION AND GROOVE WELD SIZE

The *depth of preparation* is given in the "S" position on the welding symbol for groove-type welds. See Fig. 31-13. Weld strength, or weld size for welds other than a groove weld, may also be given in this same location. Edges of metal over 1/4″ (6.4 mm) thick are always prepared in some way before making a groove weld, in order to ensure 100 percent penetration. The *groove weld size* is the depth to which a weld penetrates

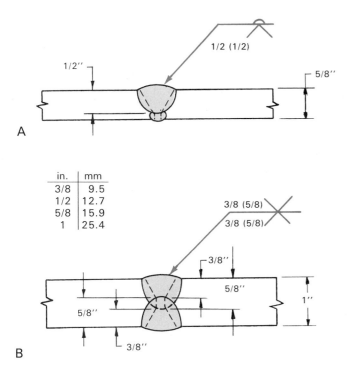

in.	mm
3/8	9.5
1/2	12.7
5/8	15.9
1	25.4

Fig. 31-13. Groove weld size and depth of preparation. At A, the groove weld size is the same as the depth of preparation, 1/2″ (12.7 mm). At B, the depth of preparation for both welds is 3/8″ (9.5 mm), but the groove weld size or depth of penetration is 5/8″ (15.9 mm).

the joint from the surface of the base metal. See Fig. 31-13. The desired depth of preparation and the depth of the weld penetration are generally determined by codes or specifications, or by a welding engineer. Sometimes, experimentation is needed to determine the correct groove weld size and depth of preparation.

The *effective throat* of a fillet weld shown in the "E" position, is the distance, minus any convexity, between the weld face and the weld root. *The effective throat can never be greater than the metal thickness.* See Fig. 31-14.

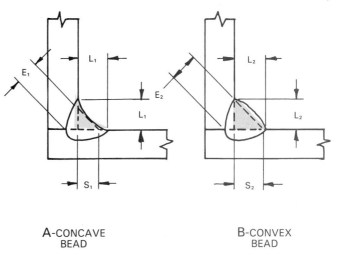

A-CONCAVE BEAD   B-CONVEX BEAD

Fig. 31-14. Fillet weld sizes and strengths. The leg size of a fillet weld (L_1 and L_2) is the length measured from the toe of the weld to the surface of the base metal. The actual size of a fillet weld (S_1 and S_2) is the length of the leg of the triangle that can be drawn within the completed weld. Note that the triangle in the convex bead is larger and therefore stronger than the triangle in the concave bead. Note also that the effective throat E_2 is larger than E_1.

The size of a fillet weld is the length of the **legs** (sides) of the triangle that can be drawn inside the cross section of the finished weld. See Fig. 31-14. The length of the fillet weld legs is given in the same "S" position as the depth of penetration for a groove weld, as shown in Fig. 31-15.

A typical fillet weld triangle will have legs of equal length. Only one dimension is given for this triangle. If the legs are unequal in length, two dimensions must be given. See Fig. 31-15. When a fillet weld with unequal legs is to be made, the weld shape is shown on the weldment drawing. The shape on the drawing will show which leg is longer and which leg is shorter. *The fillet weld size is shown to the left of the basic weld symbol. The fillet size is placed in parentheses only when two dimensions are required. The effective throat size is also shown to the left of the basic weld symbol. The effective throat size always is placed in parentheses.* See Fig. 31-16. Fillet weld leg size and effective throat size may be shown in fractions of an inch, in decimals, or in SI metric units.

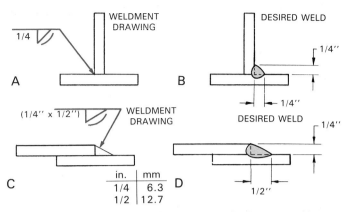

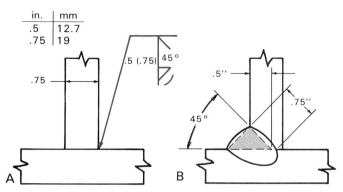

Note: No weld size dimensions or shapes are generally given on a weldment drawing. However, the shape of an unequal fillet is shown.

Fig. 31-15. Fillet weld size. At A and B, the fillet weld legs are equal, so only one dimension is necessary. At C and D, the legs are unequal. Two dimensions are given in parentheses. The weld shape for an unequal fillet weld is on the weldment drawing.

Fig. 31-16. Effective throat size. Effective throat, or the depth of the weld penetration, is shown in parentheses on the welding symbol at A. The depth of preparation of the groove weld is .5" (12.7 mm) and the effective throat of the fillet weld is .75" (19 mm).

INTERMITTENT WELDING

On many joints, it is not necessary to weld continuously from beginning to end. Where strength is not affected, and to save time and money, the joint may be made with short sections of weld. See Fig. 31-17. This type of welding is called *intermittent welding*.

When intermittent welding is done, two dimensions are given. The first is the length of each weld. The second is the pitch of the weld. This is the distance from the center of one weld to the center of the next weld, as shown in Fig. 31-17.

Intermittent welds may be made on both sides of a joint. When *chain intermittent welding* is done, the welds stop and start at the same place. The welds exactly match on the two sides of the joint. See Fig. 31-17.

When *staggered intermittent welding* is done, the welds do *not* line up on each side. The welds on one side are staggered, or offset, from the welds on the other side of the joint. See Fig. 31-17.

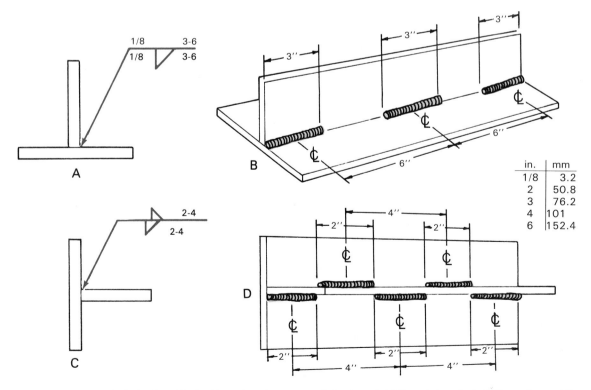

Fig. 31-17. Length and pitch of weld. A—Note the placement of the length (3") and the pitch (6") on the welding symbol. B—The weld shows a series of 3" (76.2 mm) long welds that are 6" (152.4 mm) apart from center to center. C,D—A staggered weld. Note the staggered fillet symbol at C.

MELT-THROUGH AND BACKING WELD SYMBOLS

A *melt-through symbol* is used when 100 percent penetration is required. This symbol is used on joints that are welded from one side only. See Fig. 31-18. The height of the root reinforcement may be placed to the left of the melt-through symbol.

The *backing weld symbol* is used when a stringer bead is laid on the root side of the weld to ensure complete penetration. See Fig. 31-18.

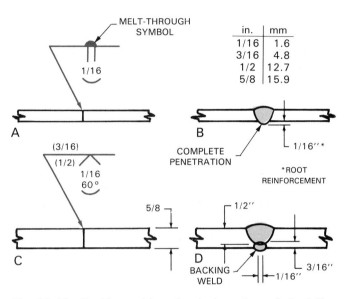

Fig. 31-18. Backing welds and melt-through symbols. A,B— The melt-through symbol is used to ensure 100 percent penetration when welding from one side only. C,D—A backing weld may be used to ensure 100 percent penetration when welding is possible on both sides of a joint.

WELD ALL-AROUND AND FIELD WELD SYMBOLS

Instructions on a welding symbol are in effect only until the weld joint makes a sharp change in direction. When a joint changes direction at a corner, a new welding symbol generally must be used.

When the same weld is made on all edges of a continuous joint, however, a *weld all-around symbol* may be used. A continuous joint may occur on a box or cylindrical part. See Fig. 31-19.

Parts joined by welding are called weldments. Some parts of a weldment may be welded in the manufacturer's shop. On large weldments like bridges and buildings, some welds must be made on the site, or "in the field."

A field weld symbol is used when welds are to be made at the construction site. See Fig. 31-19.

MULTIPLE REFERENCE LINES

When several operations are to be performed on a welded joint, a corresponding number of reference lines may be used. The reference line nearest the arrowhead indicates the first operation. The last operation on the welded joint is shown on the reference line most distant from the arrowhead. See Fig. 31-20.

PLUG AND SLOT WELDS

Sometimes, two pieces must be welded together at a point away from the edge. This is done by cutting a hole through one piece and welding the pieces together through this hole. See Fig. 31-21. The hole

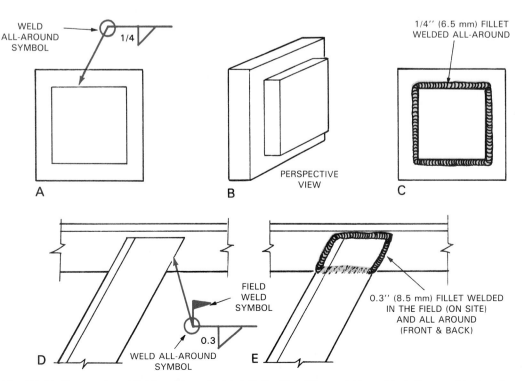

Fig. 31-19. Weld all-around and field weld symbols. A,B,C—This fillet weld is welded all-around in the shop. D,E— This 0.3" (8.5 mm) fillet weld is made in the field. It is welded all around the angle iron, both front and back.

is often round, but may be any shape. These holes may be drilled, flame-cut, or machined.

A *plug weld* is made through a round hole. A *slot weld* is made through a hole that is elongated. The sym-

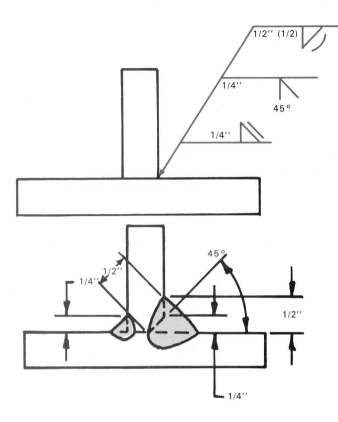

Fig. 31-20. *Multiple reference lines. When several operations are done on the same joint, a stacked (multiple) symbol is used. The weld nearest the arrow is made first.*

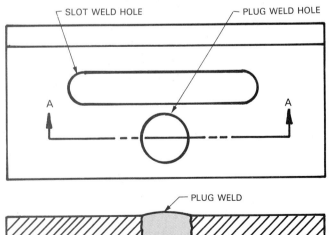

SECTION A-A

Fig. 31-21. *Plug and slot welds. The plug weld is made in a round hole. The slot weld is made in an elongated hole. Note: If no dimension is given inside the weld symbol, the plug or slot weld hole is completely filled with filler metal.*

bol for a plug weld is shown in Fig. 31-22. The diameter of the plug weld is shown to the left of the basic weld symbol. If the sides of the hole are angled, the total angle is shown below the basic weld symbol. See Fig. 31-22. The depth of the weld is shown inside the symbol. If no depth dimension is given for a plug or slot weld, the hole is completely filled.

For a slot weld, the drawing shows the length and width, and the angle of the sides. The drawing will indicate the location and spacing of the holes for both plug and slot welds. If a series of plug or slot welds is to be made, the center-to-center distance of the holes may be shown to the right of the basic weld symbol.

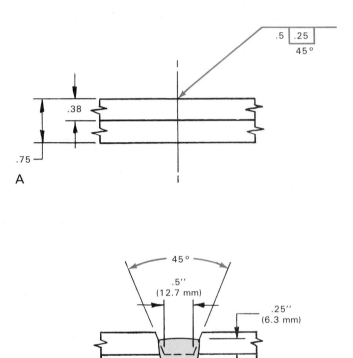

Fig. 31-22. *A plug weld. A—The weld symbol shows the depth of the weld, the hole diameter, and the angle of the hole (if it is not straight). B—A cross section view of the weld described in A.*

SPOT WELDS

A spot weld may be accomplished using resistance welding, gas tungsten arc welding, electron beam welding, or ultrasonic welding.

The *spot weld symbol* is a circle 1/4" (6.0 mm) in diameter. The circle may be on either side of the reference line or it may straddle the reference line. If the weld is accomplished from the arrow side, the spot weld symbol should be below the reference line, as in all other welding symbols. If the welding is done on both sides, as in resistance spot welding, the circle straddles the reference line. See Fig. 31-23.

Projection welding is another form of resistance welding used to produce spot welds. To indicate which

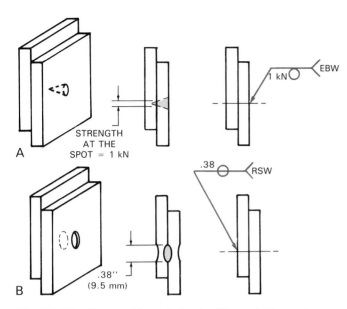

Fig. 31-23. Spot weld symbols. A—The weld is an electron beam spot weld. Its required strength is 1 kilonewton. The weld is made from the other side. B—The weld is a resistance spot weld. Its size is 3/8'' (9.5 mm). The weld is made from both sides, so the symbol straddles the reference line.

piece has the projections on it, the circle is placed above or below the reference line.

Information given for a spot weld includes size or strength, spacing, and number of spot welds.

The *weld size* or desired strength is given to the left of the weld symbol. The weld size may be given in US customary units or SI metric units. Weld strength is shown in pounds or newtons per spot. To the right of welding symbol is found the weld spacing. Centered

above or below the spot welding symbol, in parentheses, is the number of welds desired.

The welding process used is shown in the tail of the welding symbol. See Fig. 31-24 for examples of the welding symbols used for spot welding. The welding symbol may be placed in any view of a drawing.

SEAM WELDS

Seam welds may be made with a number of processes, such as electron beam, resistance, or gas tungsten arc.

The process used is shown in the tail. The size (width) of the weld, or strength of the weld, are shown to the left of the weld symbol. The strength is given in pounds per linear inch or in newtons per millimeter.

The weld symbol may straddle the reference line if welded from both sides, as in resistance seam welding. For electron beam and gas tungsten arc welding, the symbol is placed above or below the reference line. This indicates from which side of the part the weld is made. The length of the seam may be shown to the right of the weld symbol. See Fig. 31-25. For more details, see ANSI/AWS publication A2.4-93.

A periodic view of this chapter as you move through the material covered in this book and the laboratory manual will help build your print-reading ability.

REVIEW QUESTIONS

Write your answers on a separate sheet of paper.
1. _____ drawings or _____ may be used to produce welded parts.

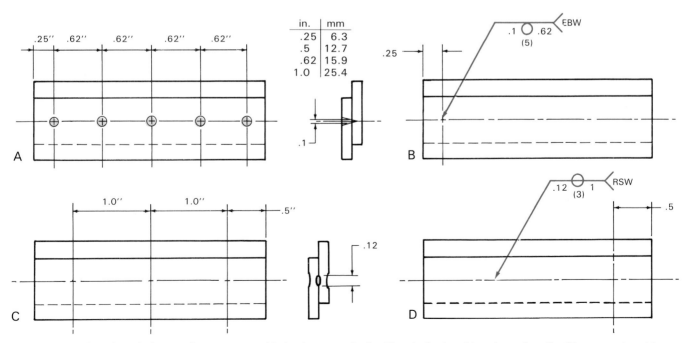

Fig. 31-24. A series of electron beam spot welds is shown at A. A—The desired weld and spacing. B—The correct weldment drawing, symbol, and dimensions. C—A resistance spot-welded part. D—The weldment drawing for this part.

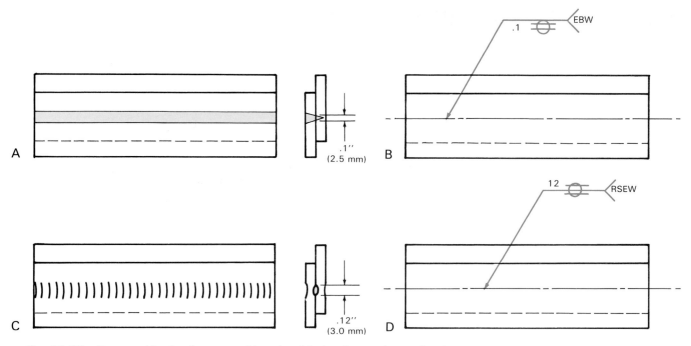

Fig. 31-25. Seam welds. A—A seam weld made with the electron beam. Its size at the fusion point is .1'' (2.5 mm). B—The weld symbol and weldment drawing for the seam weld shown at A. C—The resistance welded seam on this part results from the weldment drawing and welding symbol shown at D.

2. Almost all industrial drawings are made using a process called _____ projection.
3. The AWS welding symbol shows the type of joint to be welded. True or False?
4. List five things that the welding symbol will tell the welder about the weld to be made.
5. Could a weld all-around symbol be used in the welding symbol for the box shown in Exercise 31-1? Yes or No? Why?
6. What types of information may be shown in the tail of the welding symbol? Name two.
7. Sketch the welding symbol for the completed weld shown below:

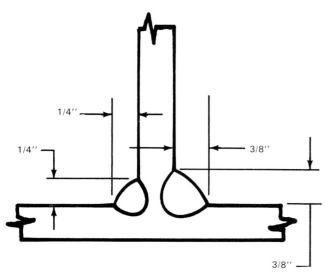

Refer to this welding symbol when answering questions 8-10:

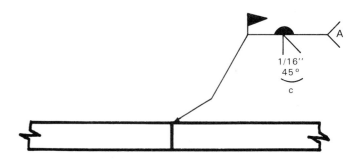

8. Which piece is cut or ground in this bevel groove weld? What is the angle of the bevel?
9. What is the shape of the face of the weld bead? How is the weld face to be finished?
10. Is this weld made in the shop or in the field (on site)? Is it welded all around? Yes or No?

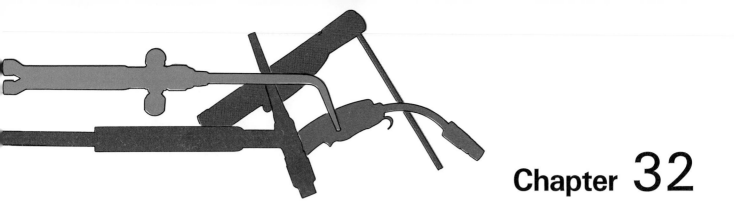

Chapter 32

Inspecting and Testing Welds

After studying this chapter, you will be able to:
☐ Describe the difference between a welding flaw and a welding defect.
☐ List the most common types of nondestructive and destructive testing done on welds.
☐ Perform several basic types of tests on welds to evaluate weld quality.
☐ Describe the methods used to prepare samples for bend tests.

TECHNICAL TERMS

Air pressure leak test, bend test, code, defect, destructive test, ductility, eddy current inspection, face bend, fillet test, flaw, free bend test, guided bend test, hardness test, impact test, indenter, liquid penetrant inspection, magnetic particle inspection, nondestructive evaluation, oscilloscope, peel test, pressure test, radiograph, root bend, side bend, tensile shear, tensile strength, tensile test, ultrasonic inspection, visual inspection, X-ray inspection.

REASONS FOR INSPECTING WELDS

All welds have flaws in them. A *flaw* is a part of a weld that is not perfect. Some flaws are so small they can be found only by examining the weld under a microscope. Other flaws are very easily seen. These include large cracks or porosity, as shown in Fig. 32-1.

A *defect* is a flaw that makes a weld unusable for the job it is intended to perform. If you are welding a handle on a garbage can lid at home, and the weld develops a small crack, it can still be used. However, if a small crack develops when welding on a gas pipeline, the crack must be repaired. In both cases, the crack is a *flaw,* but the crack is a *defect* only in the

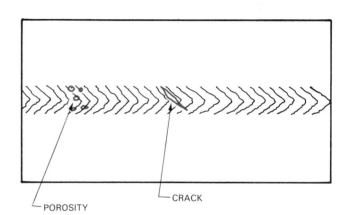

Fig. 32-1. Some flaws are seen easily when looking at a completed weld.

second case. The allowable size for a defect varies. Different applications have different requirements.

Welds must be inspected to detect flaws. Then, the flaws must be evaluated to decide if they are acceptable or if they are defects. Most welding is done to requirements of a *code* or specification. The code or specification determines how large a flaw can be before it becomes a defect. In some codes, a crack 1/32″ (.8 mm) long may be acceptable. In other codes, a crack of the same length will be a defect.

A weld is inspected to locate all defects. Once located, the defects must be repaired. A weld without defects will perform the job for which it was designed.

TYPES OF TESTS

Tests are used to determine the size and location of flaws in a weld. Tests are also used to determine the physical properties of a weld. These tests can be divided into two groups: nondestructive and destructive. *Nondestructive evaluation,* also called *inspections,* do not damage the weld or the base metal. *Destructive tests,* however, result in at least some damage to the weld. Destructive tests are usually used to determine the weld's physical properties. They often totally destroy the weld.

NONDESTRUCTIVE EVALUATION (NDE)

The most typical type of nondestructive evaluation is a *visual inspection.* Visual inspection often is sufficient to determine if a weld is a good weld. Other inspections that do not damage a completed weld include:

- Liquid penetrant inspection.
- Magnetic particle inspection.
- Ultrasonic inspection.
- X-ray inspection.
- Eddy current inspection.
- Air pressure leak test.

Nondestructive evaluation is often called NDE. It is also called NDT, which stands for nondestructive testing (a nonstandard term).

VISUAL INSPECTION

Visual inspections are performed on almost all welds. These inspections are quick and do not require much equipment.

Visual inspections are used to locate visible flaws and defects on the *surface* of the weld. Very small flaws and flaws that are inside the weld are difficult or impossible to find by this method.

A visual inspection is useful to determine the size of a weld. Gauges are available for measuring fillet sizes. See Fig. 32-2. Visual inspection also is used to check that all dimensions required by a drawing are met. A visual inspection tells a welder or an inspector if there is any undercut, overlap or other surface flaws.

Whenever you complete a weld, you can look at it to evaluate your welding skills. It also tells you if the joint is filled properly. You can look at the back side of a butt weld to see if there is complete penetration. A visual inspection tells you if the arc is too long, if the travel speed is too fast or too slow, and other information about your welding skills.

Liquid penetrant inspection

Liquid penetrant inspection is a good means of locating surface flaws. It cannot locate flaws that are inside a weld, however. This test uses a liquid dye, usually red in color for good visibility. The surface of the part must be cleaned, then the dye is applied.

If there is a flaw on the surface, some of the dye will enter the flaw. The dye is cleaned off the surface after a few minutes. However, any dye that entered a flaw is not removed by the cleaning. A developer is then applied. The developer is usually a white, dry

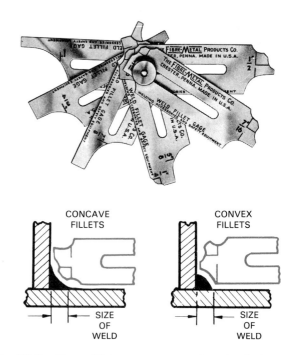

Fig. 32-2. These fillet gauges are used to check the contour and size of various convex and concave fillet welds. (Fibre Metal Products Co.)

powder. It draws out the dye, showing the location of all surface flaws in the metal or the weld. The steps followed in a liquid penetrant inspection are shown in Fig. 32-3.

Fluorescent liquids can be used instead of dyes. The surface of the part is cleaned and the fluorescent liquid is applied. After a few minutes, the liquid is cleaned off the surface. Then the developer is applied. After a short while, the part is examined under an ultraviolet lamp or "black light." The fluorescent liquid glows a light green color to show the location of all surface flaws.

Magnetic particle inspection

Magnetic particle inspection is used to locate flaws that are at the surface or very near the surface of the weld. Flaws that are deeper within the weld cannot be found using this test. Also, magnetic particle inspection can be used only on materials that are magnetic. This eliminates nonferrous metals and many stainless steels.

To examine a part using magnetic particle inspection, the surface must first be cleaned. Then, fine magnetic particles are applied to the surface. The particles can be dry or can be mixed with a liquid. Colored and fluorescent particles are available.

After the particles are applied to the surface, a strong magnetic field is applied. The field can be created by using a strong magnet or by passing an electric current through the part.

When the magnetic field is applied to the part, any flaw will act like a small magnet. The fine magnetic

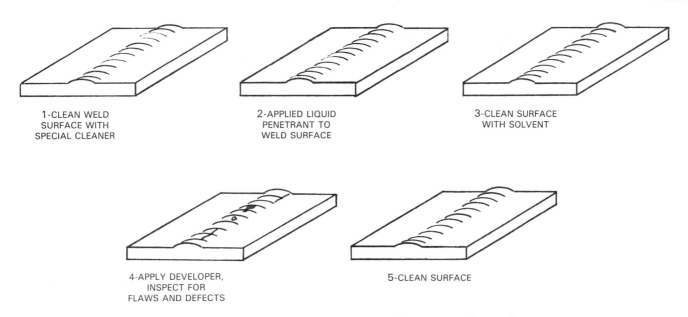

Fig. 32-3. The steps required to make a dye penetrant inspection.

particles will be drawn toward the flaw like a paper clip is drawn toward a strong maganet. These particles outline the flaw. When the magnetic field is removed, the magnetic particles remain in place. The part is then examined to locate any flaws or defects. See Fig. 32-4.

Ultrasonic inspection

Ultrasonic inspection can be used on all metals and can locate internal flaws. Sound waves are sent into the part being inspected. The waves travel to the far side of the metal and bounce back.

The sound waves are displayed on an *oscilloscope* (an electrical instrument that has a calibrated screen).

Each time the sound hits something, a pulse shows up on the oscilloscope screen. When there are no flaws in the metal, only two pulses will be seen on the screen. The first pulse is from the top surface of the metal, the second is from the bottom surface. See Fig. 32-5.

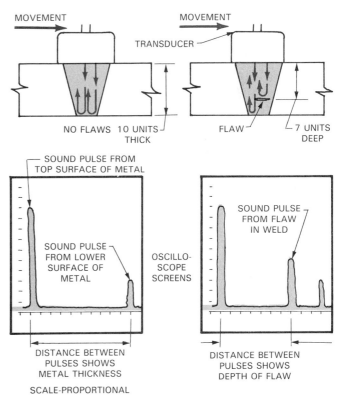

Fig. 32-5. Ultrasonic testing. The transducer sends ultrasonic waves through the test piece. The top surface shows up on the oscilloscope screen as a large pulse. Additional pulses on the screen are from the bottom surface and any flaws in the weld. The distance between the top surface, and a flaw or the bottom surface, can be measured on the screen.

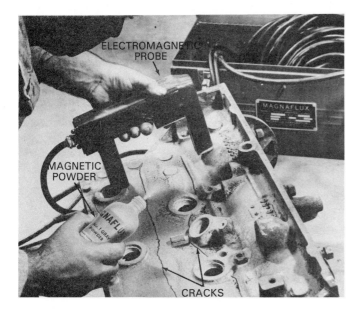

Fig. 32-4. Magnetic particle inspection being used to check an engine block for cracks. Note the dark lines, which indicate cracks. (Magnaflux Corp.)

319

If there is a flaw, another pulse will show up on the screen. The location of the pulse indicates where the flaw is located. The height of the pulse indicates the size of the flaw.

X-ray inspection

X rays are used to inspect welds in the same way that a doctor uses X rays to inspect a patient. Film is placed on one side of the part to be inspected. X rays pass through the part and expose the film. After the film is developed, any flaw or defect in the weld can be observed. The film also serves as a permanent record of the quality of the weld. Fig. 32-6 shows an X-ray view or *radiograph* of a weld.

Safe working practices are very important when using X rays or working with radioactive materials. Observe all safety procedures so neither you nor anyone working in the area is exposed to the X rays.

Other nondestructive tests

Eddy current inspection can detect porosity, cracks, slag inclusions, and lack of fusion at or near the surface of a weld. A coil carrying high-frequency alternating current is brought close to the part being inspected. This coil creates a current in the part. The current in the part is called an eddy current. Eddy currents flow in a closed circular path.

The coil is moved over the part being inspected. When a flaw is present in the part, the flow of the eddy current is interrupted. This change in current flow is measured on an oscilloscope.

An *air pressure leak test* is used to inspect welds on pipelines or storage tanks that must safely contain gases or liquids under pressure. A common testing method is to coat all welds with a soap and water solution, then fill the pipe or vessel with a gas under pressure. Carbon dioxide is often used. If a weld defect allows gas to leak out, soap bubbles will form to show its location.

Fig. 32-6. An X-ray radiograph that shows a large undersurface crack in a weld.

DESTRUCTIVE TESTS

Destructive testing is used to determine the physical properties of a weld. To do so, destructive testing requires that a completed weld be damaged in some way. Most such tests totally destroy the weld. The most common destructive tests include:
- Tensile test.
- Bend test.
- Fillet test.
- Hardness test.
- Impact test.
- Peel test (for spot welds).
- Tensile shear test (for spot welds).
- Pressure test.

Tensile test

The *tensile test* is a destructive test. This test applies a tensile (stretching or pulling) load to a prepared sample until it breaks.

After a butt weld on a plate or pipe is completed, a section is cut from the weld. This section is then prepared for a tensile test. See Fig. 32-7. Note the part of the specimen where the section has a reduced area. Fig. 32-8 shows a reduced section tensile specimen after it has been pulled to the breaking point.

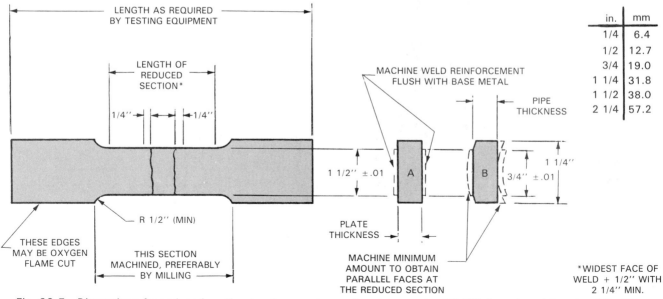

Fig. 32-7. Dimensions for reduced-section tension specimens for plates under 1'' (25.4 mm) and for pipe 6''—8'' (152 mm—203 mm) in diameter.

Fig. 32-8. A reduced-section tension specimen after it has been tested to failure. Note that it has broken outside the weld area, as it should.

A tensile test is used to determine the tensile strength and the ductility of a weld. *Tensile strength* is a measurement of the amount of force required to pull the metal apart. *Ductility* measures how much the metal stretches before breaking.

Before performing a tensile test, you must make a few measurements. Measure the thickness of the metal and the width of the reduced section. Then, mark two lines (or use a center punch to mark two points) that are exactly 2″ (50.8 mm) apart. One point should be 1″ (25.4 mm) on one side of the weld, and the other point should be about 1″ (25.4 mm) on the other side of the weld. See Fig. 32-9. The marks allow you to determine how far the metal stretched before breaking. Fig. 32-10 shows an all-weld-metal sample after it was broken in a tensile test machine.

To perform a tensile test, place the prepared section in the machine that will pull the metal apart. See Fig. 32-11. Clamp each end of the sample into the jaws of the machine. The machine will pull on each end of the sample. During the test, the metal sample will stretch and then break into two pieces. The machine will record the maximum force applied to break the sample. Write down the maximum force applied.

Fig. 32-9. A standard all-weld-metal tensile specimen. Note that the all-weld-metal sample is machined to a cylindrical shape with threads at each end. The threads are used to mount the sample into the tensile test machine. Two small punch marks are made 2″ (50.8 mm) apart before the test begins. A tool like the one shown is used to make these marks accurately.

Fig. 32-10 An all-weld-metal tension test sample mounted in a tensile test machine. Note how the diameter has been reduced in the area of the break by the metal stretching before it broke.

Remove the pieces from the jaws of the machine. Press the broken edges back together. Measure the distance between the marks that were 2″ (50.8 mm) apart before the test. Record your measurement.

Calculating tensile strength and ductility. With the data you have recorded, you can find the tensile strength and ductility of a weld. Use the following formulas:

$$Tensile\ strength = \frac{Maximum\ force\ applied\ by\ machine}{Thickness \times Width\ of\ sample}$$

$$Ductility = \frac{\begin{array}{c}(Distance\ between\ points\ after\\ test\ -\ distance\ before\ test)\end{array}}{Distance\ before\ test} \times 100$$

Tensile strength is measured in pounds per square inch (psi) or in megapascals (MPa). Ductility is measured as a percentage. For example, low carbon steel has a tensile strength of 55,000 psi (379 MPa) and a ductility of about 25 percent.

Example: A welded plate is 1/2″ (12.7 mm) thick. The reduced section is 1 1/2″ (38 mm) wide. A maximum force of 48,000 pounds (213 500 newtons) is required

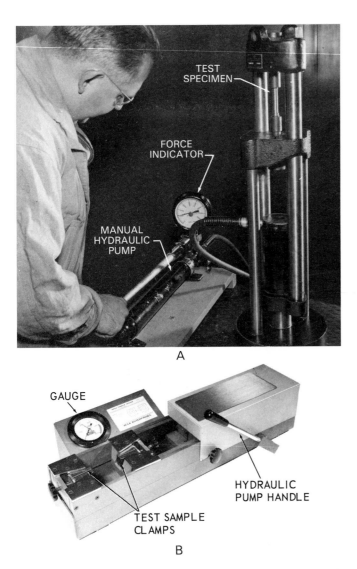

A

B

Fig. 32-11. Typical tensile testing machines used in school shops. A—A portable, hydraulic tensile tester in use. Notice the test specimen mounted in the machine. Other clamps can be installed to test samples cut from plate. B—A portable tensile testing machine. Force is applied by operating a hydraulic jack built into the tester housing. The force applied is read on the gauge. Notice the clamps. (Vega Enterprises)

to break the sample. The distance between the two points after breaking is 2 1/2″ (63.5 mm). Calculations:

$$\textit{Tensile strength} = \frac{48{,}000 \text{ pounds}}{1/2″ \times 1\,1/2″} = 64{,}000 \text{ psi}$$

$$\frac{213\,500 \text{ N}}{12.7 \text{ mm} \times 38 \text{ mm}} = 442 \text{ MPa}$$

$$\textit{Ductility} = \frac{2\,1/2″ - 2″}{2″} \times 100 = 25 \text{ percent}$$

$$\frac{63.5 \text{ mm} - 50.8 \text{ mm}}{50.8 \text{ mm}} \times 100 = 25 \text{ percent}$$

You should also visually inspect the pieces after they are broken. Note where the sample breaks. The sample should break in the base metal. If it breaks in the

weld area, look for signs of porosity, slag inclusions, or other defects.

EXERCISE 32-1 — MAKING A TENSILE TEST

1. Make a butt weld on plate or pipe, as you did for an exercise earlier in this book.
2. Remove two pieces, each 2″ (50 mm) wide. If the plate or pipe is over 1″ (25 mm) thick, the pieces should be 1 1/2″ (38 mm) wide.
3. Prepare each piece as shown in Fig. 32-7.
4. Grind the face and root of the weld flush with the base metal. All grind marks must run lengthwise on the metal samples. Also, be sure to break all sharp edges along the reduced section area.
5. Measure the thickness and width of each piece. Record the information on a separate sheet of paper. Do not write in this book.
   Thickness _____
   Width_____
6. Mark two lines or points exactly 2″ (50 mm) apart. The weld should be centered between the marks.
7. Put the piece to be tested into a tensile testing machine. Apply force until the piece breaks.
8. On your paper, record the maximum force applied.
9. Place the two pieces back together. Measure the distance between the two points. Write down the distance on your paper.
10. Calculate the tensile strength and ductility of the test piece using the formulas given in this chapter. Show your work.
    Tensile Strength _____
    Ductility_____
11. Did the piece break in the weld, next to the weld, or in the base metal area?
12. Are there any signs of welding flaws along the broken surface?

Bend test

Bend tests are used to evaluate the quality and ductility of a completed weld. The *guided bend test* is the most common type. The *free bend test* is also used. In both tests, a sample is bent into a "U" shape. In a guided bend test, the size or radius of the bend is controlled. Different sizes are used for different thicknesses and types of metal.

There are three types of bend tests: the face bend, the root bend, and the side bend. In a *face bend,* the face of the weld is on the outside of the bend. A *root bend* has the root side of a weld on the outside of the bend. See Figs. 32-12, 13, and 14. A *side bend* is used on thick material. In each test, the center of the weld must be at the center of the bend.

Bend samples can be made with the weld going across the sample or along the length of the sample. The sample is called a *transverse* sample when the weld goes

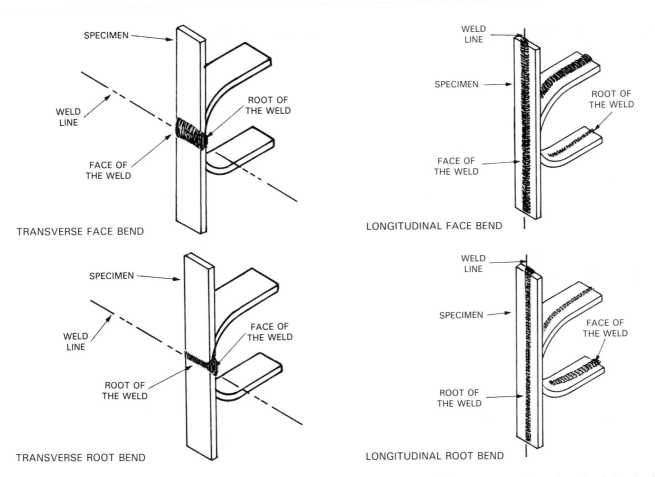

TRANSVERSE FACE BEND

LONGITUDINAL FACE BEND

TRANSVERSE ROOT BEND

LONGITUDINAL ROOT BEND

Fig. 32-12. Transverse and longitudinal bend tests. A transverse test sample is cut across the weld axis. A longitudinal test sample is cut along the weld axis.

A

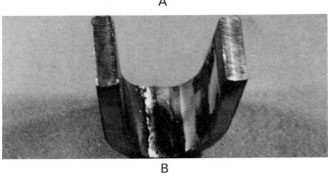

B

Fig. 32-13. A root bend sample after bending. A—The root side. B—The face side. Note that no flaws large enough to be considered defects are shown in the weld on either side.

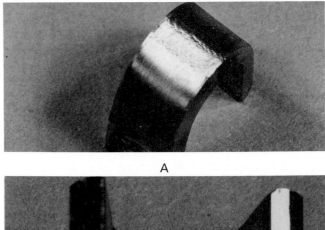

A

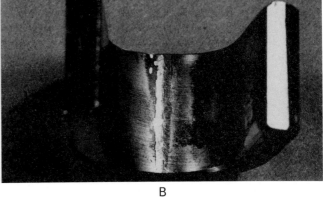

B

Fig. 32-14. A face bend sample after bending. A—The face side. B—The root side.

323

across it. A *longitudinal* sample has the weld along the length of the sample. Refer to Fig. 32-12.

Prior to bending, a section of a completed weld must be prepared. Fig. 32-15 shows the dimensions of a sample to be bent. After preparation, the sample is placed in a jig designed for bending plates. See Fig. 32-16. Fig. 32-17 shows a typical bending jig.

Force is applied to cause the sample to bend. The sample is bent until it forms a U-shape or until it breaks. If the sample breaks, it has failed the test. If the sample does not break, it is visually examined after bending. There should be no cracks or other defects 1/16" (1.6 mm) or larger in any direction on the outside bent surface. Some codes allow flaws up to 1/8"

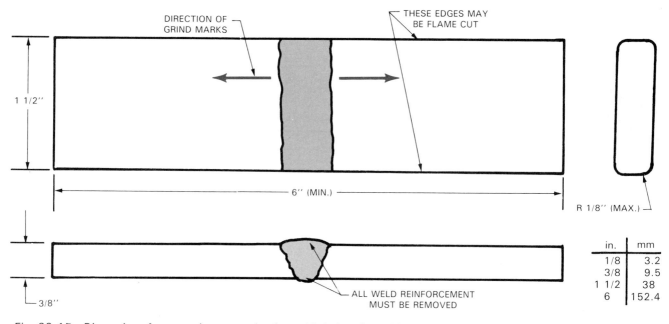

in.	mm
1/8	3.2
3/8	9.5
1 1/2	38
6	152.4

Fig. 32-15. Dimensions for preparing a sample of a welded plate for guided bend test. Remove all weld reinforcement and radius the corners. Notice the direction of the grinding marks.

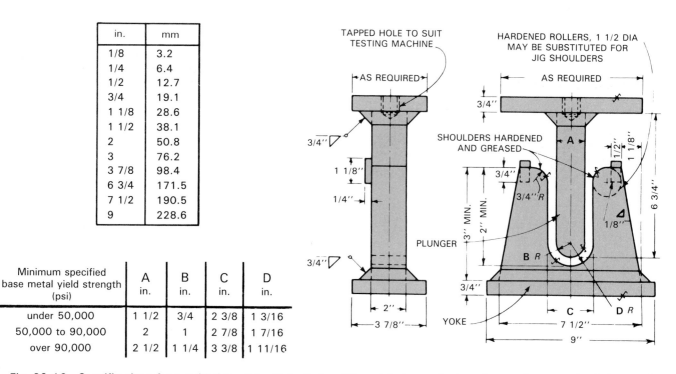

in.	mm
1/8	3.2
1/4	6.4
1/2	12.7
3/4	19.1
1 1/8	28.6
1 1/2	38.1
2	50.8
3	76.2
3 7/8	98.4
6 3/4	171.5
7 1/2	190.5
9	228.6

Minimum specified base metal yield strength (psi)	A in.	B in.	C in.	D in.
under 50,000	1 1/2	3/4	2 3/8	1 3/16
50,000 to 90,000	2	1	2 7/8	1 7/16
over 90,000	2 1/2	1 1/4	3 3/8	1 11/16

Fig. 32-16. Specifications for a guided bend jig. Note that a different jig is required as the tensile strength of the metal changes. The changing sizes for dimensions A, B, C, and D are shown in the table at lower left. The millimeter specifications are not part of the ASME-Section IX code. They are shown for reference only.

Fig. 32-17. This guided bend jig was constructed according to the dimensions given in Fig. 32-16.

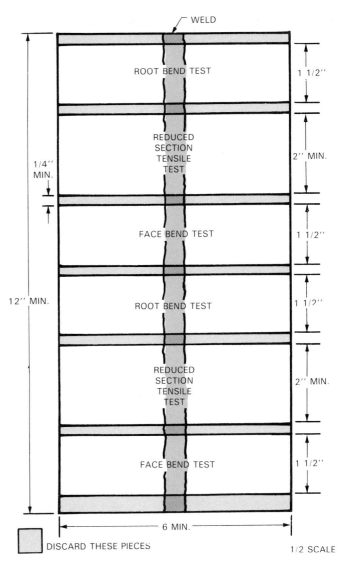

Fig. 32-18. One way to obtain bend test and tensile test samples from a test weld. You may weld a 7'' (178 mm) test piece and only perform one root bend, one face bend, and one tensile test. Different codes and specifications require different ways to cut test samples from a test weld.

(3.2 mm). Cracks that start at the corner of the sample are not counted, unless there is evidence that the flaw (porosity or lack of fusion, for example) is a result of poor welding skills. If there are no cracks larger than are allowable under code, the sample passes the bend test.

Bend test samples and tensile test samples are often taken from the same weld. This allows the strength, ductility, and quality of a welded joint to be determined. Fig. 32-18 shows how a plate may be cut to obtain samples for both bend and tensile testing.

Fillet test

The *fillet test* is used to determine the quality of a fillet weld. Two pieces of metal are tack welded to form a T-joint. Then a fillet weld is made, as shown in Fig. 32-19. The weld is stopped and restarted at the midway point of the joint. Welding for this test can be done in any position, using any process.

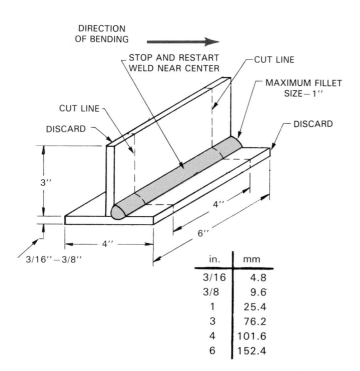

Fig. 32-19. The dimensions for a fillet weld break test specimen. The welder must stop and restart in the center of this weld. (AWS)

in.	mm
3/16	4.8
3/8	9.6
1	25.4
3	76.2
4	101.6
6	152.4

Force is applied to the welded joint until it breaks in two, or until the vertical piece is bent flat against the horizontal piece. Fig. 32-20 shows a safe way to apply the necessary force to the sample.

The weld should break along its center line. This indicates that there is good penetration into the base metal, and that the legs of the weld are equal. A weld that breaks along one edge may indicate a lack of fusion. Look at the fractured surface for signs of porosity, slag inclusions, lack of penetration at the root, or lack of fusion. Also look at the size of each leg of the weld. They should be equal. See Fig. 32-21.

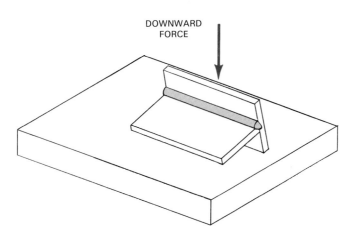

Fig. 32-20. The fillet weld specimen must be bent with a downward force, as shown. The force may be applied with a hammer, hydraulic press, or other suitable means. Safety note: The test should be made in a shielded area to protect others from injury.

To pass a fillet weld test, your weld should break through the weld metal, or the vertical piece should bend down onto the horizontal piece. The broken surface or bent surface should show no evidence of defects.

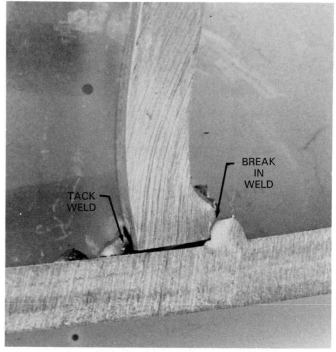

Fig. 32-21. A fillet weld that has broken. The fillet weld was well-made and broke at about the center of the weld. Note the absence of porosity or inclusions.

EXERCISE 32-3 — MAKING A FILLET TEST

1. Make a fillet weld as you did for one of the exercises earlier in this book. Stop and restart the weld in the center of the joint.
2. Apply a downward force to one piece. Continue applying force until the joint fails or until one piece is bent over onto the other piece.
3. Examine the broken or bent weld. There should be no signs of any defects.

Hardness test

This test is used to determine the hardness of a metal. Test areas include the weld metal, the heat-affected metal close to a weld, and the base metal. *Hardness tests* do not completely destroy a welded joint, like bend testing and tensile testing. They do make impressions or dents in the metal, however. Some parts can be used after hardness testing; other parts cannot.

To determine the hardness of a metal, an *indenter* is pressed into the metal by a known force. The indenter is usually a steel ball or a pointed diamond, depending upon the type of hardness test. The ball or diamond is pressed into the metal by a specified force. If the metal is soft, the indenter will be pressed into the metal more than if the metal is hard. Fig. 32-22 shows a part being hardness-tested.

Machines used for hardness testing usually have a gauge to show the hardness of the metal being tested. The most common type of tester is a Rockwell hardness testing machine. Other types include the Brinell, Vickers, Scleroscope, Knoop, and microhardness testers.

Other destructive tests

An *impact test* requires that a test sample be machined to specific dimensions. The sample is held in a clamp and broken by an impact from a large known force. The amount of energy required to break the sample is displayed on a dial.

A *peel test* is used to check the quality of spot welds. One edge of each piece is bent 90°, then force is applied to tear the metal apart. See Fig. 32-23.

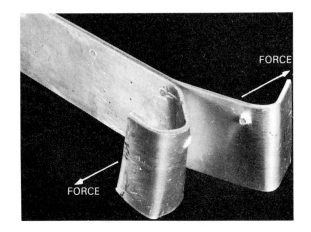

Fig. 32-23. A destructive peel test. This test is used to determine the size, strength, and proper welding machine settings for resistance spot welds.

A *tensile shear test* is also used for testing spot welds. The two pieces to be welded are staggered, so that each piece extends about 1″ (25.4 mm) past the other. The pieces of metal are spot welded together. See Fig. 32-24. Then the pieces are put into a tensile testing machine like the one shown in Fig. 32-11. Force is applied and the spot welds are torn apart. This test measures the force required to tear the welds apart, and also allows the spot welds to be examined.

Pressure tests are used to determine how much pressure a pipe or cylinder can withstand. A cap or valve is used to close each opening in the pipe, piping assembly, or cylinder. The pressure inside the closed system is steadily increased until a leak develops. See Fig. 32-25. Pressure testing may be done under water because the pipe or cylinder can explode from the high pressures applied.

A

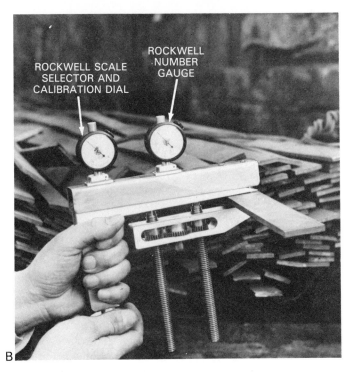

ROCKWELL SCALE SELECTOR AND CALIBRATION DIAL

ROCKWELL NUMBER GAUGE

B

Fig. 32-22. A—A laboratory type Brinell tester being used to test the hardness of a sample. B—A portable Rockwell hardness tester being used to check the hardness of steel stock.

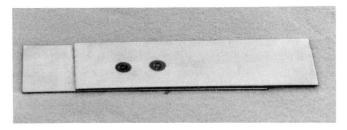

Fig. 32-24. A spot weld tensile strength specimen. The metals are overlapped as shown so they can be gripped in the tensile test machine.

Fig. 32-25. This long section of gas pipeline has a dome fitting welded at each end. Water is pumped into the pipe to fill it. The pipe is then pressurized to several thousand pounds per square inch to test all of the welds along the pipe for leaks or failure. Note the pressure relief valve, which prevents excessive pressure from developing. After the test, the domes are cut off and the pipe is welded to the next section.

REVIEW QUESTIONS

Write your answers on a separate sheet of paper.
1. Explain the difference between a flaw and a defect.
2. Why are welds inspected?
3. Which nondestructive evaluations cannot detect defects inside a weld?
4. List the five steps involved in performing a magnetic particle inspection.
5. What two values are determined from a tensile test?
6. List two things that will cause a bend test sample to fail the test.
7. Where is the root of the weld positioned on a root bend test sample after bending?
8. Where should a fillet weld test sample fail?
9. What type of indenter is pressed into a metal being hardness tested?
10. List the two types of tests used on spot welds.

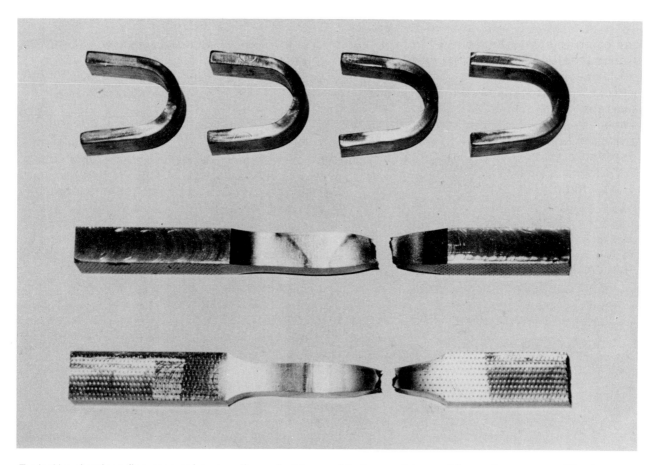

Typical bend and tensile test samples normally required for a welder to complete a welding performance qualification test.
(G.N. Fischer/Welding Journal)

Chapter 33

Welder Certification

After studying this chapter, you will be able to:
☐ Describe the use of codes and specifications to provide needed information on a required weld.
☐ Discuss the difference between a welding procedure specification and a welding performance qualification.
☐ Explain why a welder often must pass a number of welding performance qualifications.
☐ List the steps that must be followed to conform to most codes.

TECHNICAL TERMS

Contract, qualified, welding performance qualification, welding procedure qualification record, welding procedure specification.

CODES AND SPECIFICATIONS

A *contract* is an agreement between two people or organizations. Whenever a welded structure is built, a contract is signed between the organization that is buying the structure and the organization that will build it. Whether the structure is a building, a bridge, a pipeline, an item of military equipment, or something else, a contract is involved. A contract states *what* is to be built, by *when,* for *how much,* and other technical and legal matters.

How all welding will be performed is covered by the contract. Certain government agencies, technical societies and associations have written welding codes and specifications. These codes and specifications cover all the needed information for making a required weld. Thus, the contract need not list all the information for every type of weld. Instead, it can refer to specific codes and specifications that *do* list all required information on how a weld is to be made.

Codes and specifications list what can and cannot be done. They cover information on base metals, welding variables, filler metal compositions, and other details. They also cover many safety related topics. These codes and specifications help to standardize welding.

Some of the more common codes and specifications are written by these organizations:
Government Agencies:
• Federal Aviation Administration (FAA)
• Interstate Commerce Commission (ICC)
• Department of Defense — Military Specifications (MIL)
• Occupational Safety and Health Administration (OSHA)
Associations and Societies:
• American Institute of Steel Construction (AISC)
• American National Standards Institute (ANSI)
• American Petroleum Institute (API)
• American Society of Mechanical Engineers (ASME)
• American Welding Society (AWS)

Two widely used codes that deal with welding are the ASME Boiler and Pressure Vessel Code and the AWS Structural Welding Code. There are codes that cover base metal composition, electrode composition, quality of shielding gases, heat treating, and other aspects of welding. See Fig. 33-1.

WELDING PROCEDURE SPECIFICATION

Before any welding can be done on a structure, the procedure that will be used must first be approved, or *qualified.* The manufacturer specifically lists how the

Fig. 33-1. Typical AWS and ASME code and standards booklets. These booklets tell how welds are to be made and tested. They also deal with welding symbols, terms, and safety practices.

welding will be done on a *welding procedure specification (WPS)*. Many WPS's are required, since each major change that affects or could affect the quality of a weld must have its own WPS.

The manufacturer must weld sample plates using the WPS. These samples are then tested to determine the strength, ductility, hardness, and overall ability of this given set of parameters to make a high quality weld. The codes and specifications list how many and what types of tests must be performed. Many of the destructive tests covered in Chapter 32 are used to evaluate the weld samples. The most common are the tensile and bend tests. Results of these tests are listed on a *welding procedure qualification record (WPQR)*.

A welding procedure qualification record shows the properties of a weld that was made by following the WPS. When the results of the WPQR show that welding meets the code or specification, then the WPS is said to be *qualified.* Once welding procedure specifications are written and qualified, the WPS can be followed and welding can begin on the project. Both the WPS and the WPQR must be kept on file.

WELDING PERFORMANCE QUALIFICATION

After a procedure has been qualified as described in the preceding section, it can be followed to produce good quality weld joints. This assumes that a skilled welder is doing the welding.

Every welder who will make welds on a project must also be qualified. The welder's welding ability must be tested. When welders pass the test, they are said to be qualified. A qualified welder is able to make a high quality weld following a given set of welding conditions.

Welders must pass a *welding performance qualification* before making welds in accordance with a code or specification. A welding performance qualification requires the welder to weld test plates following an approved WPS. Testing of the welds is specified by the codes. The employer keeps a record of these tests. See Appendix C. Usually, face and root bends are made, along with tensile tests. On thick plate, side bends are used instead of face and root bends. Fig. 32-18 shows where test samples are taken from a welded test plate. Welds are sometimes X-rayed to find defects beneath the weld surface.

Welders often must pass a number of welding performance qualifications. A different test is required each time one of the following is changed:
- Welding process.
- Base metal composition.
- Thickness of base metal (large change).
- Welding position.
- Electrode or filler metal type.

Thus, welders would have to pass a welding performance qualification test to weld mild steel using an E6013 SMAW electrode in the horizontal welding position. They would have to pass a different test to weld using an E6013 in the overhead welding position. Fig. 33-2 shows a welder welding a test plate.

Fig. 33-2. A welder preparing a test plate for a V-groove butt weld in the horizontal welding position.

If welders do not pass a performance qualification, they are not allowed to weld using that procedure. The company doing the testing can retest welders who do not pass. They may have to receive some training before retaking a test.

In most codes, base metals that have similar properties are grouped into one category. A welder who passes a qualification test on one base metal is qualified to weld on any base metal in that category. For example, a welder who passes a test on one type of stainless steel can also weld on other stainless steels that have similar welding properties.

When a welder passes a qualification test on one thickness of material, that welder is then qualified for a range of thicknesses. Usually, a welder is qualified for a thickness up to twice the thickness of the test plate. Fig. 33-3 lists the thickness range that a welder is qualified for when completing a test plate of a given thickness.

Certain codes require welders to requalify. This means that welders must be retested to make sure they can still make quality welds. If a welder does not use a given process for three months, or if there is reason to question the welder's ability to follow the welding procedure, retesting is required. Some codes require a welder to requalify once a year.

Positions and types of joints

Some codes state that welders who pass a groove weld test are qualified to make groove welds and fillet welds. If welders pass a fillet weld test, they can make only fillet welds, not groove welds. Groove welds are considered more difficult to make.

Flat welding is considered to be the easiest position. Most welder qualifications are done in the horizontal, vertical, or overhead welding positions. Welders who pass a qualification test in one of these positions are qualified for that position and for welding in the flat position. Welders are allowed to weld only in the positions for which they have been qualified. Two separate welding performance qualifications may be required to qualify for uphill and downhill welding. Figs. 33-4 and 33-5 illustrate different positions for plate and pipe welding.

A typical welding performance qualification

Overhead welding is avoided in most plants, whenever possible. Most welding is done in the flat welding position. Some welding must be done in the horizontal and vertical positions.

Assume a company needs a welder qualified to weld 1/16′ – 1′ (1.6 mm – 25.4 mm) stainless steel in the flat, horizontal and vertical welding positions, using SMAW. One set of welding performance qualification tests that may be given to the welder is listed below:

Weld a square butt or bevel groove weld on 1/8″ (3.2 mm) stainless steel in the horizontal and vertical positions, using SMAW electrode E308.

Weld a V-groove weld in 1/2′ (12.7 mm) plate on stainless steel in the horizontal and vertical positions, using SMAW electrode E308.

The welds would be inspected visually for undercut or overlap, and to evaluate the skills of the welder. Two tensile samples, plus one face bend and one root bend sample, would be tested from each weld.

If testing shows that the welds are of high quality, the welder is considered qualified to weld in the flat, horizontal, and vertical welding positions. The welder is qualified only to weld on certain types of stainless steel, using the shielded metal arc welding process. Only certain types of covered electrodes can be used. Welding the 1/8″ (3.2 mm) material qualifies the welder to weld on thicknesses from 1/16″ – 1/4″ (1.6 mm – 6.4 mm). Welding the 1/2″ (12.7 mm) plate qualifies the welder for thicknesses from 3/16″ – 1″ (4.8 mm – 25.4 mm). See Fig. 33-3.

Summary: Welding per codes and specifications

A contract states how the welding on a particular job will be accomplished. The manufacturer is responsible for making sure that all welding meets the requirements of the codes and specifications listed in the contract. Following are the steps that are followed to conform to most codes.

1. A welding procedure specification (WPS) is written.
2. Test plates are welded, following the WPS.
3. Welded plates are tested. The results are listed on a welding procedure qualification record (WPQR).
4. The WPQR must show that the WPS does make good quality welds. Once this is done, the WPS is approved.
5. Welders are tested using a welding performance qualification. The welding performance qualification tests the ability of each welder to weld as required by an approved WPS. Welders who pass the welding performance qualification test are allowed to weld.
6. Welders are retested periodically to make sure they are still able to make high quality welds.

OTHER EMPLOYMENT CONSIDERATIONS

When reviewing applications for a welding position, an employer often considers three things: how much training, how much education, and how much experience a person has.

Welding training is obtained through industrial education classes in high schools, trade schools, community colleges, and universities. These classes usually will teach basic welding techniques. Some classes, especially those offered through a good trade school, can prepare you to take and pass welder performance qualification tests.

Employers often look to see how much education you have completed. A high school diploma is helpful in applying for a welding position.

TENSION TEST AND TRANSVERSE BEND TESTS

Thickness T of Test Coupon Welded, in. [Note (1)]	Range of Thickness T of Base Metal Qualified, in. [Note (2)]		Range of Thickness t of Deposited Weld Metal Qualified, in. [Note (2)]		Type and Number of Tests Required (Tension and Guided Bend Tests) — Note (6)			
	Min.	Max.	Min.	Max.	Tension QW-462.1[4]	Side Bend QW-462.2	Face Bend QW-462.3(a)	Root Bend QW-462.3(a)
Less than 1/16	T	2T	t	2t	2	—	2	2
1/16 to 3/8, incl.	1/16	2T	1/16	2t	2	—	2 (5)	2 (5)
Over 3/8, but less than 3/4	3/16 (8)	2T	3/16 (8)	2t	2	Note(3)	2 (5)	2 (5)
3/4 to less than 1 1/2	3/16 (8)	2T	3/16 (8)	2t when $t < 3/4$	2	4	—	—
3/4 to less than 1 1/2	3/16 (8)	2T	3/16 (8)	2t when $t \geq 3/4$	2	4	—	—
1 1/2 and over	3/16 (8)	8 (7)	3/16 (8)	2t when $t < 3/4$	2	4	—	—
1 1/2 and over	3/16 (8)	8 (7)	3/16 (8)	8 (7) when $t \geq 3/4$	2	4	—	—

NOTES:
(1) When the groove weld is filled using two or more welding processes, the thickness t of the deposited weld metal for each welding process shall be determined and used in the "Range of thickness t" column. The test coupon thickness T is applicable for each welding process.
(2) See QW-403 (.2, .3, .6, .7, .9, .10) and QW-407.4 for further limits on range of thicknesses qualified.
(3) Four side bend tests may be substituted for the required face and root bend tests.
(4) The deposited weld metal of each welding process shall be included in the tension test of QW-462.1(a), (b), (c), or (e); and in the event turned specimens of QW-462.1(d) are used, the deposited weld metal of each welding process shall be included in the reduced section insofar as possible.
(5) Applicable for a combination of welding processes only when the deposited weld metal of each welding process is on the tension side of either the face or root bend.
(6) When toughness testing is a requirement of other Sections, it shall be applied with respect to each welding process.
(7) For the welding processes of QW-403.7 only; otherwise per Note (2) or 2T, whichever is applicable.
(8) When the weld metal thickness deposited by a process is 3/8 in. or less, this minimum shall be 1/16 in. for that process.

TENSION TESTS AND LONGITUDINAL BEND TESTS

Thickness T of Test Coupon Welded, in. [Note (1)]	Range of Thickness T of Base Metal Qualified, in. [Note (2)]		Range of Thickness t of Deposited Weld Metal Qualified, in. [Note (2)]		Type and Number of Tests Required (Tension and Guided Bend Tests) - Note (5)		
	Min.	Max.	Min.	Max.	Tension QW-462.1[3]	Face Bend QW-462.3(b)	Root Bend QW-462.3(b)
Less than 1/16	T	2T	t	2t	2	2	2
1/16 to 3/8, incl.	1/16	2T	1/16	2t	2	2 (4)	2 (4)
Over 3/8	3/16 (6)	2T	3/16 (6)	2t	2	2 (4)	2 (4)

NOTES:
(1) When the groove weld is filled using two or more welding processes, the thickness t of the deposited weld metal for each welding process shall be determined and used in the "Range of thickness t" column. The test coupon thickness T is applicable for each welding process.
(2) See QW-403 (.2, .3, .6, .7, .9, .10) and QW-407.4 for further limits on range of thicknesses qualified.
(3) The deposited weld metal of each welding process shall be included in the tension test of QW-462.1(a), (b), (c), or (e); and in the event turned specimens of QW-462.1(d) are used, the deposited weld metal of each welding process shall be included in the reduced section insofar as possible.
(4) Applicable for a combination of welding processes only when the deposited weld metal of each welding process is on the tension side of either the face or root bend.
(5) When toughness testing is a requirement of other Sections, it shall be applied with respect to each welding process.
(6) When the weld metal thickness deposited by a process is 3/8 in. or less, this minimum shall be 1/16 in. for that process.

Fig. 33-3. Tables showing required tests for ASME procedure qualification. The tables show that a properly welded test coupon will qualify a welder on a range of thicknesses from T to 2T. T is the thickness of the metal welded. The QW numbers refer to parts of the ASME code.

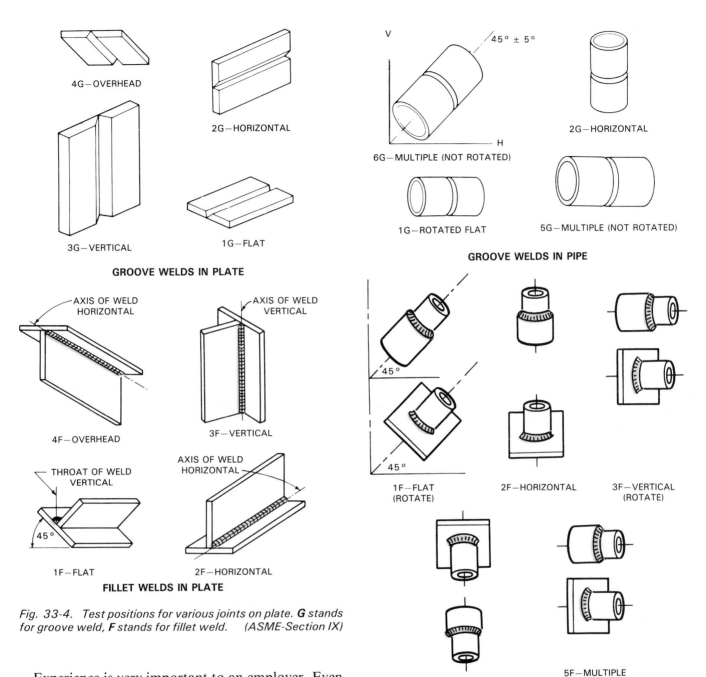

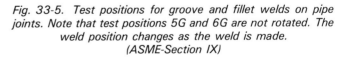

Fig. 33-4. Test positions for various joints on plate. **G** stands for groove weld, **F** stands for fillet weld. (ASME-Section IX)

Experience is very important to an employer. Even though some companies have training programs, they prefer to hire welders that are already experienced. When applying for your first welding position, you can tell the employer how many hours of training you have had.

LABOR UNIONS

On a large construction project, such as a bridge or a pipeline, the company that is building the project may not have any welders who are permanent employees. Instead of interviewing and hiring welders, the company can go to a local union for the welders it needs.

The union will send a number of qualified welders to the job site. The welders are tested by the company. Those that pass the performance tests are hired by the company to work until the project is completed.

Fig. 33-5. Test positions for groove and fillet welds on pipe joints. Note that test positions 5G and 6G are not rotated. The weld position changes as the weld is made. (ASME-Section IX)

Unions sometimes represent the employees within a company. The workers are members of the union and are also permanent employees of a company.

REVIEW QUESTIONS

Write your answers on a separate sheet of paper.
1. What agreement is signed before a welded structure is built?

2. What groups write codes and specifications?
3. List two widely used codes or specifications that deal with welding.
4. What document must a manufacturer prepare that specifically lists how welding will be performed?
5. On what document are the results from the test plates recorded?
6. What test is given to determine a welder's skill in following the welding procedure?
7. Name three destructive tests that are usually performed on a welder's test plates to determine the quality of the weld.
8. If a welder welds a test plate that is 1″ (25.4 cm) thick, what range of thickness is that welder qualified to weld? Refer to Fig. 33-3.
9. When qualified to weld on one type of material, what other type of material can the welder weld on?
10. What three things does an employer often consider when a person applies for a job?

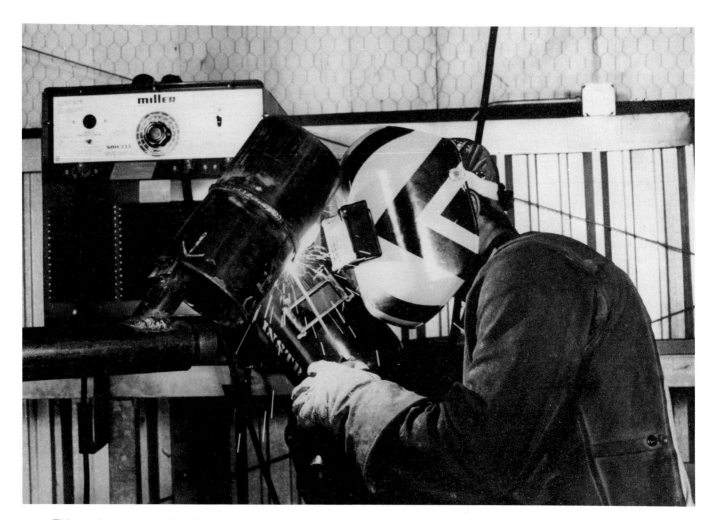

This student is using SMAW to weld a V-groove butt weld on a pipe in the inclined (6G) position. The pipe will be cut into test coupons and the welds tested to qualify the welder to weld pipe.   (Miller Electric Mfg. Co.)

Form A.7.1

SUGGESTED
WELDING PROCEDURE SPECIFICATION (WPS)

Identification _____

Date _____ Revision _____

Company name _____

Supporting PQR no.(s) _____ Type - Manual ( ) Semi-Automatic ( )

Welding process(es) _____ Machine ( ) Automatic ( )

Backing: Yes ( ) No ( )

Backing material (type) _____

Material number _____ Group _____ To material number _____ Group _____

Material spec. type and grade _____ To material spec. type and grade _____

Base metal thickness range: Groove _____ Fillet _____

Deposited weld metal thickness range _____

Filler metal F no. _____ A no. _____

Spec. no. (AWS) _____ Flux tradename _____

Electrode-flux (Class) _____ Type _____

Consumable insert: Yes ( ) No ( ) Classifications _____

Shape _____

Position(s) of joint _____ Size _____

Welding progression: Up ( ) Down ( ) Ferrite number (when reqd.) _____

PREHEAT: **GAS:**

Preheat temp., min _____ Shielding gas(es) _____

Interpass temp., max _____ Percent composition _____

(continuous or special heating, where Flow rate _____

applicable, should be recorded) Root shielding gas _____

POSTWELD HEAT TREATMENT: Trailing gas composition _____

Trailing gas flow rate _____

Temperature range _____

Time range _____

Tungsten electrode, type and size _____

Mode of metal transfer for GMAW: Short-circuiting ( ) Globular ( ) Spray ( )

Electrode wire feed speed range: _____

Stringer bead ( ) Weave bead ( ) Peening: Yes ( ) No ( )

Oscillation _____

Standoff distance

Multiple ( ) or single electrode ( )

Other _____

| Filler metal | | | | | Current | | | |
Weld layer(s)	Process	Class	Dia.	Type & polarity	Amp range	Volt range	Travel speed range	
								e.g., Remarks, comments, hot wire addition, technique, torch angle, etc.

Approved for Production by _____

Employer

Note: Those items that are not applicable should be marked N.A.

(American Welding Society)

Appendix B

Form A.7.2 **SUGGESTED** Page 1 of 2
WELDING PROCEDURE QUALIFICATION RECORD (WPQR)

WPS no. used for test _____ Welding process(es) _____

Company _____ Equipment type and model (sw) _____

JOINT DESIGN USED (2.6.1)

WELD INCREMENT SEQUENCE

Single ( ) Double weld ( )

Backing material _____

Root opening _____ Root face dimension _____

Groove angle _____ Radius (J-U) _____

Back gouging: Yes ( ) No ( ) Method _____

POSTWELD HEAT TREAMTENT (2.6.6):

Temp. _____

Time _____

Other _____

GAS (2.6.7)

Gas type(s) _____

Gas mixture percentage _____

Flow rate _____

Root shielding gas _____ Flow rate _____

EBW vacuum ( ) Absolute pressure ( )

BASE METALS (2.6.2)

Material spec. _____ To _____

Type or grade _____ To _____

Material no. _____ To material no. _____

Group no. _____ To group no. _____

Thickness _____

Diameter (pipe) _____

Surfacing: Material _____ Thickness _____

Chemical composition _____

Other _____

ELECTRICAL CHARACTERISTICS (2.6.8)

Electrode extension _____

Standoff distance _____

FILLER METALS (2.6.3)

Weld metal analysis A no. _____

Filler metal F no. _____

AWS specification _____

AWS classification _____

Flux class _____ Flux brand _____

Consumable insert: Spec. _____ Class. _____

Supplemental filler metal spec. _____ Class. _____

Non-classified filler metals _____

Consumable guide (ESW) Yes ( ) No ( )

Supplemental deoxidant (EBW) _____

Transfer mode (GMAW) _____

Electrode diameter tungsten _____

Type tungsten electrode _____

Current: AC ( ) DCEP ( ) DCEN ( ) Pulsed ( )

Heat input _____

EBW: beam focus current _____ Pulse freq. _____

Filament type _____ Shape ___ Size _____

Other _____

TECHNIQUE (2.6.9)

Oscillation frequency _____Weave width _____

Dwell time _____

String or weave bead _____ Weave width _____

Multi-pass or single pass (per side) _____

Number of electrodes _____

Peening _____

Electrode spacing _____

POSITION (2.6.4)

Position of groove _____ Fillet _____

Vertical progression: Up ( ) Down ( )

PREHEAT (2.6.5)

Preheat temp., actual min _____

Interpass temp., actual max _____

Arc timing (SW) _____ Lift ( )

PAW: Conventional ( ) Key hole ( )

Interpass cleaning: _____

Pass no.	Filler metal size	Amps	Volts	Travel speed (ipm)	Filler metal wire (ipm)	Slope induction	Special notes (process, etc.)

Note: Those items that are not applicable should be marked N.A.

Form A.7.2   **Page 2 of 2**

TENSILE TEST SPECIMENS: SUGGESTED WELDING PROCEDURE QUALIFICATION RECORD   WPQR No.

Type: _____ Tensile specimen size: _____

Groove ( )   Reinforcing bar ( )   Stud welds ( )

Tensile test results: (Minimum required UTS _____ psi)

Specimen no.	Width, in.	Thickness, in.	Area, in.^{2}	Max load lbs	UTS, psi	Type failure and location

GUIDED BEND TEST SPECIMENS - SPECIMEN SIZE: _____

Type	Result	Type	Result

MACRO-EXAMINATION RESULTS: Reinforcing bar ( )   Stud ( )

1. _____   4. _____
2. _____   5. _____
3. _____

SHEAR TEST RESULTS - FILLETS:
1. _____   3. _____
2. _____   4. _____

IMPACT TEST SPECIMENS

Type: _____ Size: _____

Test temperature: _____

Specimen location: WM = weld metal; BM = base metal; HAZ = heat-affected zone

Test results:

Welding position	Specimen location	Energy absorbed (ft.-lbs.)	Ductile fracture area (percent)	Lateral expansion (mils)

IF APPLICABLE   **RESULTS**

Hardness tests: ( ) Values _____   Acceptable ( )   Unacceptable ( )

Visual (special weldments 2.4.2) ( )   Acceptable ( )   Unacceptable ( )

Torque ( ) psi   Acceptable ( )   Unacceptable ( )

Proof test ( ) Method _____   Acceptable ( )   Unacceptable ( )

Chemical analysis ( )   Acceptable ( )   Unacceptable ( )

Non-destructive exam ( ) Process _____   Acceptable ( )   Unacceptable ( )

Other _____   Acceptable ( )   Unacceptable ( )

Mechanical Testing by (Company) _____ Lab No. _____

We certify that the statements in this Record are correct and that the test welds were prepared, welded, and tested in accordance with the requirements of the American Welding Society Standard for Welding Procedure and Performance Qualification (AWS B2.1).

Qualifier: _____

Date: _____

(American Welding Society)

Appendix C

Form A.7.3
SUGGESTED
WELDING PERFORMANCE QUALIFICATION TEST RECORD

Name _____ Identification _____ Welder ( )   Operator ( )

Social security number: _____ Qualified to WPS no. _____

Process(es) _____ Manual ( )   Semi-Automatic ( )   Automatic ( )   Machine ( )

Test base metal specification _____ To _____

Material number _____ To _____

Fuel gas (OFW) _____

AWS filler metal classification _____ F no. _____

Backing: Yes ( )   No ( )   Double ( ) or Single side ( )
Current: AC ( )   DC ( )   Short-circuiting arc (GMAW) Yes ( ) No ( )
Consumable insert: Yes ( )   No ( )
Root shielding: Yes ( )   No ( )

TEST WELDMENT   **POSITION TESTED**   **WELDMENT THICKNESS (T)**

GROOVE:
Pipe   1G ( ) 2G ( ) 5G ( ) 6G ( ) 6GR ( ) Diameter(s) _____ (T) _____
Plate   1G ( ) 2G ( ) 3G ( ) 4G ( )   (T) _____
Rebar   1G ( ) 2G ( ) 3G ( ) 4G ( )   Bar size _____ Butt   ( )
                                                 Spliced butt   ( )

FILLET:
Pipe ( ) 1F ( ) 2F ( ) 3F ( ) 4F ( ) 5F ( ) Diameter _____ (T) _____
Plate ( ) 1F ( ) 2F ( ) 3F ( ) 4F ( )   (T) _____
Other (describe) _____

Test results: Remarks _____

Visual test results   N/A ( )   Pass ( )   Fail ( )
Bend test results   N/A ( )   Pass ( )   Fail ( )
Macro test results   N/A ( )   Pass ( )   Fail ( )
Tension test   N/A ( )   Pass ( )   Fail ( )
Radiographic test results N/A ( )   Pass ( )   Fail ( )
Penetrant test   N/A ( )   Pass ( )   Fail ( )

QUALIFIED FOR:
PROCESSES
GROOVE: **THICKNESS**
Pipe   1G ( ) 2G ( ) 5G ( ) 6G ( ) 6GR ( ) (T) Min ____ Max ____ Dia ____
Plate 1G ( ) 2G ( ) 3G ( ) 4G ( )   (T) Min ____ Max ____
Rebar 1G ( ) 2G ( ) 3G ( ) 4G ( )   Bar size   Min ____ Max ____

FILLET:
Pipe   1F ( ) 2F ( ) 4F ( ) 5F ( )   (T) Min ____ Max ____
Plate 1F ( ) 2F ( ) 3F ( ) 4F ( )   (T) Min ____ Max ____
Rebar 1F ( ) 2F ( ) 3F ( ) 4F ( )   Bar size   Min ____ Max ____

Weld cladding ( )   Position(s) _____ T Min ____ Max ____ Clad Min ____

Consumable insert ( )   Backing type ( )
Vertical Up ( )   Down ( )
Single side ( )   Double side ( )   No backing ( )
Short-circuiting arc ( )   Spray arc ( )   Pulsed arc ( )
Reinforcing bar - butt ( ) or Spliced butt ( )

The above named person is qualified for the welding process(es) used in this test within the limits of essential variables including materials and filler metal variables of the AWS Standard for Welding Procedure and Performance Qualification (AWS B2.1).

Date tested _____   Signed by _____
                                          Qualifier

(American Welding Society)

METRIC—INCH EQUIVALENTS

INCHES FRACTIONS	INCHES DECIMALS	MILLIMETERS	INCHES FRACTIONS	INCHES DECIMALS	MILLIMETERS
	.00394	.1	15/32	.46875	11.9063
	.00787	.2		.47244	12.00
	.01181	.3	31/64	.484375	12.3031
1/64	.015625	.3969	1/2	.5000	12.70
	.01575	.4		.51181	13.00
	.01969	.5	33/64	.515625	13.0969
	.02362	.6	17/32	.53125	13.4938
	.02756	.7	35/64	.546875	13.8907
1/32	.03125	.7938		.55118	14.00
	.0315	.8	9/16	.5625	14.2875
	.03543	.9	37/64	.578125	14.6844
	.03937	1.00		.59055	15.00
3/64	.046875	1.1906	19/32	.59375	15.0813
1/16	.0625	1.5875	39/64	.609375	15.4782
5/64	.078125	1.9844	5/8	.625	15.875
	.07874	2.00		.62992	16.00
3/32	.09375	2.3813	41/64	.640625	16.2719
7/64	.109375	2.7781	21/32	.65625	16.6688
	.11811	3.00		.66929	17.00
1/8	.125	3.175	43/64	.671875	17.0657
9/64	.140625	3.5719	11/16	.6875	17.4625
5/32	.15625	3.9688	45/64	.703125	17.8594
	.15748	4.00		.70866	18.00
11/64	.171875	4.3656	23/32	.71875	18.2563
3/16	.1875	4.7625	47/64	.734375	18.6532
	.19685	5.00		.74803	19.00
13/64	.203125	5.1594	3/4	.7500	19.05
7/32	.21875	5.5563	49/64	.765625	19.4469
15/64	.234375	5.9531	25/32	.78125	19.8438
	.23622	6.00		.7874	20.00
1/4	.2500	6.35	51/64	.796875	20.2407
17/64	.265625	6.7469	13/16	.8125	20.6375
	.27559	7.00		.82677	21.00
9/32	.28125	7.1438	53/64	.828125	21.0344
19/64	.296875	7.5406	27/32	.84375	21.4313
5/16	.3125	7.9375	55/64	.859375	21.8282
	.31496	8.00		.86614	22.00
21/64	.328125	8.3344	7/8	.875	22.225
11/32	.34375	8.7313	57/64	.890625	22.6219
	.35433	9.00		.90551	23.00
23/64	.359375	9.1281	29/32	.90625	23.0188
3/8	.375	9.525	59/64	.921875	23.4157
25/64	.390625	9.9219	15/16	.9375	23.8125
	.3937	10.00		.94488	24.00
13/32	.40625	10.3188	61/64	.953125	24.2094
27/64	.421875	10.7156	31/32	.96875	24.6063
	.43307	11.00		.98425	25.00
7/16	.4375	11.1125	63/64	.984375	25.0032
29/64	.453125	11.5094	1	1.0000	25.4001

ADVANTAGES/DISADVANTAGES OF VARIOUS WELDING PROCESSES USING CONSUMABLE ELECTRODES

PROCESS	ADVANTAGES/DISADVANTAGES
SMAW	**Advantages** • Low initial cost. • Flexibility. • Usable in all positions. • Portability. • Many types of filler metal with special characteristics available. **Disadvantages** • Requires slag removal.
GMAW	**Advantages** • Higher deposition rates than SMAW. • Relative flexibility. • Adaptable to mechanized, robotic, or automatic welding method. • Doesn't require slag removal. **Disadvantages** • Needs special power source to be usable in all positions. • Needs external gas supply and wire feeder.
FCAW	**Advantages** • Higher deposition rates than SMAW and GMAW. • Adaptable to mechanized, robotic, or automatic welding methods. • Relative flexibility. • Usable in all positions without special power source. **Disadvantages** • Needs external gas supply for most electrodes. • Requires wire feeder. • Requires slag removal.
SAW	**Advantages** • Very high deposition rate when mechanized. • High-quality, low-cost process when mechanized. **Disadvantages** • Can be used only for flat and horizontal position welding. • Needs large capital investment. • Requires slag and flux removal.

PERCENT OF WELDING FILLER METAL THAT ACTUALLY BECOMES PART OF A COMPLETED WELD JOINT

FILLER METAL FORM/PROCESS	PERCENT DEPOSITED IN WELD
Bare solid wire	
Electroslag	95-99
Gas metal arc	90-95
Gas tungsten arc	99
Submerged arc	95-99
Covered electrodes (SMAW)	
14'' length	55-65
18'' length	60-70
28'' length	65-75
Flux-cored electrodes	
Flux-cored arc	80-85

VARIABLES THAT AFFECT THE COST OF WELDING

Type of:
electrode or filler wire
joint
shielding
weld

Size of:
electrode or filler wire
weld

Other variables:
arc voltage
electrode deposition efficiency
flux consumption ratio
power source efficiency
shielding gas flow rate
total time to complete welding
weld current

Expense of:
edge preparation
electric power
filler metal
finishing
flux
inspection
labor
overhead
set-up
shielding gas

Glossary of Technical Terms

A

Abrasion: worn condition produced by rubbing.

Abrasion resistance: quality of a material that allows it to resist abrasive wear.

ABS: acrylonitrile butadiene styrene, a plastic widely used for automotive body panels and other components.

Acetone: a colorless and extremely flammable liquid used to fill the air pockets of the porous material inside an acetylene storage cylinder. The acetylene is absorbed into the acetone for safe storage.

Acetylene (C_2H_2): a colorless fuel gas composed of two parts carbon and two parts hydrogen. When burned in the presence of oxygen, acetylene produces one of the highest flame temperatures obtainable.

Acetylene generator: a device used to produce acetylene fuel gas in a controlled environment. Calcium carbide is placed in a hopper and fed into water at a controlled rate to chemically produce acetylene.

Active gases: gases that will combine with the weld metal if given the chance. Carbon dioxide and oxygen are active gases.

Actuator: a device used to move part of a machine. Robots use electric and hydraulic actuators.

Air pressure leak test: inspection method used for pipelines and storage tanks. Welds are coated with a soap solution, then the vessel is filled with gas under pressure. Bubbles show any leaks due to defective welds.

Air-fuel gas torch: a torch burning a combination of air and a fuel gas, such as propane, to provide heat for soldering.

Alloy: a pure metal that has had additional metal or metallic elements added to it while molten. An alloy has mechanical properties that are different (usually improved) from those of the pure metal.

Alternating current (AC): type of electricity in which the direction of electron flow reverses at regular intervals.

Ammeter: instrument used to measure electrical current flow in amperes.

Ampere: unit of measurement of electrical current. One ampere of current will flow through a conductor that has a resistance of one ohm at a potential (electrical pressure) of one volt.

Annealing: softening metals by heat treatment.

ANSI: American National Standards Institute.

Applicator: a device used to add flux to a joint being soldered.

Arc: flow of electricity across a gaseous space (air gap).

Arc blow: wandering of a welding arc from its normal path because of magnetic forces.

Arc crater: nonstandard term for weld pool.

Arc length: distance between the electrode and the base metal. In SMAW, this distance should be approximately equal to the electrode diameter.

Arc spraying (ASP): surfacing process using electric arc between two electrodes of surfacing material and a pressurized gas to propel the vaporized material onto the base metal surface.

Arc stud welding (SW): arc welding process in which studs, nails, or other fasteners are used as the electrode while they are welded to a surface. Once the base metal melts, the special stud arc gun presses the fastener into the surface.

Arc welding: a group of welding processes used to melt and weld metal using the heat of an electric arc, with or without filler metal.

Argon (Ar): an inert gas found in the atmosphere and used as a shielding gas in some welding operations.

Arrow: on a welding symbol, a line with an arrow point that touches the line to be welded.

Arrow side: area under the welding symbol's reference line; the side of the weld touched by the arrow.

Automatic welding: welding with equipment that requires no (or only occasional) observation of the

341

welding, and no manual adjustment of the welding controls.

AWS: American Welding Society.

AWS electrode specifications: strict manufacturing specifications developed by the American Welding Society for electrodes and fluxes. These specifications are published and revised every five years.

Axis of weld: imaginary line through the center of the weld metal, from the beginning to the end of the weld.

B

Backfire: short "pop" of the torch flame, followed by extinguishing of the flame or continued burning of the gases.

Background current: a relatively low amperage used to maintain the arc during pulsed spray transfer GMAW.

Backhand welding: moving the weld in the direction opposite that to which the gas flame or electrode is pointing.

Backing: material placed beyond the root opening to control penetration and prevent a hot shortness hole. Backing may be machined to control the shape of the penetration bead.

Backing ring: metal ring placed inside a pipe before butt welding to ensure complete weld penetration and a smooth inside surface.

Backing weld: weld placed on the root side of a weld to aid in the control of penetration.

Backward arc blow: arc blow that occurs at the end of the joint as the magnetic field tries to stay in the metal. This causes the filler metal to blow toward the center of the joint.

Base metal: metal to be welded, cut, or brazed.

Basic weld symbol: symbol used to describe the shape of the welding joint. It is used as part of the welding symbol.

Bead: the shape of the finished joint when fusion welding is done. Usually a filler metal is added to form a bead while welding.

Bell-mouthed kerf: a kerf that is widened at the bottom as a result of using too much oxygen while cutting.

Bend test: a test described by code in which a sample weld plate is bent under specific conditions until it fails.

Bevel: angling the metal edge where welding is to take place.

Bevel angle: the angle formed between the prepared edge of one piece and a plane perpendicular to the surface of that piece.

Bird's nest: a tangle of welding wire in a GMAW wire drive mechanism.

Braze welding: making an adhesion groove, fillet, or plug connection above 840°F (450°C). The brazing material does not flow in the joint by capillary action.

Brazement: assembly joined by brazing.

Brazing: making an adhesion connection with a minimum of alloy that melts above 840°F (450°C). The brazing material flows between the parts by capillary action.

Brazing rod: filler metal rod used in brazing and braze welding.

Bridge: the process of filling wide gaps in poorly fitted assemblies.

Brinell hardness test: an accurate measurement of a metal's hardness, made with an instrument that presses a hard steel ball into the surface of the metal under standard conditions.

Brittleness: quality of a material that causes the material to develop cracks when it is deformed (bent) only slightly.

Buildup: deposited surfacing materials.

Buildup process: applying surfacing materials to a part so that its worn surface is returned to its original dimensions.

Bullet-shaped bead: a completed stringer bead of even width that shows evenly spaced, bullet-shaped ripples.

Burning: see **Flame cutting**.

Butt joint: assembly in which two pieces joined are in the same plane, with the edge of one piece touching the edge of the other.

Buttering: process used to form a transition layer when welding dissimilar metals. One or more layers of easily welded material are applied to the surface of a part with poor welding characteristics.

C

Cable: see **Lead**.

Cadmium (Cd): a chemical element sometimes used in soldering alloys. It is a toxic material.

Calcium carbide (CaC$_2$): a chemical compound combined with water in an acetylene generator to produce acetylene fuel gas.

Cap pass: see **Cover pass**.

Capacitor: a component in electrical circuits used to store an electrical charge.

Capillary action: in brazing, the process that occurs when the filler metal is drawn between tightly fitted, mating surfaces.

Carbon dioxide (CO$_2$): a chemical compound used as a shielding gas. CO$_2$ is often used when welding carbon steel with GMAW.

Carbon steel: a steel made by adding small, controlled amounts of carbon to pure iron.

Carburizing flame: an oxyfuel gas flame that occurs when there is too much fuel gas in the mixture and not enough oxygen to burn it.

Carrying a pool: the process of creating a weld pool of molten base metal and carrying it along a line.

CC: see **Constant current**.

Cell: see **Robot cell**.

Chain intermittent welding: intermittent welding on two sides of a joint arranged so that the welds on the two sides are opposite each other.

Check valve: valve designed to allow a gas or fluid to

flow in only one direction, used to prevent the reverse flow of gases through the torch, hoses, and/or regulators.

Chemical composition: a list of the different metals or elements that are combined to produce a certain metal or metal alloy.

Chemical corrosion: eating away of a metal surface as a result of chemical reaction.

Chemical flux cutting: an oxyfuel gas cutting process which uses a chemical flux to help in cutting certain materials.

Chemical properties: properties that determine how a material will withstand the effects of the environment.

Chipping: the process of cleaning slag from a welding bead, by striking it with a chipping hammer.

Chipping goggles: eye protection worn while chipping slag from a welding bead.

Chipping hammer: cleaning tool with a sharp pointed pick at one end of its head, used to remove slag from welding beads.

Circuit: the path of electron flow from the source through various components and back to the source.

Circuit breaker: a device designed to shut off the flow of current in a circuit if the amperage becomes too high.

Cladding: process used to apply surfacing materials that will improve the corrosion or heat resistance of a part.

Clearance: gap or space between adjoining or mating surfaces.

Clockwise: rotation to the right, when facing the object being rotated; in the same direction as movement of a clock's hands.

Closed arc: an arc enclosed by the heavy covering material on a SMAW electrode.

Coarse amperage range: one of several ranges that provide electrical current output on a welding machine. Coarse amperage ranges usually extend over 50A or more.

Coated electrode: see **Covered electrode**.

Coating: a relatively thin layer of material applied to a surface to prevent corrosion, wear, or temperature scaling.

Code: a written system of figures, symbols, rules, regulations, and procedures used to regulate the welding industry.

Cold welding: a process for fusing metal parts by using high pressure. No outside heat is applied.

Collet: a sleeve used in a GMAW gun or GTAW torch to hold the electrode tightly enough to ensure electrical contact with the welding power supply.

Collet body: the part of the torch or gun that accepts the collet.

Combustible: capable of being easily ignited; flammable.

Complete fusion: fusion that has taken place over all the fusion faces and between all adjoining weld beads.

Compressive strength: greatest stress developed in material under compression.

Computer memory: electronic components of a computer that store the programs and other information that has been entered.

Concave bead: a weld bead with a surface that curves inward (toward the root of the weld).

Cone: the visible inner flame shape of a neutral or nearly neutral flame.

Constant current (CC): term used for arc welding machines that produce a nearly constant current even though the arc gap voltage may vary.

Constant potential: an alternate term for constant voltage.

Constant voltage (CV): term used for arc welding machines that produce a nearly constant voltage even when the current changes.

Constricting: reducing in size or diameter, as in a constricting orifice.

Consumable electrode: an electrode that is melted and becomes part of the weld.

Contact tube: device that transfers current to a continually fed electrode.

Contactor switch: switch on a welding machine used for GTAW that allows the welder to choose whether current, shielding gas, and water flow can be controlled remotely with a foot pedal or thumb switch while welding.

Contamination: an undesirable substance in or around a weld.

Continuous weld: completing the weld in one operation, without stopping.

Contour: the shape of the face of the weld.

Contract: a legal agreement between two parties.

Contraction: the shortening of metal, usually upon cooling.

Controller: an electronic device (usually a computer) with a memory that can be programmed to signal automatic equipment when and where to move.

Convex bead: a weld bead with a surface that curves outward (away from the root of the weld).

Corner joint: junction formed by the edges of two pieces of metal meeting at an angle (usually a right angle, or 90 degrees).

Corrosion: chemical and electrochemical reaction of a metal with its surroundings, causing the metal to deteriorate.

Corrosion resistance: the ability of a material to withstand corrosive attack.

Counterclockwise: rotation to the left, when facing the object being rotated; in the direction opposite movement of a clock's hands.

Cover pass: the final pass that forms the face of the weld.

Cover plates: replaceable pieces of clear glass or plastic, used to protect the more costly filter lenses or plates of welding goggles and helmets.

Covered electrode: in arc welding, a metal rod with a covering of materials that aid in the welding process.

Crack: break or separation in rigid material, such as

weld metal, that runs in more or less one direction.

Crater: depression in the face of the weld. Usually found at the termination of the weld.

Creating a continuous weld pool: creating a pool with molten base metal and moving it along a line, maintaining a consistent width.

Crescent-shaped motion: a curved side-to-side motion of the welding torch or electrode holder while, at the same time, moving forward. This creates a wider bead.

C-shaped weld pool: a crescent-shaped weld pool which indicates that the base metal surfaces have melted enough to create a well-fused weld.

Cup: see **Nozzle**.

Current: the flow of electrons through an electrical circuit.

Current switch: switch on a welding machine used for GTAW that allows the welder to choose whether current can be adjusted with a foot pedal or thumb switch while welding.

Cutting machine: motor-driven device with one or more oxyfuel gas cutting torches mounted on it, used to make complex, long, or multiple cuts.

Cutting oxygen orifice: center orifice in the cutting tip of an oxyfuel gas cutting torch.

Cutting tip: part of an oxyfuel gas cutting torch from which gases are released for combustion.

Cutting torch: in oxyfuel gas cutting, the device that controls and directs the fuel gases and oxygen needed for cutting and removing metal.

Cutting torch attachment: device that converts a welding torch body into a cutting torch, eliminating the need for a separate cutting torch.

CV: see **Constant voltage**.

Cycle: set of events that repeat in a specific order; one complete reversal of alternating current.

Cylinder: a thick-walled container holding, under high pressure, a supply of gas used for welding.

Cylinder pressure gauge: gauge on a pressure regulator that shows pressure of the gas in the cylinder.

Cylinder valve: device that can be opened or closed to control the flow of gas from a cylinder.

D

Dash number: an AWS-assigned number that indicates how an FCAW electrode is to be used.

Defect: an imperfection that (because of its size, shape, location, or makeup) reduces the useful service of a part.

Degree: one unit of the temperature scale (°F or °C are most common).

Density: the weight of a particular metal per unit volume.

Deoxidizer: a substance added to molten metal to remove either free or combined oxygen.

Deoxidizing: the process of removing oxygen from molten metal with a deoxidizer; in metal finishing, removing oxide films with chemical or electrochemical processes.

Deposition rate: weight of material applied in a unit of time, usually expressed in lbs/hr or kg/hr.

Destructive test: testing method that involves applying stress to a sample until it fails. Destructive testing is used to determine how large a discontinuity can be before it is considered a flaw.

Dewar flask: a pressurized container with an insulated double wall, used to store liquid oxygen.

Dimensions: the size information shown on a mechanical drawing of a part or assembly.

Direct current (DC): electric current that flows in only one direction.

Direct current electrode negative (DCEN): arc welding method in which direct current flows from the electrode (cathode) to the workpiece (anode).

Direct current electrode positive (DCEP): arc welding method in which direct current flows from the workpiece (cathode) to the electrode (anode).

Direct current reverse polarity (DCRP): see **Direct current electrode positive**.

Direct current straight polarity (DCSP): see **Direct current electrode negative**.

Downhand welding: see **Flat position welding**.

Downhill: a vertical weld made from the top of the joint to the bottom.

Downslope: the downward curve of a volt-ampere graph.

Drafting machine: a device used by drafters to simply and accurately draw parallel, perpendicular, and diagonal lines.

Drag: the offset distance between the actual and theoretical exit points of the cutting oxygen stream, measured on the exit side of the material.

Drag angle: the angle between the electrode and the workpiece when backhand welding.

Drag welding: welding method in which the heavy covering of a covered electrode is dragged across the surface of the workpiece to maintain a constant arc length.

Drive rolls: the rolls in a wire drive unit that are directly driven by the unit's drive motor.

Droopers: a term for constant current welding machines, also called droop curve machines because of the voltage versus amperage curve they produce.

Ductility: the ability of a material to be changed in shape without cracking or breaking.

Duty cycle (arc welding): percentage of time in a 10-minute period that a welding machine can be used at its rated output without overloading.

E

Eddy current inspection: changes in the flow of an induced eddy current, visible on an oscilloscope, indicate the location of a crack or other weld defect.

Edge joint: joint formed when two pieces of metal are lapped, with at least one edge of each piece at an edge of the other.

Edge preparation: shaping of the edges of the joint prior to welding.

Edge weld: a weld produced on an edge joint.

Effective throat: the minimum distance, minus any convexity, from the weld face to the root of the weld.

Elbow (joint): one of several axes (straight lines about which rotation can take place) on a typical robot.

Electric arc spraying: see **Arc spraying**.

Electric motor-driven carriage: a type of automatic cutting machine.

Electrode: terminal point to which electricity is brought to produce the arc for welding. In many electric arc welding processes, the electrode is melted and becomes part of the weld.

Electrode covering: flux materials combined into a thick clay-like mixture and then applied to the electrode wire in a very exact thickness.

Electrode drying oven: equipment in which an adequate supply of SMAW electrodes can be kept under ideal conditions.

Electrode extension: the length of unmelted electrode extending beyond the end of the contact tube in GMAW, FCAW, SMAW, and GTAW.

Electrode holder: device in which an electrode is held for welding.

Electrode identification system: a system of identification numbers for all welding electrodes, developed by the American Welding Society.

Electrode lead: the electrical conductor between the welding machine and the electrode holder.

Electrode tip: in arc welding processes, the extreme end of the electrode. The arc extends from the electrode tip to the base metal. Proper shaping of the tip is especially important in GTAW.

Electrode tip diameter: dimension that determines the size of the spot weld in resistance spot welding.

Electrogas welding (EGW): welding process that uses one or more torches (GMA or FCA) with a shielding gas and a set of water-cooled shoes that move up the joint to contain the molten weld metal as the weld progresses.

Electron: a fundamental part of an atom. It carries a small negative electrical charge.

Electron beam cutting (EBC): cutting process in which a focused stream of electrons is used to cut metals.

Electron beam welding (EBW): welding process in which a focused stream of electrons heats and fuses the material being welded.

Electronic pattern tracer: an electronic eye device used to follow a pattern (usually black on white) of a part to be cut. May be used to direct one or more cutting torches.

Electroslag welding (ESW): welding process in which one or more arcs are established between continuously fed metal electrodes and the base metal. As the weld progresses vertically, moving water-cooled shoes contain the molten metal, flux, and slag.

Element: chemical substance that cannot be broken down into simpler substances by chemical action.

Elongation: the percentage increase in a specimen's length when that specimen is stressed to its yield strength.

Equal pressure torch: see **Positive pressure torch**.

Expansion: the increase in the dimensions of a piece of metal as it is heated.

Explosion welding (EXW): welding process that uses the powerful shock waves from an explosive charge to cause metal flow and resulting fusion.

Expulsion weld: resistance spot weld that squirts (expels) molten metal from the weld area.

F

Face bend: a test performed by bending a weld sample with the face of the weld on the outside of the bend.

Face of the weld: the outer surface of the weld bead on the side of the weld.

Face reinforcement: the distance from the top of the weld face to the surface of the base metal.

Fatigue: a condition in which repeated stress below the tensile strength of a material (usually a metal) will cause the material to crack.

Feed rate: the speed, in a unit of time, at which an electrode or filler metal rod is fed to the arc crater.

Field weld symbol: a symbol added to the basic AWS welding symbol to indicate that a weld is to be made at the job site ("in the field"), rather than in a fabricating shop.

Filler metal: metal or alloy to be added to the base metal to make welded, brazed, or soldered joints.

Filler passes: intermediate passes used to fill the joint in a multiple pass weld.

Filler rod: see **Welding rod**.

Fillet test: test used to evaluate fillet welds. The vertical piece of the weldment is bent until it cracks or touches the horizontal piece.

Fillet weld: inside corner weld made at the intersection of two surfaces that form approximately a right (90 degree) angle.

Fillet weld size: the length of the legs of the triangle inscribed into the cross section of a completed fillet weld.

Filter lenses or plates: lenses in welding goggles or plates in helmets with optical properties that protect the welder's eyes from infrared, ultraviolet, and visible radiation.

Fire watch: safety practice in which a person is designated to observe welding operations to prevent fires started by stray sparks or heat from welding activities.

Fitting: threaded connectors on the ends of fuel gas hoses, used to join them to regulators and torches.

Fixture: device used to hold a weldment in proper alignment for welding.

Flame cutting: see **Oxyacetylene gas cutting**.

Flame spraying (FLSP): a thermal metal spraying process in which an oxyfuel gas flame is the heat source for melting the coating material. Compressed gas may be used to propel the coating material onto the base material.

Flange joint: joint formed when the edge of one or more pieces of the base metal are bent and form a flange(s).

Flared: curved outward.

Flare-groove joint: joint formed when the flanged edges of one or both pieces of base metal are placed together to form a single-flare-bevel or double-flare-V-groove.

Flash: the impact of electric arc rays against the human eye. Also, the surplus metal thrown out at the seam of a resistance weld.

Flash goggles: goggles worn under their helmets by arc welders to protect their eyes from flashes from the rear.

Flashback: an extremely dangerous occurrence where the torch flame moves into or beyond the mixing chamber of the torch.

Flashback arrestor: a device usually installed between the torch and welding hose to prevent the flow of a burning fuel gas and oxygen mixture from the torch into the hoses, regulators, and cylinders.

Flat (1G) position: welding done with the weld axis and base metal surface nearly horizontal.

Flat bead: a relatively flat bead contour used when the bead is to be ground or machined.

Flaw: a discontinuity in a weld that is within the accepted limit.

Flow meter: device that controls the amount of gas that goes to the welding torch.

Flow rate: the volume of a liquid or gas flowing through a hose or pipe, usually measured in cfh or Lpm.

Fluorescent dye: a dye that gives off or reflects light when exposed to short wave radiation.

Fluorescent penetrant: a penetrating fluid to which fluorescent dye has been added to improve visibility of any discontinuities in a weld.

Flux: material used to prevent, dissolve, or facilitate removal of oxides and other undesirable surface substances.

Flux-cored arc welding (FCAW): welding method in which heat is supplied by an arc between the base metal and a hollow, flux-filled electrode.

Flux-cored electrode: hollow metal electrode filled with a flux material, usually used in FCAW or ESW.

Foot pedal: foot-operated rheostat used for remote control of the output of an arc welding machine.

Force gauge: gauge used to measure applied force, such as the pressure being exerted by electrodes in resistance spot welding.

Forehand welding: welding in the same direction as the flame or electrode is pointing.

Forge welding (FOW): weld made by heating the parts in a forge, then applying pressure or striking blows with enough force to make them fuse together.

Forward arc blow: arc blow that occurs at the begin-ning of the joint as the magnetic field tries to stay in the metal. This causes the filler metal to blow toward the center of the joint.

Free bend test: bending a test specimen, usually in a vise, without using a fixture or guide.

Freeze: refers to a welding rod or electrode that becomes stuck in the weld pool.

Friction welding (FRW): welding process in which frictional heat is created by revolving one part against another under very heavy pressure.

Fuel gases: gases that support combustion, such as acetylene, propane, and butane.

Full anneal: application of the heat-treating (annealing) process to full thickness of the workpiece.

Fuse: device designed to prevent current flow in an electrical circuit if the amperage becomes too high.

Fusible plug: a steel plug used to prevent acetylene cylinders from exploding in a fire. The plug is filled with a metal that melts at approximately 212°F (100°C).

Fusion: the intimate mixing or combining of molten metals.

Fusion welding: any type of welding that uses fusion as part of the process.

G

Galvanized metal: metal coated with zinc to prevent rust. The zinc coating gives off toxic fumes when heated excessively.

Gas-cooled torch: GTAW torch used for light duty work. These torches are not recommended for use with over 200 A.

Gas lens: a series of fine stainless steel wire screens that makes the shielding gas exit the nozzle in a column. This allows the electrode to stick out farther, so the welder can see the weld better.

Gas metal arc welding (GMAW): arc welding process that uses a continuously fed consumable electrode and a shielding gas. Sometimes called MIG welding.

Gas tungsten arc welding (GTAW): arc welding process that uses a tungsten electrode and a shielding gas. The filler metal is added using a welding rod.

Gauntlet gloves: gloves with cuffs that extend above the wrist for protection.

Generator: a device that produces electricity; also, a device that produces acetylene.

Globular transfer: in GMAW, the movement of molten metal in large droplets from the consumable electrode across the arc.

Globule: a large droplet of molten metal.

GMAW: see **Gas metal arc welding**.

Gouging: the process of cutting a groove in the surface of a piece of metal, using an oxyfuel gas or arc cutting outfit.

Grain size: an important factor in determining the mechanical properties of a material. Coarse-grained materials are brittle and have low ductility; fine-grained materials are less brittle and are more ductile.

Grit blasting: mechanical cleaning method in which abrasive particles are propelled at high velocity against the metal surface.

Groove angle: the total angle formed between the groove face on one piece and the groove face on the other piece.

Groove face: the surface formed on the edge of the base metal after it has been machined, bent, or flame cut.

Groove joint: joint formed when there is a designed space in the form of an angled or shaped groove between the pieces being joined.

Groove weld: weld made with filler material that is fused into a joint that has had base metal removed to form a V, U, or J shape at the edges to be joined.

GTAW: see **Gas tungsten arc welding.**

Guided bend test: bending a test specimen in a specific way, using a specially designed fixture.

H

Hand truck: a two-wheel cart with a means of securing cylinders to it, such as chains or straps.

Hardfacing: surfacing process in which hard materials are applied to the surface of a part to reduce wear or loss of materials by impact, abrasion, or corrosion.

Hardness: the ability of a metal to resist plastic deformation; the same term may refer to stiffness or temper, resistance to abrading or scratching.

Hardness test: a measurement of hardness made by forcing a steel ball or pointed diamond under a given force into the surface of a material.

Heat: molecular energy of motion.

Heat-affected zone (HAZ): the area of the base metal around the joint that has been changed (mechanically or in microstructure) by welding, brazing, or soldering.

Heliarc welding: welding process that uses two tungsten electrodes to create an arc. Sometimes used incorrectly to refer to GTAW.

Helium (He): an inert, colorless gas used as a shielding gas in welding.

Helmet: protective hood that fits over the welder's head and includes a filter plate through which the welder can safely observe the electric arc.

Hertz: unit of frequency equal to one cycle per second.

High frequency: frequency considerably more rapid than the 60 hertz frequency of normal electric current.

Hold time: the time that force is maintained on resistance welding electrodes after the current is turned off.

Horizontal: in a plane parallel to the earth's surface.

Horizontal (2G) position: weld performed on a nearly horizontal weld axis and a nearly vertical surface.

Horn spacing: space between the parts of the resistance welding machine that hold the electrodes, measured when the electrodes are touching.

Hose: a flexible rubber tube used to convey gases from a pressure regulator to a welding torch.

Hot hardness: the ability of a metal to retain its strength at high temperatures.

Hot pass: the second pass in a multiple-pass welding joint.

Hot shortness: a weakness of the metal that occurs in the hot forming range (above the recrystallization temperature).

Hydraulic: a term that refers to a device actuated by the use of a fluid to transmit and/or multiply force.

Hydrogen (H): a chemical element used as a fuel gas in oxyfuel gas welding.

I

ID (inside diameter): the interior size of a pipe or tube, measured at its widest point.

Ignition temperature: the temperature at which a material will burn if enough oxygen is present.

Impact strength: the ability of a material to withstand impact or hammering forces without cracking or breaking.

Impact test: a test that carefully measures how materials behave under heavy loading, such as bending, tension, or torsion. Charpy or Izod tests, for example, measure energy absorbed when breaking a specimen.

Impurities: undesirable elements or compounds in a material.

Inch switch: a switch on a welding machine that is used to slowly feed consumable electrode wire through a combination cable to a GTAW torch.

Inclusions: foreign matter introduced to, and remaining in, welds or castings.

Incomplete fusion: failure of weld metal to fuse completely with base metal or the preceding bead.

Indentation: depression left on the surface of base metal after a spot, seam, or projection weld is made.

Indenter: in a hardness test, the ball or diamond that is pressed into the surface being tested.

Inductance: in the presence of a varying current in a circuit, the magnetic field surrounding the conductor generates an electromagnetic force in the circuit itself. If a second circuit is adjacent to the first, the changing magnetic field will cause (*induce*) voltage in the second circuit. An important application of this principle is the step-down transformer used in welding machines.

Inert gases: shielding gases, such as argon and helium, that do not react with the weld.

Infrared rays: heat rays produced by either an arc or welding flame.

Injector-type torch: torch used with a low-pressure acetylene generator. Acetylene is drawn in by traveling through the injector, and mixed with oxygen in the mixing chamber. The mixed gases flow to the tip of the torch where they are burned.

Inorganic: being or composed of material that was never living; mineral, as compared to plant or animal.

Inorganic fluxes: soldering fluxes that do not contain carbon. They are very corrosive, so they are not used on electrical or electronic parts.

Input power: electric power required (220V or 440V, for example) to operate a given welding machine.

Inside corner joint: joint made by welding along the inside of the intersection (usually a 90° angle) of two pieces of base metal.

Inspection: the process of examining welds for suitability, without damaging or destroying them.

Insulation: a material that will not permit the flow of electricity, used as a covering on wires, cables, and electrode holders.

Intermittent welding: the process of joining two pieces with welds that are not continuous; leaving gaps between welds.

Interpass heating: heating or reheating a joint between the passes needed to complete the weld.

J

Jack: a socket on the front of a welding machine that accepts a connector on the end of the electrode or work lead.

Jig: a fixture or template used to accurately position and hold a part during welding or machining.

Joint: the point or line at which two pieces come together and are fastened by welding or other means.

Joint geometry: the shape and dimensions of a (weld) joint, in cross section, prior to welding.

Joint penetration: the depth that a weld extends into the joint from the surface.

K

Kerf: the slot or opening produced in the metal when cutting.

Keyhole: an enlarged root opening that looks like an old-fashioned keyhole.

Keyhole welding: a welding technique in which concentrated heat penetrates the workpiece, leaving a hole at the leading edge of the weld. As the heat source moves on, the keyhole is continually formed and filled.

Kilopascal (kPa): one thousand pascals (Pa is the SI metric unit of pressure).

Kinetic energy: the energy in a moving object.

L

Lack of fusion: a weld defect resulting from failure of the weld metal and base metal to mix (fuse) completely.

Lap joint: a joint formed by overlapping the edges of two pieces of base metal.

Laser: a device that emits a beam of coherent light.

Laser beam cutting (LBC): process that uses the energy of a laser beam to cut material.

Laser beam welding (LBW): process that uses the energy of a laser beam to fuse materials.

Layer: one level of a thick welding joint made up of one or more passes.

Laying a bead: forming a line of fused weld metal along a line or a joint of the weldment.

Lead: wire that carries electricity from a power source to the electrode or ground clamp; also, a metallic element (symbol: Pb) that is a major part of some solders.

Leading edge: forward edge of the weld pool.

Leathers: protective clothing worn by a welder, especially when welding out of position.

Leg of a fillet weld: the shortest distance from the toe of the weld to the point where the pieces of base metal touch.

Lens: see **Filter lenses or plates.**

Leverage: the mechanical advantage provided by using a lever and a properly placed fulcrum.

Liner: a flexible tube placed inside the combination cable through which a consumable wire electrode passes on its way to the GMAW torch.

Liquefaction: a process that liquefies air and then separates the gases in it at their various boiling points.

Liquid penetrant inspection: a nondestructive method of checking a weld for flaws, using a special liquid and dye.

Liquidus: the temperature at which the filler metal becomes completely liquid.

Low hydrogen electrodes: electrodes used for SMAW that have little or no hydrogen in them.

Lugs: heavy-duty electrical terminals that are cylindrical at one end and flat on the other end. A hole is drilled in the flat end.

M

Machinability: the ability of a part to be machined or ground to size.

Magnetic field: field created around a wire or electrode whenever electricity travels through that wire or electrode.

Magnetic particle inspection: procedure used to check a weld for flaws, using a liquid that contains magnetic particles. The particles are drawn into the flaws when a magnetic field is applied to the weldment.

Mandrel: solid, pointed shaft that is forced through a nearly molten metal rod to form seamless tubing.

Manifold: an assembly of pipes that delivers gas from several cylinders into a single pipe for distribution to several work stations.

Manual welding: welding in which time, distance, speed and other variables are controlled by the person making the weld.

MAPP: trade name for stabilized methylacetylene-propadiene fuel gas.

Mechanical drawings: parts or assembly drawings made by mechanical means, either traditional "board" drafting or CAD (computer-aided drafting).

Mechanical properties: description of a material's behavior when force is applied to determine that material's suitability for mechanical use. Examples:

tensile strength, hardness, modulus of elasticity, elongation, fatigue limit.

Megapascal (mPa): one million pascals. (Pa is the SI metric unit of pressure).

Melt-through symbol: symbol used on the welding symbol to indicate 100 percent penetration on a weld made from one side of the base metal.

Metal inert gas welding (MIG): see **Gas metal arc welding (GMAW).**

Metal powder cutting (POC): process in which a metal powder (usually iron or aluminum) is fed into the flame to improve cutting effectiveness.

Metal-to-metal wear resistance: the ability of a material to resist wear from metal-to-metal contact.

Metal transfer: movement of metal from one surface to another (as in *metal transfer wear*).

Metal transfer wear: wear that occurs when metal leaves one surface and fuses or sticks to another surface.

Metric system: see **SI metric system.**

MIG: see **Gas metal arc welding.**

Mixing chamber: part of the welding torch in which gases come together and are mixed before combustion.

Motor-driven beam-mounted torch: a type of automatic cutting machine that uses an electronic tracer to follow the edge of the pattern. The tracer may control one or more gas cutting torches mounted on the beam.

Motor-driven magnetic tracer: a relatively inexpensive automatic cutting machine that works best with shapes that are not too complex. The magnetic tracer follows a steel pattern. One torch is generally carried on this type of cutting machine.

MPS: methylacetylene-propadiene (stabilized) fuel gas.

Multiple-pass weld: a welding joint requiring many passes.

Mushroomed: see **Flared.**

N

Natural gas: a naturally occurring mixture of hydrocarbon gases used as a fuel gas in some applications.

ND:yag: neodymium-doped yttrium aluminum garnet, a material used to make one type of laser.

Neutral flame: flame resulting from combining oxygen and a fuel gas in perfect proportions.

Newton (N): the unit of force in the SI metric system.

Nonconsumable electrode: an electrode that does not melt and become part of the weld.

Nondestructive evaluation (NDE): a means of testing for defects that does not damage or destroy the weld.

Non-petroleum-based: made from materials that do not contain petroleum or petroleum products.

Nontransferred arc: in plasma arc welding or spraying, an arc established between the electrode and the restricted nozzle. The work is not in the circuit.

Normalizing: heat treating process to eliminate internal stresses and create uniform grain size. Cooling is more rapid than in the annealing process.

Nozzle: a ceramic or metal cup used on a torch to direct the shielding gas to the weld area.

O

Occupation: a person's career or job. The status of an occupation usually depends on the amount of education a person obtains and his or her experience in the chosen field.

OD (outside diameter): the exterior width of a pipe or tube, measured at its widest point.

Off time: in resistance welding, the time between repeating cycles, when the electrodes are off the work.

Ohmmeter: an instrument for measuring electrical resistance.

Ohms: the unit of measurement for resistance to the flow of electricity through a circuit.

Open arc: a visible arc.

Open circuit voltage: the voltage in the welding machine circuit when it is on, but the arc is not struck.

Operator controls: in robotics, the controls used by an operator to start a programmed operation.

Organic: originating in once-living matter; animal or plant, not mineral.

Organic fluxes: fluxes that contain carbon. They are corrosive during the soldering operation, but become noncorrosive when the soldering is completed and can be washed away with water.

Orifice: a precisely bored hole through which gases flow. May be in a regulator, torch, or torch tip.

Orifice gas: the gas that surrounds the electrode in plasma arc welding and cutting. It becomes ionized in the arc to form the plasma.

Orthographic projection: a method of developing a mechanical drawing, based on viewing an object as if it were in a transparent cube.

Oscillating: moving back and forth in an uninterrupted manner.

Oscilloscope: an instrument that shows electrical impulses on a calibrated screen.

OSHA: The Occupational Safety and Health Administration, the federal agency responsible for safety in the workplace.

Other side: the area of the welding symbol above the horizontal reference line. Also, the side of the welding joint opposite the side touched by the welding symbol arrow.

Outfit: all the *welding* equipment needed to make a weld.

Out-of-position welding: welds made in the vertical, horizontal, or overhead welding positions.

Outside corner joint: joint made by welding along the outside of the intersection (usually a 90° angle) of two pieces of base metal.

Overhead (4G) position: weld made on the underside of a joint, with the face of the weld in an approximately horizontal welding position.

Overheating: applying excessive heat, and thus damaging the properties of the metal. When the original properties cannot be restored, the damage is referred to as *burning*.

Overlap: a condition in which the bead is not fused

into the base metal at the toe of the bead.

Oxidation: the process in which oxygen combines with a material to form a chemical compound called an oxide.

Oxidation resistance: the ability of a material to resist the formation of oxides. Metal oxides occur when oxygen is combined with metal.

Oxidized: combined with oxygen.

Oxidizing flame: the flame produced by an excess of oxygen in the torch mixture. The excess oxygen tends to burn the molten metal.

Oxyacetylene cutting (OFC-A): a process that uses an oxygen and acetylene flame for heat and a jet of oxygen to oxidize and cut the molten metal.

Oxyacetylene welding (OAW): a process in which oxygen and acetylene are combined and burned to provide the heat for welding.

Oxyfuel gas cutting (OFC) underwater: a process used for ship repair and similar applications. Oxygen and a fuel gas are burned to provide the necessary heat for cutting.

Oxyfuel gas welding (OFW): a group of welding and cutting processes and methods that use heat produced by a gas flame.

Oxygen (O_2): a colorless, odorless gas making up about one-fifth of the earth's atmosphere. Used in OFW and OFC to support combustion.

Oxygen arc cutting underwater: a process that uses the arc from a hollow electrode to heat the metal, and pressurized oxygen flowing through the electrode to make the cut.

P

Parallel: term describing the relationship of two lines or surfaces that, if extended, will never touch.

Parameters: welding conditions or variables.

Parts positioner: a device that holds the weldment and rotates or otherwise positions it for welding operations. For robotic welding, the positioner may be positioned by a controller and computer program.

Pascal (Pa): the unit of measurement for pressure in the SI metric system.

Pass: see **Weld pass.**

Peak current: during pulsed GMAW, the higher current that is used for welding. The lower *background* current maintains the arc and allows the weld to cool.

Peel test: a destructive test in which a resistance welded lap joint is mechanically separated by peeling one piece away from the other.

Penetrant: a liquid applied to the surface of a weld to locate discontinuities. It enters the cracks and makes them visible.

Penetration: the depth of fusion of the weld below the surface.

Percent heat control: on a resistance welding machine, the control that permits fine adjustment of current within the amperage limits of the tap switch setting.

Petroleum: a natural hydrocarbon compound that is processed into lubricants (grease and oil) and liquid fuels such as gasoline, heating oil, and kerosene.

Petroleum-based: term used to describe lubricants. Flammable lubricants derived from petroleum must not be used to lubricate any part of a welding or cutting outfit.

Phosgene gas: a highly toxic (poisonous) gas that can be released by the action of heat on cleaning chemicals that contain chlorinated hydrocarbons.

Photons: units of light energy emitted by a laser.

Physical properties: the physical characteristics used to identify or describe a metal, such as color, melting temperature, or density.

Pinch force: in short circuiting GMAW metal transfer, the magnetic force that squeezes off the droplet of the molten electrode metal.

Pipe: a hollow cylinder with a relatively thick wall; pipe size is identified by its inside diameter and schedule number.

Pipe schedule number: a one-, two-, or three-digit number that classifies pipe. Schedule numbers are determined by a combination of the pipe's inside diameter and its wall thickness.

Plasma: a temporary physical state assumed by a gas after it has been exposed to, and reacted to, an extremely high temperature.

Plasma arc cutting (PAC): a process using an electric arc and fast-flowing ionized gases to cut metal.

Plasma arc welding (PAW): a process in which an electric arc ionizes a gas, creating a plasma that generates the heat for welding.

Plasma spraying (PSP): a thermal spraying process that uses a nontransferred plasma arc to melt and propel a coating material onto a base material.

Plastic: soft and easily shaped; a metal that is in a plastic state is almost molten.

Plastics: synthetically produced nonmetallic compounds that can be shaped or molded into desired form.

Plastic state: in welding, the almost molten state of heated metal pieces being joined.

Plastic welding: process in which heated air is used to soften and fuse plastic materials.

Plug weld: weld made through a hole in one piece of metal that is lapped over another piece.

Polarity: direction of the flow of electrons in a closed direct current welding circuit. When the electrons flow from the electrode to the base metal, the polarity is DCEN (direct current electrode negative) or DCSP. When the current flows from the base metal to the electrode, the polarity is DCEP (direct current electrode positive), or DCRP.

Polyethylene: one of the most extensively used thermoplastic resins.

Polypropylene: hard, tough thermoplastic used for molded articles and fibers (especially rope).

Polyurethane: plastic resin used to form tough surface

coatings; also used as a casting or potting resin to protect electronic components.

Polyvinyl chloride: widely used vinyl resin used to make pipes and other rigid molded products.

Porosity: the presence of gas pockets or voids in weld metal.

Positioner: see **Parts positioner**.

Positive pressure torch: a torch in which the working pressure of the oxygen and fuel gas is high enough to force the gases into the mixing chamber.

Post flow: shielding gas flow that continues for a short time after the weld current stops.

Post flow adjustment: control on a welding machine that allows variation in the length of time that post flow continues.

Pounds per square inch (psi): the unit of measurement for pressure in the US conventional system.

Preheating: a process that heats the metal to a specified temperature prior to a surfacing or welding operation. Also, the process of heating base metal to its kindling temperature before cutting.

Preheating orifice: openings in a cutting torch tip through which oxygen and fuel gas for the preheating flame are supplied.

Pressure: force exerted on a given area, expressed in psi or pascals.

Pressure regulators: device used to reduce the cylinder pressure of a gas to a usable (working) pressure for welding.

Pressure roll: in gas metal arc welding, the drive roll on the wire feeder that applies pressure to electrode wire.

Pressure test: procedure used to determine the maximum pressure that a cylinder or tank can hold. The test consists of forcing a gas or fluid into the cylinder or tank under increasing pressure until the vessel fails.

Primary circuit: in a transformer, the circuit and windings connected to the input power supplied by the electrical utility.

Process: an operation used to produce a product.

Prods: a pair of hand-held electrodes used to magnetize a part for a magnetic particle inspection.

Program: a series of step-by-step directions and process parameters (times, pressures) set on the automatic controls of a welding machine or placed in the memory of computer/controller.

Projection: bump stamped into one piece to be resistance welded. A weld nugget forms at the point where the projection touches the surface of the second part.

Projection welding (PW): resistance spot welding process in which current is concentrated at points where projections on adjacent pieces are in contact.

Prototype parts: the first models of parts that later may be mass-produced.

PSI: see **Pounds per square inch**.

Pull gun: a GMAW welding gun that pulls the electrode wire through the cable.

Pulsed current: similar to direct current except that it has two current levels—a high current period followed by a low current period.

Pulsed spray transfer: a GMAW process in which the current is pulsed to take advantage of the spray mode of metal transfer, but with current values below the spray transition current.

Pulses: regular alternations of level or intensity, as in pulsed current.

Purge switch: switch on a welding machine used to manually control the flow of shielding gas.

Purging: the process of passing the correct gas through a regulator, torch, and hose to clean out any air or undesirable gas that may be in the system.

Push-pull setup: a system formed when a GMAW pull gun pulls the electrode wire through the cable and the wire feeder pushes the wire.

Q

Qualified: tested and approved. See **Welding performance qualification**.

Quenching: rapid cooling of metal in a heat-treating process as a means of hardening it.

Quick-connect terminal: a heavy-duty electrical terminal that easily connects and disconnects from a welding machine.

R

Radiation: see **Thermal radiation**.

Radiograph: an image on film produced by X rays.

Range switch: the coarse adjustment lever on a welding machine.

Rated output current: the maximum current flow a welding machine can produce.

Reading the bead: The process of visually inspecting the weld bead to determine whether the weld was made properly.

Rectified current: alternating current that has been made to flow in one direction only through use of a rectifier.

Rectifier: an electronic device (such as a diode) that acts like a one-way valve as current flows through it. It converts AC (alternating current) to DC (direct current).

Reducing flame: an oxyfuel gas flame with a slight excess of fuel gas.

Reducing gas: a gas that removes oxygen from the atmosphere by combining with the oxygen.

Reference line: the horizontal line drawn on a welding symbol. All information about the weld is positioned above or below this line.

Regulator: device used to control the volume and pressure of a welding or shielding gas as it flows from the cylinder to the torch.

Regulator adjusting screw: a screw that controls the working pressure of gas delivered by the regulator.

Remote contactor control: receptacle on the welding machine where the wire feed control mechanism is plugged in.

Repeatability: term used to describe a robot's ability to make welds in the same place, and of the same quality, on part after part in production.

Residual stress: stress that remains in a body (such as a weldment) after the external forces or thermal gradients have been removed.

Resistance: the property of a material that causes the flow of current in a circuit to be retarded.

Resistance seam welding (RSEW): process that usually uses a round, rotating ("wheel") electrode to make a continuous seam. A seam can also be formed with overlapping spot welds.

Resistance spot welding (RSW): welding overlapping pieces of metal together in small spots between two electrodes.

Resistance welding: process that uses the resistance of metals to electrical flow as a source of heat for welding.

Restarting the arc: establishing a new arc after changing electrodes, or after losing the arc due to faulty welding technique.

Restricted nozzle: in plasma arc welding, the nozzle through which orifice gas flows to form the plasma.

Reversed polarity: see **Direct current electrode positive.**

Robot: device that uses a computer program to direct its movements as it completes a series of welds or other operations.

Robot cell: three-dimensional space enclosing the robot and its working area, usually fenced off for safety of employees.

Robot controller: the computer and associated devices that direct a robot in carrying out the program developed for it.

Robotic welding system: industrial production operation using one or more robots, controllers, positioners, and other equipment to make a programmed series of welds on a weldment.

Robotic welding work station: location where a single robotic welder makes a specific set of welds.

Rockwell C hardness test: generally used for hard materials, this test uses a special piece of equipment to force a pointed diamond into the metal surface.

Root bend: a test performed by bending a weld sample with the root of the weld on the outside of the bend.

Root face: the distance from the root of the joint to the point where the bevel angle begins.

Root of the weld: the points where the root surface of the weld intersect the base metal. Also known as the "weld root."

Root opening: the distance between the two pieces at the root of the weld.

Root pass: the first weld pass made into the root of the joint.

Root penetration: the depth to which weld metal extends into the root of a welding joint.

Root reinforcement: the distance that the penetration projects from the root side of the joint.

Rosin: substance used as a welding flux. It is derived from the sap of pine trees.

Rosin fluxes: these fluxes are noncorrosive, and are recommended on electrical and electronic parts.

Running a bead: the process of making a weld bead.

RWMA: the Resistance Welder Manufacturers Association.

S

Safety cap: forged steel cap that should be screwed over the cylinder valve to protect it when the cylinder is stored or moved.

Safety valve: a device that prevents a gas cylinder from exploding when exposed to high temperatures. The valve includes a disc that ruptures under increased pressure.

Sagging: the sinking or downward curving of metal (as in a weld pool), due to gravity or pressure.

Secondary circuit: in a welding machine transformer, the electrical circuit and winding connected to the electrodes.

Self-shielding electrode: an electrode that produces its own shielding gas and does not require additional shielding gas.

Semi-automatic system: system in which some functions are performed manually and others by a programmed mechanical device.

Sequence: order in which operations or events take place.

Servomotor: small electric motor capable of precisely controlled movement.

Shielded metal arc welding (SMAW): a welding process in which the base metal is heated to fusion temperature by an electric arc created between a covered metal electrode and the base metal.

Shielding gas: a gas, usually inert, that is used to blanket the weld area and prevent contamination from the air.

Short arc: see **Short circuiting transfer.**

Short circuit: in the short circuiting transfer process, the condition that occurs when the electrode touches the base metal, causing metal from the electrode to enter the weld.

Short circuiting transfer: a gas metal arc process that uses relatively low voltage. The arc is constantly interrupted and restarted as the molten electrode shorts out against the base metal.

Shoulder (joint): one of several axes (straight lines about which rotation can take place) on a typical robot.

SI metric system: measuring system used in most countries. It uses such units as millimeters, kilograms, liters, newtons, and pascals. See **US conventional system.**

Side bend: a test performed by bending a weld sample to the side toward the thickness of the metal. The side bend is used on thick material.

Single-stage regulator: a regulator that reduces cylinder pressure to working pressure in one stage (step).
Size of the weld: see **Fillet weld size.**
Sketches: simple drawings of parts. Sketches are made without drafting tools.
Slag: the hard, brittle metal that covers a finished shielded metal arc, flux cored arc, and submerged arc welding beads; metal oxides and other materials that form on the underside of a flame or arc cut.
Slag inclusions: nonfused, nonmetallic substances in the weld metal.
Slope: the downward curve of the volt-ampere diagram for an arc welding machine.
Slope adjustment: changing the slope of the volt-amp curve by decreasing the maximum short circuit current. This reduces spatter during short circuiting transfer.
Slot weld: a weld similar to a plug weld, but made through an elongated (rather than round) hole.
SMAW: see **Shielded metal arc welding.**
Socket connection: preformed fitting soldered in place to connect lengths of thin-wall tubing.
Solder: filler metal used in soldering.
Soldering: a group of welding processes that join materials by heating them to the soldering temperature and by using a filler metal having a liquidus below 840°F (450°C) and below the solidus of the base metals.
Soldering alloy: combinations of various metals such as aluminum, antimony, cadmium, copper, indium, lead, nickel, silver, tin, or zinc.
Solid state welding: a group of welding processes that weld metals at temperatures below the melting point of the base metal, without the addition of filler metal. Examples are friction, explosion, and ultrasonic welding.
Solidify: to become solid or hard.
Solidus: the temperature at which the metal or alloy begins to melt. The *liquidus* is the temperature at which a metal becomes completely liquid.
Sonotrode: the rapidly vibrating "sound electrode" that causes materials to bond together in an ultrasonic weld.
Spark lighter: device that creates a spark to ignite the oxyfuel gas torch flame.
Spatter: scattering of molten metal droplets over the surface near an arc weld.
Specification: see **Welding procedure specification.**
Speed welding tip: tip for a welding torch that allows one-handed welding of plastics.
Spool: the drum, mounted on the wire drive mechanism, that contains the electrode wire for gas metal arc welding.
Spot weld: see **Resistance spot weld.**
Spray transfer: a gas metal arc process which has an arc voltage high enough to continuously transfer the electrode metal across the arc in small globules.
Squeeze time: in resistance welding, the time, measured in cycles, that the electrodes are under a force to clamp the parts and ensure good electrical flow.
Staggered intermittent welds: intermittent welds, made on both sides of a joint, that are offset from each other.
Stainless steel: alloy steel, containing chromium, that resists corrosion and oxidation (rusting).
Step-down transformer: electrical device used in welding machines that reduces voltage and increases amperage in its secondary circuit.
Straight polarity: see **Direct current electrode negative.**
Strain: the reaction of an object to stress.
Strength: the ability of a material to withstand applied loads without failure.
Stress: the load imposed on an object.
Stress relieving: heat treating process that involves even heating to a temperature below the material's critical temperature, followed by a slow, even cooling.
Strike an arc: the act of touching the electrode to the workpiece, then withdrawing it a distance sufficient to maintain the electrical flow across the arc.
Stringer bead: bead made by moving a torch or electrode holder along the weld without any side-to-side motion.
Submerged arc welding (SAW): process in which the electric arc is submerged under a heavy layer of flux granules.
Surfacing: the application by welding, brazing, or thermal spraying of a layer(s) of material to a surface to obtain desired properties or dimensions, as opposed to making a joint.

T

Tack weld: a small weld used to temporarily hold pieces in alignment.
Tack welding tip: pointed tip for a plastic welding torch, used to tack weld plastic weldments.
Tail: the "V" shape, drawn at one end of a welding symbol, in which special notes are placed.
Tank: a thin-walled container for liquids or gases. Tank walls are thinner than the walls of cylinders used for pressurized gases.
Tap: one of the several electrical contacts available on the controls of a resistance welding machine. Each tap provides a different range of amperage for the electrodes.
Tap settings: the various tap positions available on the resistance welding machine.
Teach pendant: an electronic control used to program a robot to perform a series of actions; the sequence of movements is placed in computer memory as a program.
Teflon® liner: smooth seamless tubing used inside a cable to make electrode wire feed more smoothly to a GMAW welding gun.
Tempering: heat treating process in which metal is heated to a temperature just below its melting point.

This heating decreases the metal's hardness, while increasing its toughness.

Tensile shear: amount of force required to pull a spot weld apart in a direction perpendicular to the weld axis or "in shear."

Tensile strength: measurement of the amount of force required to pull metal apart.

Tensile test: a test in which a specially prepared sample is pulled until it fails. The test determines the weld's ability to withstand forces that would pull it apart.

Terminals: the physical connectors on a welding machine for attaching the workpiece lead and the electrode lead.

Thermal radiation: heat rays given off by a welding arc or oxyfuel gas flame, or by the heated base metal.

Thermal spraying: a process in which a material (metallic or nonmetallic) is heated and sprayed onto a surface.

Thermoplastic: a plastic material that can be formed or reformed by applying heat; damaged articles can be repaired by welding.

Thermosetting: a plastic material that, once formed, cannot be reformed by heating. Damaged articles cannot be repaired by welding.

Thoria: a metal used in GTAW electrodes.

Throat depth: distance measured from the center of the electrodes to the frame of the resistance spot welding machine while the electrodes are closed (in contact).

Throat of a fillet weld: distance from the weld root to the weld face; also known as the "actual throat."

Thumb switch: switch mounted on a GTAW torch to control the amperage; so-called because it is usually operated by the welder's thumb.

TIG: see **Gas tungsten arc welding (GTAW).**

Tip: the end of the torch where the fuel gas mixture burns, producing a high-temperature flame. Also used to refer to the end of the electrode in the spot welding process.

Tip nut: large threaded nut used to hold a tip in the cutting torch head.

T-joint: joint formed by two pieces of base metal placed at approximately a 90° angle to one another to form at "T" shape.

Toe of the weld: the point where the weld bead contacts the base metal surface.

Torch: device used to control and mix the fuel gas and oxygen and to direct the gas flame to the welding, brazing, cutting, or soldering area. Also, the assembly that holds the electrode in gas metal arc welding and gas tungsten arc welding.

Torch angle: the angle to the base metal or weld axis at which the torch is held while welding.

Torch body: main portion of the welding torch, to which fuel gas hoses are connected and the torch tip is attached.

Torch tip: the part of the end of the torch where the fuel gas and oxygen are ignited.

Torch valves: valves that control the flow of oxygen and fuel gas into the torch.

Toughness: the ability of material to withstand all types of stresses without tearing or breaking.

Toxic: poisonous.

Track: a straight or curved guide path for an electric motor-driven carriage that carries a cutting torch.

Transferred arc: arc that is established and maintained between the electrode and the workpiece.

Transformer: electrical device used to reduce or increase voltage. See **Step-down transformer.**

Transition current: the amount of current required to convert from globular transfer to spray transfer.

Tube: a hollow cylinder similar to a pipe, but generally with thinner walls; size of a tube is specified by its outside diameter.

Tungsten inclusions: in GTAW, a weld defect caused by getting tungsten from the electrode in the weld.

Tungsten inert gas (TIG): see **Gas tungsten arc welding (GTAW).**

Two-stage regulator: regulator reduces cylinder pressure to working pressure in two stages (steps).

U

Ultimate tensile strength: the greatest tensile stress that a material can withstand.

Ultrasonic: vibrations generated at frequencies above the range of human hearing.

Ultrasonic inspection: a nondestructive evaluation method in which ultrasonic waves are passed through the material being inspected. Echo patterns will locate any discontinuities.

Ultrasonic transducer: device used to generate ultrasonic vibrations that, in turn, vibrate the sonotrode to perform ultrasonic welds.

Ultrasonic welding: process used to weld metal or other material through use of ultrasonic vibrations.

Ultraviolet rays: light rays with very short wavelengths that are given off by electric arcs and welding flames. Filter lenses in welding goggles and helmets screen out such rays, which can be harmful to the eyes.

Under load: in arc welding, term used to describe state of the welding machine when current is flowing through the welding circuit (while welding or with the electrode holder touching the table or workpiece).

Undercut: a depression at the toe of the weld — the weld metal is below the level of the base metal.

Underfill: a condition in which the molten metal sags downward and does not fill the top part of the welding joint.

Unstable arc: an arc that has not established itself in the ionized space between the electrode and the workpiece. An unstable arc may wander or stop.

Uphill weld: a vertical weld made from the bottom of the joint to the top.

Upset: the term used to describe the enlarged weld where the parts meet in a friction weld.

Upslope: the rising line on a current/time program graph.

US conventional system of measurement: measuring system that uses pounds, feet, and other traditional units. See **SI metric system**.

V

Vacuum: a condition in which atmospheric pressure in a closed vessel has been decreased. This is done by pumping out air.

Vapor degreaser: device used to clean metal with chemical vapors before applying surfacing materials.

Vertical: perpendicular to the earth's surface.

Vertical down weld: nonstandard term for downhill weld.

Vertical (3G) position: weld performed with the weld axis and base metal surfaces nearly vertical.

Vertical up weld: nonstandard term for uphill weld.

Visual inspection: determining quality of a weld by looking at it and comparing it to welds of known high quality.

Volt: unit of measure of electrical pressure.

Voltage: electrical pressure in a circuit; the force that causes current to flow.

Voltage drop: voltage loss that occurs when electricity travels a long distance from the welding machine.

Voltmeter: instrument for measuring the voltage in a circuit.

W

Waist (joint): one of several axes (straight lines about which rotation can take place) on a typical robot.

Warped: twisted out of shape; in metals, usually a result of improperly applied heat.

Water-cooled torch: A torch that has a continuous flow of water passing through the torch body to remove heat. This type of torch is used to carry currents over 200 A. See **Gas-cooled torch**.

Weaving bead: bead formed by moving a torch or electrode holder from side-to-side as the pass progresses along the welding joint.

Weld: the blending or mixing of two or more metals or nonmetals by heating them until they are molten and flow together.

Weld all-around symbol: circle drawn on the welding symbol, indicating that the described weld is to be made all around the part.

Weld axis: an imaginary line running through the center of a completed weld.

Weld bead: one thickness of filler metal that is added to a welding joint.

Weld face: the outer surface of a weld on the side from which the weld is made.

Weld groove: a cut, ground, or machined surface on the workpiece, designed to provide space for welding.

Weld metal: the fused portion of the base metal, or fused base metal and filler metal.

Weld nugget: the weld metal in a resistance spot, seam, or projection weld.

Weld pass: one bead along a welding joint.

Weld pool: the small molten volume of metal under the torch flame or electrode, prior to its solidification as weld metal.

Weld reinforcement: weld metal extending beyond the upper and lower surfaces of the base metal.

Weld schedule: see **Welding sequence**.

Weld size: see **Joint penetration**.

Weld time: the time, measured in cycles, when current flows to make a resistance weld.

Weld toe: see **Toe of the weld**.

Welder: person who performs welding activities. (Use of "welder" to describe a welding machine is incorrect.)

Welding: the process of making a weld on a joint.

Welding gun: in GMAW, the device used to hold the consumable electrode.

Welding machine: a device that provides and controls the proper voltage and current for a welding task.

Welding outfit: the welding machine and other equipment required to actually create a weld.

Welding performance qualification test (WPQT): test of a welder's ability to weld a joint as directed by the welding procedure specification.

Welding positions: the position of the weld axis and weld face determines whether a weld is made in the flat, horizontal, vertical, or overhead position.

Welding procedure: method by which a weld is to be made, as outlined in a *Welding procedure specification*.

Welding procedure qualification record (WPQR): document containing the actual welding variables used to produce an acceptable weld. The procedure qualification record is used to qualify a welding procedure specification.

Welding procedure specification (WPS): document that lists in detail the specifics of the job: the base metal to be welded, the filler metals to be used, the preheat or post-welding treatment to be used, the metal thickness, and all other variables for each welding process. All items in the specification are identified as essential or nonessential.

Welding rod: welding filler metal, usually packaged in straight lengths. Unlike an electrode, a welding rod is not used to conduct electricity.

Welding sequence: the order in which components (parts) of a structure are welded.

Welding station: a work area that contains the fuel gas welding or cutting outfit or welding machine, booth, ventilation, and all required supplies to perform welds.

Welding symbol: the symbol (designed by the American Welding Society) that appears on a drawing of a weldment. The symbol describes the joint preparation, the weld, and other considerations.

Weldment: an assembly of parts joined by welding.

Wheel electrode: rotating electrode used in resistance seam welding.

Whip motion: rapid movement of the electrode or flame away from and back to the weld pool or arc crater. The motion allows time for the weld metal to cool.

Wire brush: a brush with bristles of metal wire, used to clean the weld bead between passes.

Wire feeder: in GMAW, the device that continuously feeds consumable electrode wire to the welding gun.

Wire tension control knob: device used to control the pressure applied to the drive rolls of a wire feeder.

Work booth: arc welding work area shielded from the view of workers without eye protection by solid walls, canvas curtains, or filtered, transparent plastic curtains.

Working pressure gauge: gauge on a pressure regulator that shows pressure of the gas being supplied to the torch.

Working volume: the three-dimensional space in which a robot moves while performing programmed tasks.

Workpiece: the object or assembly being welded.

Workpiece lead: the electrical cable that connects the base metal to the welding machine.

Wrist (joint): one of several axes (straight lines about which rotation can take place) on a typical robot.

Wrought iron: an easily welded or forged iron containing about 0.2 percent carbon.

X

X ray: a stream of high-energy photons; common term for a photographic image made through the use of X rays. See **Radiograph**.

X-ray inspection: the use of X rays to check a weld for flaws or defects.

Y

Yield strength: the point at which, when a metal is being stretched, it takes a permanent set and will not return to its original dimensions when the stretching force is released.

Z

Zirconia: zirconium dioxide, a compound used in some GTAW electrodes for AC welding. It permits easier arc starting than pure tungsten.

Acknowledgments

The authors wish to express their appreciation to the following companies for their cooperation in furnishing photographs and technical information used in the preparation of this book:

Advanced Robotics Corporation
American Welding Society
Anchor Swan Corporation
Automation International, Inc.
Bernard, Dover Industries Company
Broco, Inc.
Century Manufacturing Company
CK Worldwide, Inc.
CONCOA (Controls Corp. of America)
Craftsweld Equipment Corporation
Cronatron Welding Systems, Inc.
Detroit Testing Machine Company
Dockson Corporation
E.H. Wachs Company
ESAB Welding & Cutting Products
FANUC Robotics North America, Inc.
Fischer Engineering Company
Fusion, Inc.
Goss, Inc.
Handy & Harman/Lucas Milhaupt, Inc.
Henry Ford Museum & Greenfield Village
Hercules Welding Products
Highland Fabricators
Hobart Brothers Company
Jackson Products, Inc.

James Morton, Inc.
Jewel Mfg. Company
Kamweld Products Company, Inc.
Kioke Aronson Machine, Inc.
Laramy Products Company, Inc.
Lenco - NLC, Inc.
Linde Div., Union Carbide Corp.
LORS Machinery, Inc.
Magnaflux Corporation
Magnatech Limited Partnership
Miller Electric Mfg. Company
Modern Engineering Co., Inc.
Nederman, Inc.
Nelson Stud Welding Div., TRW, Inc.
Pontiac Motor Division
Pressed Steel Tank Company, Inc.
PTR — Precision Technologies, Inc.
Red-D-Arc, Ltd.
Rego/Engineered Controls, Intl. Inc.
Rexarc, Inc.
Sciaky, Inc., Subsidiary of Phillips Service Industries
Seelye, Inc.
Smith Equipment, Div. of Tescom Corp.
Sonobond Ultrasonics, Inc.

Stoody Company
Strippit, Inc., Unit of IDEX Corp.
The Aluminum Association
The American Society of Mechanical Engineers
The Fibre-Metal Products Company
The Gage Co., Taylor Engineering Co.
The Lincoln Electric Company
The Nippert Company
The Ohio Nut & Bolt Company
Thermacote-Welco Company
Thermadyne Industries, Inc.
Tuffaloy Products, Inc.
Tweco/Arcair, Div. of Thermadyne Industries, Inc.
U.S. Steel
Uniweld Products, Inc.
Vega Enterprises
Veriflow Corporation
Victor Equipment Co., Division Thermadyne Industries, Inc.
Wall Colmonoy Corporation
Weldcraft Products, Inc.
Welding & Fabrication Data Book
Welding Design & Fabrication

The authors wish to thank the following instructors and their students for their cooperation in creating many of the required photographs:

Clyde Hall
Welding Instructor
Washtenaw Community College
Ann Arbor, Michigan

James Etchill, Robert Solmose,
 and Ronald Watkins
Welding Instructors
Breithaupt Vocational/Technical Center
Detroit Public Schools
Detroit, Michigan

William Scrimger
Welding Instructor
Southwest Oakland Vocational Education Center
Wixom, Michigan

Most photographs in this book not otherwise credited were taken by Phillip Slaughter of Romulus, Michigan, Nathaniel Garcia of Ann Arbor, Michigan, or the coauthor, William Bowditch.

Index